EMBRYOS
Color Atlas of Development

Edited by
Jonathan B. L. Bard

**MRC Human Genetics Unit
Western General Hospital
Edinburgh**

M WOLFE

Copyright © 1994 Mosby–Year Book Europe Limited except illustrations in Chapter 12 copyright ©1994 M.A. England.
Published in 1994 by Wolfe Publishing, an imprint of Mosby–Year Book Europe Limited
Printed in Singapore by Imago Productions Pte., Ltd.
ISBN 0 7234 1740 7

All rights reserved. No part of this publication may be reproduced, stored in a retrieval system, copied or transmitted, in any form or by any means, electronic, mechanical, photocopying, recording or otherwise without written permission from the Publisher or in accordance with the provisions of the Copyright Act 1956 (as amended), or under the terms of any licence permitting limited copying issued by the Copyright Licensing Agency, 33–34 Alfred Place, London, WC1E 7DP.

Any person who does any unauthorised act in relation to this publication may be liable to criminal prosecution and civil claims for damages.

Permission to photocopy or reproduce solely for internal or personal use is permitted for libraries or other users registered with the Copyright Clearance Center, provided that the base fee of $4.00 per chapter plus $.10 per page is paid directly to the Copyright Clearance Center, 21 Congress Street, Salem, MA 01970. This consent does not extend to other kinds of copying, such as copying for general distribution, for advertising or promotional purposes, for creating new collected works, or for resale.

For full details of all Mosby–Year Book Europe Limited titles please write to Mosby–Year Book Europe Limited, Lynton House, 7–12, Tavistock Square, London WC1H 9LB, England.

A CIP catalogue record for this book is available from the British Library.

Contents

Preface	4
Introduction	5
1. *Arabidopsis* Gerd Jürgens and *Ulrike Mayer* Institut für Genetik und Mikrobiologie, University of Munich, Munich D8000, Germany	7
2. *Dictyostelium discoideum* Robert R. Kay and *Robert H. Insall* MRC Laboratory of Molecular Biology, Hills Rd., Cambridge, CB2 2QH, U.K.	23
3. The Sea Urchin *Jeff Hardin* Dept. Zoology, University of Wisconsin, Madison, WI 53706, U.S.A.	37
4. *Caenorhabditis elegans,* the Nematode Worm *Ian A. Hope* Dept. Pure and Applied Biology, University of Leeds, Leeds LS2 9JT, U.K.	55
5. Molluscs *Jo A.M. van den Biggelaar, Wim J.A.G. Dictus* and *Florenci Serras* Zoological Laboratory, State University of Utrecht, Utrecht, Netherlands	77
6. The Leech *David A. Weisblat* Dept. Zoology, University of California, Berkely, CA 94720, U.S.A.	93
7. *Drosophila* *Maria Leptin* Max Plank Institut für Entwicklungsbiologie, Spemannstrasse 35, 7400 Tubingen, Germany	113
8. The Zebrafish *Walter K. Metcalfe* Institute of Neuroscience, University of Oregon, Eugene, OR 97403, U.S.A.	135
9. *Xenopus* and Other Amphibians *Jonathan M.W. Slack* Dept. Zoology, Oxford University, Oxford, OX1 3PX, U.K.	149
10. The Chick *Claudio D. Stern* Dept. Human Anatomy, Oxford University, Oxford, OX1 3QX, U.K.	167
11. The Mouse *Jonathan B.L. Bard* and *Matthew H. Kaufman* MRC Human Genetics Unit, Western General Hospital, Edinburgh, EH4 2XU, U.K. Dept. Anatomy, Edinburgh University, Edinburgh, EH8 9AG, U.K.	183
12. The Human *Marjorie A. England* Dept. Anatomy, Medical Sciences, University of Leicester, Leicester, LE1 9HN, U.K.	207
Index	221

Preface

This book was written to help developmental biologists and molecular geneticists working on one embryo understand research being done on others, but it has two other purposes. One is to highlight the wide range of developmental and morphogenetic problems that are currently being investigated, many of which still lack even the beginnings of a molecular solution, and the other is to display the sheer wonder at the way in which the complexity and richness of the functioning organism are generated by a modest egg. I therefore hope that developmental biologists from undergraduates onwards will not only find the book useful but will also enjoy browsing through its pages and letting the embryos speak of their mysteries through their pictures.

This book is a joint effort and I am grateful to my collaborators who have set aside their research work in order to show how their embryos develop and what they can do. In addition, I particularly appreciate the generosity of our many colleagues who have allowed us to use their photographs: embryology is a visual subject and pictures are more powerful than words can ever be in bringing the subject to life. I would also like to thank Vernon French for his comments and advice, Tracy Cooper, my editor, for her support, enthusiasm and help, and Penny Bourke who had the difficult task of integrating the text and pictures to produce a book that was both easy to read and pleasing to look at.

Jonathan Bard

Introduction

This book summarizes the development and uses of the most important embryos currently being studied and the range is very wide, mainly because different lines of work have always required different types of organism[1]. Experimentalists whose work involved surgical manipulations have tended to use amphibian and chick embryos because they are large and robust, or sea urchins because they are transparent. Geneticists, on the other hand, need organisms that are quick to breed and easy to mutate and orignally chose *Drosophila* but now they also use *C. elegans*, zebrafish and mice where mutations can be made to order. Developmental biologists interested in lineage analysis have, however, mainly opted for invertebrates such as leeches, nematodes and molluscs because their development is largely mosaic. And so the list of working organisms has lengthened.

Over the years, this diversity of choice and approach had one unfortunate consequence: even though the field was not very large, it began to lose its sense of cohesion. Scientists working with one organism naturally tried to keep up to date with research being done on others because it was both interesting and collegiate to do so. They did not, however, have to keep too close an eye on what was happening elsewhere because results from one embryo could, in general, only be applied to related species. As there was no experimental interest that was common to all developing systems and no unifying body of knowledge underpinning the subject, signs of fragmentation began to appear.

The advent of molecular genetics has counteracted this tendency to balkanization for the surprising reason that the new methodology is essentially embryo-free, with its concerns and techniques being much the same for embryos across the phyla, even the kingdoms. Moreover, because the approach starts with nucleotide sequences, sensible and objective comparisons can be made, even among different phyla.

The first reward of the new technology has been the discovery of a host of molecules that regulate developmental processes, and these molecules are in turn providing direct evidence for a key assumption of four or five generations of embryologists, that the ontogeny of very different organisms is based on similar molecular mechanisms. This success does, however, mean that contemporary developmental biologists have to know about a much wider range of embryos than did their predecessors. If it is reported, say, that a transcription factor is present as the *Drosophila* wing forms, anyone working with another embryo will inevitably wonder whether it expresses a homologous gene and, if so, how its field of activity relates to *Drosophila* wing development. For this reason, molecular genetics is forcing members of a greatly enlarged developmental biology community to take a much more immediate interest in one another's work than used to be required.

The ability to make comparisons among different embryos does, however, require a background knowledge of development that is not easy to come by and it is still much simpler to check a cDNA library for the presence of a new molecule than to delve into the development of an alien embryo. Hence this book, and, given the nature of contemporary work, it was not difficult to choose which embryos should be included[2]. It was, however, harder to know whether it should focus on the molecular basis of development or on the overt processes of embryogenesis. On the whole, the latter option has been chosen, partly because it defines the subject and partly because it is hard to know which aspects of the current flood of molecular data will prove to be important. The processes of embryogenesis, in contrast, are timeless.

It may seem ironic that, while this book is intended to facilitate the study of the molecular basis of development, it pays relatively little attention to the area and, instead, concentrates on how embryos develop and why they are used. This approach with its pictorial emphasis[3] does, however, have unexpected benefits for it highlights the wide range of activities taking place in embryos and so reminds us of how little of development we yet understand and how great are the problems that we still need to solve.

Jonathan Bard

---

[1] For historical references, see
 Bard, J.B.L. (1990). *Morphogenesis: the Cellular and Molecular Processes of Developmental Anatomy*. Cambridge University Press.
 Gilbert, S. (1991). *Developmental Biology*. Sinauer Press.
[2] *D. discoideum* does not of course have an embryo, but this slime mould is such an important developmental model that it has honorary embryonic status.
[3] Still photographs can never do justice to the dynamics of embryogenesis so it is worth noting that a time-lapse video of early development entitled *A Dozen Eggs* has been compiled by Rachel Fink and distributed for the American Society of Developmental Biologists by Sinauer Associates (Sunderland, MA 01375, USA. American format). Most of the embryos discussed in this book are shown on the video.

1. *Arabidopsis*

Gerd Jürgens and *Ulrike Mayer*

Introduction

Animals and plants have evolved different developmental strategies. Animals essentially establish their body organization during embryogenesis even if the adult form originates from metamorphosis. Plant embryogenesis, by contrast, merely produces a primary body organization, as expressed in the seedling, to which new structures are continually added from terminal growth zones, the meristems of the shoot and the root. The adult plant thus looks very different from the mature embryo or the seedling. Since postembryonic development is so impressive in flowering plants, the fact that the essential features of the body organization are already present in the seedling is often overlooked. In addition, the importance of pattern formation in the embryo may also have been obscured by the striking plasticity of plant development: cells other than the zygote can undergo embryogenesis in culture, and fertile plants can be regenerated from disorganized tissue (callus).

Flowering plants are a group of closely related species which share a common body plan. This is reflected in the body organization of the seedling, which is remarkably uniform across species, although dicot and monocot types can be distinguished. The analysis of a plant model system like *Arabidopsis thaliana* is therefore likely to provide insights into mechanisms of development that also apply to other flowering plants even though work on *Arabidopsis* did not contribute to the historic development of plant embryology. The early descriptive studies involved a fairly large number of species, including the favourite teaching object, *Capsella bursa-pastoris* (shepherd's purse), while somatic embryogenesis has been most extensively studied in *Daucus carota* (carrot). With the recent advances in molecular-biological techniques, however, combined genetic and molecular approaches are increasingly being used to analyse plant developmental processes and it is in this context that *Arabidopsis* has become the experimental organism of choice.

Arabidopsis as an experimental organism

The wall cress, *Arabidopsis thaliana*, is a member of the mustard family (Brassicaceae). This 'common lab weed' has a number of favourable features that facilitate genetic analysis. Due to its small size (**1.1**), *Arabidopsis* can be grown in the laboratory at a density of 10,000 plants/m^2, completing its life cycle in about 6 weeks when the plants are grown at 25°C under continuous illumination. *Arabidopsis* has hermaphroditic flowers which are naturally self-fertilizing, and this mode of reproduction makes it very easy to isolate recessive mutants, while the crosses necessary for the genetic characterization of mutants can also be created by removing the immature stamens from flowers to be cross-pollinated. Since *Arabidopsis* continues to flower for several weeks, progenies produced by selfing or crossing can be obtained from different flowers of the same parent plant. This is important in complementation studies involving recessive lethal mutants. Each flower produces 20–50 seeds, and a single plant can produce several hundreds or even thousands of seeds.

While higher ploidy numbers are fairly common in plants, *Arabidopsis thaliana* is a truly diploid species, and mutations in most genes are thus likely to cause recognizable phenotypes. More than 100 genes identified by mutant alleles have been localized on the genetic map which consists of five linkage groups representing the haploid set of chromosomes

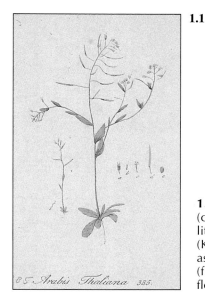

1.1 *Arabidopsis thaliana* (called *Arabis* in this lithography from *ca.* 1840). (Key: **a**, **b** and **c**, different aspects of flower; **d**, silique (fruit); **e**, seed; **f**, young flowering plant.)

(Koornneef, 1990). The chromosomes of *Arabidopsis* are fairly small and cannot be used for cytological mapping of genes, unlike in *Drosophila*. For molecular cloning of genes, this disadvantage is being overcome with the establishment of

RFLP maps, and there has now been substantial progress in constructing a physical map of the genome (see later).

Arabidopsis is thus a particularly suitable organism for investigating the molecular basis of plant development through the identification and analysis of mutants (for reviews on technical and other matters, see Redei, 1970; Meyerowitz, 1987, 1989; Koncz *et al.*, 1992). In this chapter, we will describe the normal development of the plant and then focus on the progress being made in the isolation and analysis of mutants that affect development, particularly those that alter pattern formation processes in the embryo.

Normal development

Being a short-lived weed, *Arabidopsis* has an abbreviated version of the life cycle in which the vegetative phase is reduced to a minimum; in other respects, its life cycle is representative of the flowering plants and includes the 'alternation of generations' (**1.2**). Its life cycle thus has three main phases: gamete formation (about 1 week), embryogenesis (about 2 weeks), and the period between seed germination and flower formation (about 3 weeks). Embryogenesis also has three phases: an early period between fertilization and the heart stage when the primary body organization is laid down, a period when primordia grow and cells differentiate, and a final desiccation phase. The mature embryo lies within the seed, the unit of dispersal, which is resistant to unfavourable conditions and the dormant seed is therefore the state in which wild-type and mutant strains are maintained in laboratories.

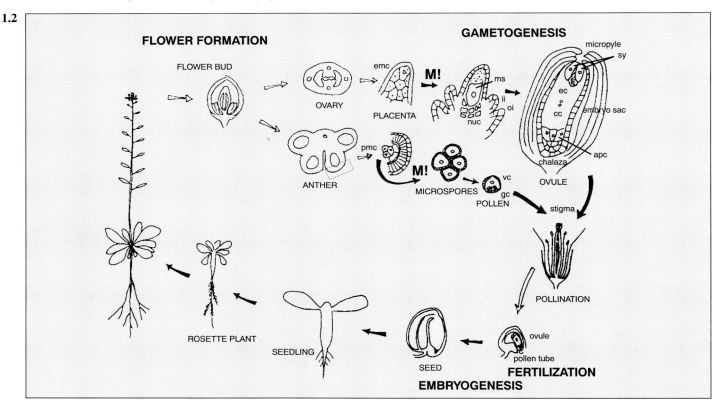

1.2 Life cycle of *Arabidopsis*. (Key: **apc**, antipodal cells; **cc**, central cell; **ec**, egg cell (stippled); **emc**, embryo sac mother cell; **gc**, generative cell, giving rise to two sperm cells; **ii**, inner integument; **M!**, meiosis; **ms**, functional megaspore, giving rise to embryo sac; **nuc**, nucellus; **oi**, outer integument; **pmc**, pollen mother cell; **sy**, synergids; **vc**, vegetative cell.)

Meiosis to fertilization

In contrast to animal meiosis, which directly produces the gametes, the meiotic products of flowering plants divide mitotically to produce few-celled haploid gametophytes. The male gametophyte (pollen) comprises a large vegetative cell which encloses two tiny sperm cells. The female gametophyte (embryo sac), which lies within an ovule, often consists of seven cells: an egg cell and two synergids at the micropylar end, three antipodal cells at the other end, and a diploid central cell with a large vacuole (**1.3**). The polarized egg cell has its nucleus near the future apical end and a vacuole at the other end which abuts the micropylar end of the ovule. The micropyle is formed by the integuments which, together with other diploid maternal tissues, surround the embryo sac. Rows of ovules are attached to two lines of thickened tissue named placenta which run down the length of the ovary (**1.4**) and these ovules are supplied with nutrients by the mother plant via vascular connections.

The mature anthers deposit their pollen by touching the papillate surface of the stigma. Following germination of the pollen grain, the large vegetative cell is turned into the pollen tube which grows down the style to reach the ovary, transporting the sperm cells at its tip. Since the pollen tubes greatly outnumber the ovules to be fertilized, pollen tube growth is a race, and delivery of the sperm cells into the embryo sac occurs on a first-come, first-served basis. This competition is called 'certation' if the outcome is influenced by the pollen genotype. A successful pollen tube grows through the micropyle into one of the two synergids where the sperm cells are released. As the synergid degenerates, one sperm fuses with the egg to give the diploid zygote and the other fertilizes the diploid central cell which will form the triploid endosperm ('double fertilization') that will nourish the growing embryo during the later stages of embryogenesis (the two thus differ in gene dosage). Fusion of the nuclei completes fertilization.

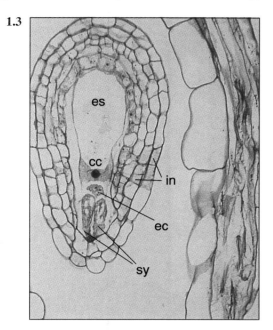

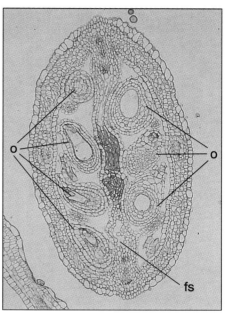

1.3 Section through ovule showing the micropylar end of mature embryo sac. (Key: **cc**, central cell; **ec**, egg cell; **es**, embryo sac; **in**, integuments; **sy**, synergid; ×300.)

1.4 Section through silique showing rows of ovules and the false septum. (Key: **o**, ovules; **fs**, false septum; ×100.)

The stages of embryogenesis

Although there is no cell movement in plant development, the embryo alters its morphology as it grows. Shape changes are brought about by position-dependent mitosis, orientation of cell-division planes and directional cell expansion (Lyndon, 1990). It is therefore possible to stage embryos by their cell numbers and shapes, although this becomes difficult in the later period. Nevertheless, quadrant and octant stages of early embryos, and the globular, heart, torpedo and bent-cotyledon stages of later ones can readily be identified (Johri, 1984). Furthermore, additional stages such as the dermatogen stage have been introduced, while other stages have been subdivided (Mansfield and Briarty, 1991). To unify the system, we have put forward a sequence of 20 stages based on morphological criteria (*Table 1.1*)[1]. Times are in percentages of embryogenesis (0% is fertilization, 100% is the mature-embryo stage) which takes about 9 days at 25°C. Although *Arabidopsis* embryos are comparatively small and develop fast, other plants go through similar stages (Johri, 1984). In *Arabidopsis*, the egg cell measures about 20μm in diameter, while the mature embryo nearly fills a seed that is almost 500μm in size. **1.5–1.23** show the stages of embryogenesis (histological sections).

---

[1] For practical purposes, one would like to correlate stages of embryogenesis with stages of flower development. However, the morphological changes of the flower cannot be sufficiently resolved in time to match specific developmental stages of the embryo (Müller, 1961, 1963; Smyth *et al.*, 1990). Müller's stages roughly correlate with those given here in the following way: B4 includes Stages 1–2 (or 3); B5, Stages 3 (or 4)–5; B6, Stage 6; B7.1, Stages 7–10; B7.2, Stages 11–16; B7.3, Stages 17–18; B7.4–B7.5, Stage 19; and B8–B9, Stage 20.

Table 1.1 Developmental stages of embryogenesis.

Stage No.	Designation	Time[a] (%)	Number of Cells[b]	Specific Features
1	Zygote	0	(1)	
2	Elongated zygote	2	(1)	
3	One-cell	5	1 + (1)	One embryonic cell
4	Two-cell	8	2 + (1)	Two embryonic cells
5	Quadrant	11	4 + (1)	Four embryonic cells
6	Octant	14	8 + (1)	Upper and lower tiers of embryonic cells
7	Dermatogen	17	17	Outer epidermal layer; hypophysis
8	Early globular	20	33	Lens-shaped derivative of hypophysis; suspensor complete
9	Mid globular	23	approx. 65	Elongate centre cells (vascular primordium); expansion of lower tier
10	Late globular	29	approx. 110	First vertical division of the two hypophyseal derivatives
11	Triangular	30	approx. 150	Apical surface flat; second vertical division of hypophyseal derivatives
12	Early heart	32	approx. 250	Cotyledonary primordia bulging; two layers of hypophyseal derivatives
13	Mid heart	37	approx. 500	Central root cap initials produce first layer of central root cap cells
14	Late heart	41	approx. 1000	Lateral root cap initials; central root cap cells divide
15	Early torpedo	44	approx. 2000	Cotyledonary primordia nearly parallel to axis; central root cap initials divide
16	Mid torpedo	47	approx. 4000	Straight embryo of maximum extension; central root cap initials produce second layer of central root cap cells
17	Late torpedo	57	approx. 8000	Cotyledons bending away from axis; endodermal layer of root cortex visible
18	Bent cotyledon	69	approx. 12000	Embryo 'boomerang-shaped'; shoot meristem bulging
19	Mature embryo	100	15000–20000	Cotyledonary tips at level of root end; central root cap initials produce third layer of central root cap cells before cessation of cell division
20	Desiccation	(120)		Embryo shrinks and turns pale yellow

[a] % embryogenesis (0% fertilization, 100% mature-embryo stage); approximate values.

[b] The zygote not only gives rise to the embryo but also to the extra-embryonic suspensor, and cells also contributing to the latter are given in brackets; from stage 9 on, cell numbers are estimates.

Stages 1–3: the zygote and the first cell division

Within a few hours of fertilization (Stage 1, **1.5**), the egg cell elongates about three-fold in the apical-basal axis (Stage 2, **1.6**) before undergoing the first division (Mansfield and Briarty, 1991). This directional expansion of the zygote is associated with a re-organization of the microtubular cytoskeleton (Webb and Gunning, 1991). The zygote subsequently divides perpendicular to the axis about a quarter of the way down from the top, producing a smaller apical and a larger basal cell (Stage 3, **1.7**). The apical cell will give rise to all of the embryo except for its very basal end. The basal cell will produce by repeated transverse divisions a file of 7–9 cells of which all but the uppermost one form the extra-embryonic suspensor. While the embryo is still one-celled, the basal cell divides once or sometimes twice (Stage 3).

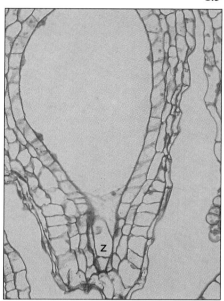

1.5 Zygote (Stage 1). (Key: **z**, zygote; ×300.)

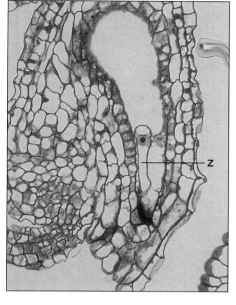

1.6 Fully elongated zygote (Stage 2). (Key: **z**, zygote; ×300.)

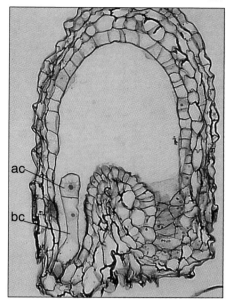

1.7 One-cell stage (Stage 3). (Key: **ac**, apical cell; **bc**, basal cell; ×300.)

Stages 4–6: the two-cell to the octant stage

The apical cell undergoes three rounds of 'pseudo-cleavage' divisions to produce the octant-stage embryo which is only slightly larger than the apical cell (Mansfield and Briarty, 1991). The cell volume shrinks with each of these divisions until a nearly constant nucleo-cytoplasmic ratio is reached. The first two rounds of division occur longitudinally, at right angles to one another, while the third division is transverse, separating two tiers, each of four cells. In Stage 4 (**1.8**), the embryo consists of two cells, with two or three cells being derived from the basal cell. A round of division later, the quadrant of embryonic cells is attached to a file of four cells (Stage 5, **1.9**). The embryonic cell group then expands slightly before transverse divisions generate the octant stage (Stage 6, **1.10**). The basal cell may have produced a file of four or five cells, with the largest cell, which also has the largest vacuole, lying at the micropylar end. The uppermost derivative of the basal cell begins to join the group of embryonic cells.

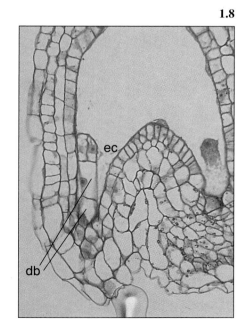

1.8 Two-cell stage (Stage 4). (Key: **ec**, embryonic cells; **db**, derivatives of basal cell; ×300.)

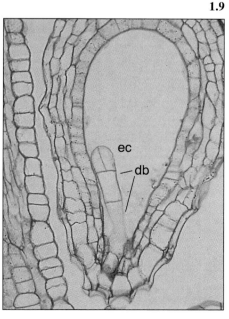

1.9 Quadrant stage (Stage 5). (Key: **ec**, embryonic cells; **db**, derivatives of basal cell; ×300.)

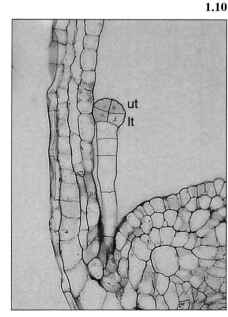

1.10 Octant stage (Stage 6). (Key: **lt**, lower tier; **ut**, upper tier of embryonic cells; ×300.)

Stage 7: dermatogen stage

The octant stage gives way to the dermatogen stage by the next round of cell divisions, which occur tangentially to produce an outer layer of eight epidermal precursor cells (protoderm) and an inner group of eight cells (**1.11**). These divisions are not always synchronous, and the cells on one side sometimes divide before the cells on the other side (Mansfield and Briarty, 1991). The uppermost derivative of the basal cell divides transversely, and the daughter cell adjacent to the embryonic cell mass becomes the hypophysis which will give rise to the basal end of the embryo. The other daughter cell and the remainder of the file of basal cell derivatives become the extra-embryonic suspensor, which at Stage 7 consists of five or six cells. In histological sections, the hypophysis stains more lightly than the other cells of the embryo and more closely resembles the suspensor cells.

Stage 8: early-globular stage

The embryo has enlarged, due to another round of cell division. The epidermal cell layer consists of 16 cells, produced by divisions along the surface. The eight inner cells have divided nearly parallel to the apical-basal axis (**1.12**), and their centrally located daughter cells become the primordium of the vascular tissue (procambium).

By the end of Stage 8, the hypophysis has, by asymmetric transverse division, produced two daughter cells (**1.12**). The smaller upper cell is lens-shaped with its convex apical surface abutting the lower end of the embryonic cell mass, which includes cells of all three tissue primordia. The larger lower cell touches the epidermal cell layer laterally and, at its basal surface, the uppermost suspensor cell. The suspensor is now complete, consisting of 6–8 cells.

Stage 9: mid-globular stage

The mid-globular embryo consists of about 65 cells. The inner cells derived from the lower tier of the octant stage have divided horizontally. The cells of the vascular primordium are of unequal size, with the basal ones being more elongated (**1.13**). The cell group derived from the lower tier of the octant stage has started to expand in the apical-basal axis faster than the upper tier. Although the two groups were of equal height in Stage 8, the lower tier derivatives will account for two-thirds of the embryo length at Stage 11.

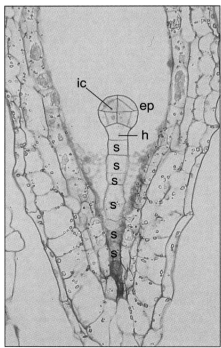

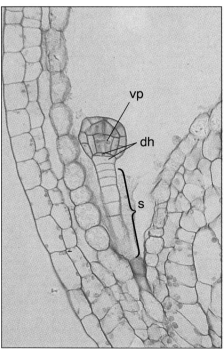

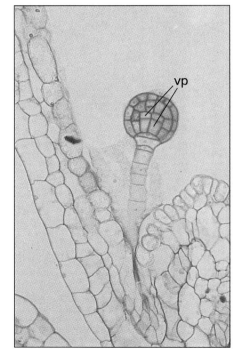

1.11 Dermatogen stage (Stage 7). (Key: **ep**, epidermal primordium (protoderm); **h**, hypophysis; **ic**, inner cells; **s**, suspensor; ×300.)

1.12 Early-globular stage (Stage 8). (Key: **dh**, derivatives of hypophysis; **s**, suspensor; **vp**, vascular primordium; ×300.)

1.13 Mid-globular stage (Stage 9). (Key: **vp**, vascular primordium; ×300.)

Stage 10: late-globular stage

The cell number has increased to about 110. The upper half of the embryo starts to widen, due to predominantly vertical cell divisions, although the apical surface is still curved (**1.14**). The provascular cells are more elongated than the surrounding cells, and the lower tier continues to expand in the apical-basal axis. At the basal end, the daughter cells of the hypophysis have divided vertically.

Stage 11: triangular stage

The transition from the globular to the heart stage is associated with a change from radial to bilateral symmetry and this reflects the cellular changes which lead to the emergence of the cotyledonary primordia. The apical surface of the late-globular embryo initially flattens, and the embryo attains a triangular shape in longitudinal section (**1.15**). This stage is brief as the continuing cell divisions in the upper half of the embryo cause bulging of the incipient cotyledonary primordia, and so form the early-heart stage. The vascular primordium is now four cells wide. The derivatives of the hypophyseal cell undergo another vertical division at right angles to the preceding one, resulting in two layers of four cells.

Stage 12: early-heart stage

By now, the cotyledonary primordia have grown above the level of the shoot pole where vertical divisions occur in the epidermal layer, while horizontal divisions underneath produce two subepidermal layers in the future shoot meristem (**1.16**). Cell divisions expand the ground tissue of the future hypocotyl which is now two layers wide on each side. At the root end, there are two layers of four cells, each derived from the hypophysis.

1.14

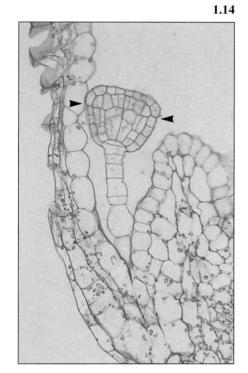

1.15

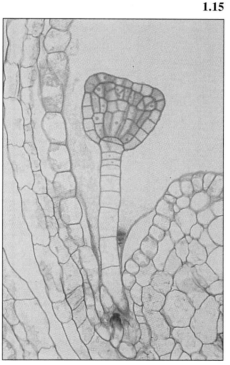

1.16

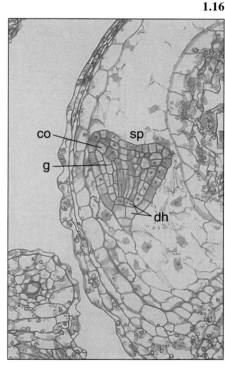

1.14 Late-globular stage (Stage 10): arrowheads mark the boundary between the derivatives of the upper and the lower tier (×300).

1.15 Triangular stage (Stage 11, ×300).

1.16 Early-heart stage (Stage 12). (Key: **co**, emerging cotyledonary primordia; **dh**, derivatives of hypophysis; **g**, ground tissue; **sp**, shoot pole; ×300.)

Stage 13: mid-heart stage
This stage marks the end of the first phase of embryogenesis in that the primordia of the major seedling structures are recognizable (**1.17**). From now on, the cotyledonary primordia grow rapidly by cell division, and the axis of the embryo expands relative to its width. The vascular primordium begins to extend into the bases of the cotyledonary primordia, while, at the root end, the cells in the lower layer of hypophyseal derivatives divide horizontally to produce a third layer.

Stage 14: late-heart stage
The growing cotyledonary primordia are still oriented away from the axis. At the basal end, periclinal cell divisions occur in the outer layer adjacent to the hypophyseal derivatives, producing the (outer) initials of the lateral root cap and the (inner) initials of the root epidermis (**1.18**). The four cells in the bottom layer of hypophyseal derivatives divide once and there are now 16 cells derived from the hypophysis. The embryo turns green at the end of this stage.

Stage 15: early-torpedo stage
The distal parts of elongating cotyledonary primordia come to lie parallel to the axis, and the hypocotyl expands further (**1.19**). The cells of the ground tissue are almost as wide as they are long, although, in contrast, the vascular cells are narrower and more elongated. The vascular primordium, which is six cells wide, extends further into the cotyledonary primordia. At the basal end, the initials of the lateral root cap have given off a layer of daughter cells. The derivatives of the hypophyseal cell have increased to 20, as a result of vertical divisions in the middle layer.

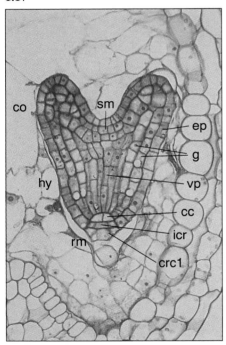
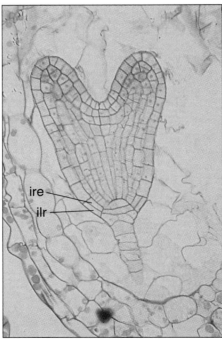
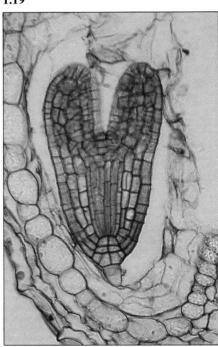

1.17 Mid-heart stage (Stage 13). (Key: **cc**, central cells (of root meristem); **co**, cotyledonary primordia; **crc1**, first layer of central root cap cells; **ep**, epidermis; **g**, ground tissue; **hy**, hypocotyl; **icr**, initials of central root cap; **rm**, incipient root meristem; **sm**, incipient shoot meristem; **vp**, vascular primordium; ×300.)

1.18 Late-heart stage (Stage 14). (Key: **ilr**, initials of lateral root cap; **ire**, initials of root epidermis; ×300.)

1.19 Early-torpedo stage (Stage 15, ×300).

Stage 16: mid-torpedo stage

The embryo is now straight and maximally extended (**1.20**). The growing cotyledons have approached each other at their tips, and the hypocotyl has further expanded in the axis. The shoot meristem consists of three cell layers abutting the vascular primordium. The files of vascular cells have been extended by cell divisions and the ground tissue is now three cells wide. At the basal end, the root cap forms a contiguous layer of surface cells, the central portion of which is derived from the hypophysis. The middle layer of hypophyseal derivatives has divided horizontally, producing another layer of future root cap cells, which increases the number of derivatives to 40. In the embryonic root, a file of four lateral root cap cells lies alongside a file of epidermis cells. The initials of the ground tissue produce the first endodermal cells, and the pericycle cells appear in the vascular primordium.

Stage 17: late-torpedo stage

Following further elongation, the cotyledonary primordia bend away from the axis (**1.21**). At the basal end, the endodermal layer of the root cortex is now clearly visible.

Stage 18: bent-cotyledon stage

The growing cotyledonary primordia turn towards the hypocotyl, giving the embryo a 'boomerang' shape (**1.22**). The precursor cells of the vascular strands appear in the cotyledonary primordia, and the shoot meristem starts to bulge out.

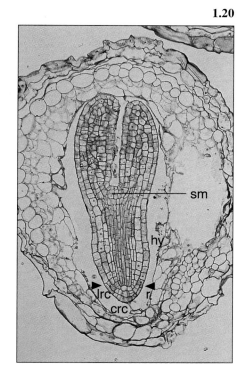
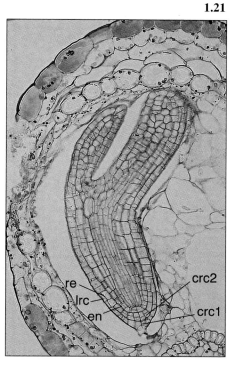
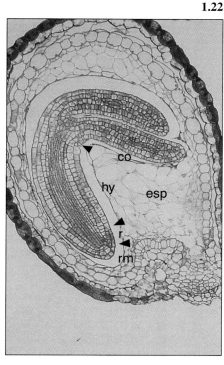

1.20 Mid-torpedo stage (Stage 16). (Key: arrowheads mark boundary between hypocotyl **hy** and embryonic root **r**; **crc**, central root cap; **lrc**, lateral root cap; **sm**, shoot meristem; ×200.)

1.21 Late-torpedo stage (Stage 17). (Key: **crc1** and **crc2**, the first and second layer of central root cap cells; **en**, endodermis; **lrc**, lateral root cap; **re**, root epidermis; ×200.)

1.22 Bent-cotyledon stage (Stage 18). (Key: arrowheads mark boundaries of cotyledonary primordia **co**; **hy**, hypocotyl; **r**, embryonic root, and **rm**, root meristem; **esp**, endosperm; ×150.)

Stage 19: mature-embryo stage
The tips of cotyledonary primordia have reached the level of the root pole, and the mature embryo almost completely fills the ovule, while the endosperm has disappeared (**1.23**). At the basal end, the second layer of hypophyseal derivatives divides once more to produce another layer of root cap cells. There are now five cell layers derived from the hypophysis, with the uppermost layer of four cells having been mitotically inactive. These central cells may therefore correspond to the quiescent centre of the root meristem. When cell division ceases, the mature embryo consists of 15,000–20,000 cells.

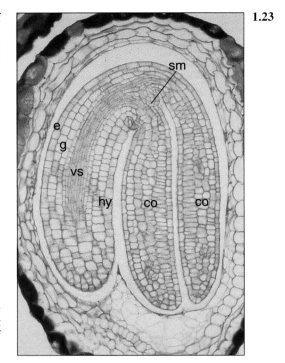

1.23 Mature-embryo stage (Stage 19). (Key: **co**, cotyledons; **e**, epidermis; **g**, ground tissue; **hy**, hypocotyl; **sm**, shoot meristem; **vs**, vascular tissue.) Root end of embryo is out of plane of sectioning; ×200).

Stage 20: desiccation stage
The embryo and the seed enter dormancy, which is characterized by shrinkage due to loss of water and pale yellow pigmentation of the embryo, while the hardening seed coat turns yellowish-brown.

Seed germination to the adult plant

Postembryonic development starts with the germination of the seed from which the embryo emerges as the seedling. The meristems of the shoot and the root become mitotically active early on when the seedling mainly lives on stored substances and before the cotyledons turn green. The root meristem forms the main root of the plant, while, at some distance from the growing root tip, lateral root primordia are initiated from the internal layer of pericycle cells.

The shoot meristem initially produces vegetative parts, mostly leaf primordia. In *Arabidopsis*, the leaves are arranged in a rosette, due to the internodes' lack of elongation. About 2 weeks after germination, the shoot meristem switches to reproductive growth, producing flower primordia. How this transition, called induction of flowering, is brought about is not known, although a dozen genes have been identified by mutant alleles which delay flowering (Koornneef *et al.*, 1991). Induction of flowering not only causes elongation of the stem above the rosette level ('bolting'), but also involves a change in the organization of the shoot meristem itself.

The inflorescence meristem of *Arabidopsis* is indeterminate and produces a succession of flowers along the growing main stem, with the oldest flower being farthest away from the meristem. As a consequence, different stages of embryogenesis are observed in different fruits (siliques) along the stem, while, within any one silique, the embryos develop nearly synchronously. The indeterminacy of the inflorescence meristem is genetically controlled (Alvarez *et al.*, 1992).

Each flower primordium produces a well-defined structure with a limited number of pattern elements arranged in concentric whorls. Four different whorls can be distinguished: the outermost one consists of four leaf-like sepals; the next contains four petals; this is followed by the whorl of six stamens; and the centre is occupied by two fused carpels. The genetic basis of floral patterning has been studied extensively in both *Arabidopsis* and *Antirrhinum,* and common principles have been discerned (reviewed by Coen and Meyerowitz, 1991).

Analysis of embryogenesis

The body organization of the seedling develops during embryogenesis. Due to the lack of cell migration and the nearly invariant pattern of cell division, the structures of the seedling can be traced back to cell groups in the early embryo by histological study of increasingly younger embryos (see *The stages of embryogenesis*, page 9). The apical-basal pattern, including the meristems of the shoot and root, cotyledons, hypocotyl and embryonic root (radicle), arises over the first thirteen stages as the axis of polarity appears to be partitioned into successively smaller regions. The development of the radial pattern perpendicular to the axis seems to be confined to a shorter period, with the three main tissues (epidermis, ground tissue and vascular tissue) formed between the dermatogen and triangular stages (Stages 7–11).

Similarly, the bilateral symmetry of the seedling is presaged by the emerging cotyledonary primordia at the early-heart stage (Stage 12). However, such descriptive studies do not tell us anything of the mechanisms by which the primary body organization is established in plant embryos.

In animal embryogenesis, the generation of the body pattern has been analysed by two complementary approaches: experimental studies and genetic dissection. While the former mainly investigates 'system properties' such as regulative capability, regional autonomy, induction or cell interaction, the latter ultimately aims at the molecular analysis of developmental genes and their interaction through the isolation and characterization of relevant mutants. Essentially similar approaches have been developed for the analysis of plant embryogenesis.

Experimental studies: in vitro *work*

Plant embryos are less accessible to experimental manipulation than animal embryos, mainly because their development occurs within an ovule which is, in turn, enclosed within a fruit. Attempts have been made to overcome this difficulty by *in vitro* techniques, taking advantage of the plasticity of plant development. These techniques, which were not originally devised for *Arabidopsis*, but have been adapted for its use, include *in vitro* culture of excised embryos, somatic embryogenesis and regeneration of plants from callus (disorganized tissue). Although these techniques have great potential, they have not yet contributed much to our understanding of pattern formation in the plant embryo.

Developing embryos removed from their fruit and transferred to small volumes of culture medium can give rise to normal seedlings provided that they are not too immature (Wu *et al.*, 1992). While heart-shaped and older embryos do well, octant embryos rarely produce viable plantlets but tend to form callus. Once the appropriate culture conditions have been worked out, explanted early embryos can be subjected to the sorts of experiments routinely done on animal embryos and it will thus be possible to undertake cell-ablation and fate-mapping studies.

Somatic embryogenesis, where adult tissue is induced to become embryonic, provides another experimental system for studying the conditions of embryogenesis. So far, however, experimental manipulatation has been limited to older carrot embryos and here bisected upper halves will regenerate entire embryos (Schiavone and Racusen, 1991). While somatic embryos of carrot can be derived from single cells or pro-embryogenic masses (reviewed by Van Engelen and De Vries, 1992), only pro-embryogenic masses have been found to produce somatic embryos in *Arabidopsis* (Marton and Browse, 1991; Wu *et al.*, 1992). It should also be mentioned that somatic embryos provide an excellent assay system for identifying and purifying factors that promote normal development (Van Engelen and De Vries, 1992).

Another *in vitro* technique uses appropriate culture conditions to regenerate shoots and roots from callus, disorganized tissue that often forms when plant material is explanted and cultured. In *Arabidopsis*, callus has been derived from various sources that include leaf discs, root explants, protoplasts and cultured zygotic embryos (for references to earlier work, see Sangwan *et al.*, 1991, and Wu *et al.*, 1992). One of the main applications of callus culture has been the generation of transgenic plants which can be used for the molecular isolation or identification of plant genes (see later). In the future, callus culture may gain more importance for investigating the basis of abnormal development in mutant embryos because regeneration produces plants via a non-embryogenic pathway. If, for example, callus from a mutant embryo were to give a normal plant, the mutation would clearly be embryo-specific (provided that the primary defect is not caused by the lack of a diffusible molecule that can be supplemented by the medium). As a side effect, homozygous mutant plants would produce 100% mutant embryos and so provide homozygous material for molecular studies (see later). If, on the other hand, plants were unable to regenerate from mutant callus, then the mutant gene might well be required throughout development or repeatedly during the life cycle. Although these sorts of experimental studies may eventually contribute to our understanding of plant embryogenesis, the combined approach of genetic dissection and molecular analysis looks more promising at present.

Genetic dissection: developmental mutants

Developmental processes are dissected genetically for two different purposes. First, mutant phenotypes resulting from the inactivation of individual genes provide information about the way the system responds to the removal of a single component. Second, in cases where the structure or function of their products is not known, identifying relevant genes by their mutant phenotypes provides the only starting point for their molecular analysis.

In self-fertilizing species like *Arabidopsis*, mutations are predominantly induced by exposing seeds to chemical mutagens like EMS (ethyl methanesulfonate) which mainly causes point mutations. Plants developing from mutagenized seeds are mosaics containing genetically distinct cell clones. Clones derived from subepidermal cells of the shoot meristem give rise to the generative cells of flowers, and, on average, every other flower then produces offspring; about a quarter of these are homozygous mutants which can be recognized by their phenotype. If the mutant is lethal or sterile, the mutant strain is established from the normal-looking siblings from the same silique because about two-thirds of them are heterozygous and will again produce mutant offspring by selfing.

Newly-isolated mutants are genetically characterized by mapping and by complementation testing. Mapping against known RFLP markers is a prerequisite for molecular cloning of the gene by chromosomal walking (see later). Complementation tests are used to determine whether two or more mutants with similar phenotypes represent alleles of the same gene. If this were the case, the mutant phenotype would be likely to result from inactivation of the gene. If the mutants are lethal or sterile, one has to make sure that the plants used in the test crosses are heterozygous. This is done by analysing selfed progeny of the plants involved in the cross along with their progeny derived from the cross.

The largest class of mutants made by EMS, by X-irradiation and by T-DNA insertion are embryonic lethals and a wide range of mutant phenotypes has been observed (for review, see Meinke, 1991). Embryonic lethals are, however, a heterogeneous class of mutants, and it is difficult to distinguish developmental from 'house-keeping' genes by phenotypic criteria. By contrast, embryonic pattern mutants that do not die during embryogenesis but alter the body organization of the seedling are likely to play important roles in embryogenesis (for discussion of this problem, see Jürgens *et al.*, 1991).

Pattern mutants: a case study

Of about 250 putative pattern mutants that have been isolated from EMS-mutagenized seeds, 73 representing nine complementation groups have been studied in some detail. By phenotypic criteria, mutations in these genes affect three different aspects of the seedling organization: apical-basal pattern, radial pattern and shape (Mayer *et al.*, 1991). The analysis of developing mutant embryos demonstrated that embryogenesis can become abnormal at a very early stage.

Four of the genes so far identified appear to be involved in the partitioning of the apical-basal axis (*gurke*, *fackel*, *monopteros* and *gnom*) and their mutant phenotypes are characterized by deletions of pattern elements at the seedling stage (**1.24–1.33**), which correspond by position with defects in the early embryo (**1.34–1.43**). In formal terms, the pattern deletions define three regions, apical, central and basal, which may correlate with the subdivisions of the octant-stage embryo (**1.44**). As mutations in different pairs of genes affect different regions, it seems that the genes regulate pattern in either a combinatorial or hierarchical way. Indeed, one mutant phenotype indicates that the normal gene may be active at the one-cell stage, or even before the asymmetric division of the zygote. In summary, the apical-basal pattern deletion mutants support the idea that the apical-basal axis is partitioned into three regions around the octant stage. So far, no mutants have been described that reveal any later subdivisions of the axis.

Radial pattern defects have been produced by mutations in two genes (*knolle* and *keule*) which both affect the epidermis (**1.24–1.33**). These defects have been traced back to the early-globular stage: in the case of *keule*, the primordium of the epidermis is bloated, while, for *knolle*, no distinction can be made between the cells of the epidermal primordium and the inner cells (**1.34–1.43**). These two genes seem to be involved in the formation and/or the normal development of the epidermal primordium.

Shape mutants do not affect pattern formation but change the overall shape of the seedling. While mutations in three genes (*fass*, *knopf* and *mickey*) are known to produce this effect, mutations in the *fass* gene specifically affect the shape of cells. The primordia of such seedling structures cannot therefore be recognized by their characteristic cell shape or arrangement of cells (**1.24–1.33**, **1.34–1.43**). Although the mutant cells do not undergo position-specific cell changes during embryogenesis, the mutant embryos do acquire bilateral symmetry and differentiate to give the elements of the seedling pattern. This suggests that cell shape change normally occurs in response to positional cues rather than playing a major role in pattern formation.

The pattern mutants analysed so far support the notion that a specific group of genes is involved in developmental decisions during plant embryogenesis. In retrospect, this may not seem surprising, given the precedent in *Drosophila* (see chapter 7). In order to learn more about the developmental roles of these genes, it will be necessary to analyse them at the molecular level.

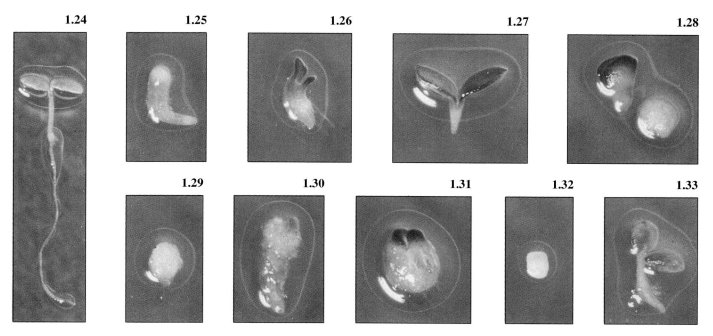

1.24–1.33 Seedling phenotypes: **1.24**, wild-type; **1.25–1.28**, apical-basal pattern deletion mutants: **1.25**, *gurke*; **1.26**, *fackel*; **1.27**, *monopteros*; **1.28**, *gnom*; **1.29, 1.30**, radial pattern defect mutants: **1.29**, *knolle*; **1.30**, *keule*; **1.31–1.33**, shape mutants: **1.31**, *fass*; **1.32**, *knopf*; **1.33**, *mickey*. (Reprinted by permission of *Nature*, **353**, 402–407. Copyright 1991 Macmillan Magazines Ltd.)

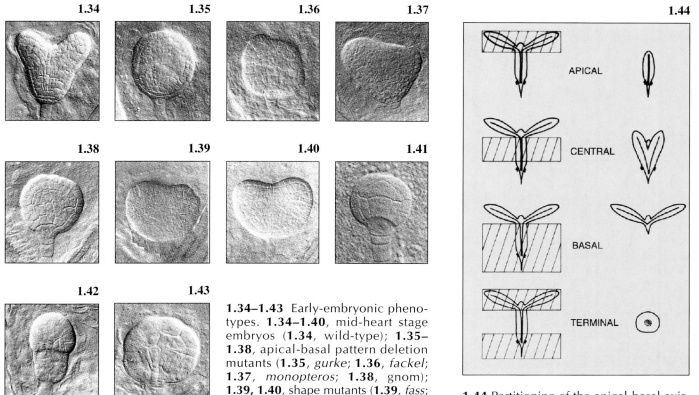

1.34–1.43 Early-embryonic phenotypes. **1.34–1.40**, mid-heart stage embryos (**1.34**, wild-type); **1.35–1.38**, apical-basal pattern deletion mutants (**1.35**, *gurke*; **1.36**, *fackel*; **1.37**, *monopteros*; **1.38**, *gnom*); **1.39, 1.40**, shape mutants (**1.39**, *fass*; **1.40**, *knopf*); **1.41–1.43**, early-globular embryos (**1.41**, wild-type); **1.42, 1.43**, radial pattern defect mutants (**1.42**, *knolle*; **1.43**, *keule*). (Reprinted by permission of *Nature*, **353**, 402–407. Copyright 1991 Macmillan Magazines Ltd.)

1.44 Partitioning of the apical-basal axis. Pattern deletions caused by mutations in individual genes are schematically represented as shaded parts of the wild-type seedling pattern on the left and as mutant patterns on the right. (Reprinted by permission of *Nature*, **353**, 402–407. Copyright 1991 Macmillan Magazines Ltd.)

Molecular basis of embryogenesis

Ultimately, we would like to know the function of developmental genes in embryogenesis and how their expression is regulated in space and time. In *Arabidopsis*, there are two options for cloning genes solely defined by mutant phenotype and map position. One approach called 'gene tagging' or insertion mutagenesis is widely spread for analysing flowering plants: when insertion of T-DNA from *Agrobacterium* or a transposable element disrupts a gene function resulting in a mutant phenotype, the foreign DNA can be used as a molecular probe to isolate flanking plant DNA that corresponds to the gene. The T-DNA tagging approach has been quite successful in *Arabidopsis* (for review, see Feldmann, 1991). T-DNA is also used as a vector for generating transgenic plants in order to identify genes molecularly: relevant segments of wild-type DNA are recognized by their ability to suppress the phenotype of a mutant allele (see, for example, Yanofsky *et al.*, 1990).

The other gene cloning approach which involves RFLP mapping and chromosomal walking was specifically proposed for *Arabidopsis* because of its special genome organization among experimental plants (Meyerowitz, 1989). The genome size has been estimated to be about 100,000 kb (Hauge *et al.*, 1991); by comparison, the *Caenorhabditis* genome is similar in size, while that of *Drosophila* is nearly twice as large. More importantly, the *Arabidopsis* genome appears to be largely devoid of interspersed repetitive sequences. At present, a number of starting points for chromosome walks are available: these are based on collections of RFLP markers (Chang *et al.*, 1988; Nam *et al.*, 1989), mapped YAC clones corresponding to about one third of the genome (Hwang *et al.*, 1991), and 'contigs' of ordered cosmid clones covering about 90% of the genome (Hauge *et al.*, 1991). In consequence, we may soon be able to clone genes simply by localizing mutant alleles on the complete physical map.

So far, the T-DNA tagging approach has been more successful than the alternative strategy of RFLP mapping and chromosomal walking. There are, for instance, several embryonic-lethal mutants whose phenotype co-segregates with the T-DNA inserts (Errampalli *et al.*, 1991), and the first floral patterning gene to be cloned had been tagged by T-DNA insertion (reviewed by Coen and Meyerowitz, 1991). Nonetheless, if a specific gene is to be cloned, RFLP mapping and chromosomal walking may be more reliable because T-DNA insertion is one or two orders of magnitude less frequent than EMS mutagenesis and tends to be non-random. On the other hand, while EMS-induced mutant alleles of any gene can readily be mapped against RFLP markers, chromosomal walking from the nearest RFLP to the gene and its molecular identification are rather laborious.

Once a developmental gene has been cloned by one or another strategy, a number of obvious experiments can be done. Chief among them is the study of gene expression patterns, which involves *in situ* hybridization to tissue sections of both wild-type and mutant individuals. This level of analysis has only been reached with floral patterning, and the results have led to insights into mechanisms of flower development (Coen and Meyerowitz, 1991). The analysis of pattern formation in the embryo is heading in the same direction, but the genes involved are only now being cloned.

The future

Arabidopsis has become the organism of choice for the genetic analysis of plant-specific processes. Considering the ease with which mutants can be isolated on a large scale, it is conceivable that many problems in plant developmental biology will soon be dissected by mutation analysis. With the recent progress in establishing the physical map of the genome, molecular cloning will soon become fairly straightforward and the analysis of developmental genes can be taken to the molecular level. The main emphasis is then likely to shift in two directions: the isolation of ever more 'interesting' mutants and the analysis of cloned genes which will include studies of expression patterns and DNA-protein interactions. In the longer term, attention will shift to the molecular basis of plant cell behaviour and the functional role of gene products that do not directly affect gene regulation. Cell behaviour will thus become accessible to analysis and this may well be an area where there are considerable differences between plants and animals.

References

Alvarez, J., Guli, C.L., Yu, X.-H. and Smyth, D.R. (1992). Terminal flower: a gene affecting inflorescence development in *Arabidopsis thaliana*. *Plant J.*, **2**, 103–116.

Chang, C., Bowman, J.L., DeJohn, A.W., Lander, E.S. and Meyerowitz, E.M. (1988). Restriction fragment length polymorphism linkage map for *Arabidopsis thaliana*. *Proc. Natl. Acad. Sci.*, **85**, 6856–6860.

Coen, E.S. and Meyerowitz, E.M. (1991). The war of whorls: genetic interactions controlling flower development. *Nature*, **353**, 31–37.

Errampalli, D., Patton, D., Castle, L., Mickelson, L., Hansen, K., Schnall, J., Feldmann K. and Meinke, D. (1991). Embryonic lethals and T-DNA insertional mutagenesis in *Arabidopsis*. *Plant Cell*, **3**, 149–157.

Feldmann, K.A. (1991). T-DNA insertion mutagenesis in *Arabidopsis*: mutational spectrum. *Plant J.*, **1**, 71–82.

Hauge, B.M., Hanley, S., Giraudat, J. and Goodman, H.M. (1991). Mapping the *Arabidopsis* genome. In *Molecular Biology of Plant Development., Symp. Soc. Exp. Biol.*, **45**, 45–56.

Hwang, I., Kohchi, T., Hauge, B.M., Goodman, H.M., Schmidt, R., Cnops, G., Dean, C., Gibson, S., Iba, K., Arondel, V., Danhoff, L. and Somerville, C. (1991). Identification and map position of YAC clones comprising one third of the *Arabidopsis* genome. *Plant J.*, **1**, 367–374.

Johri, B.M. (1984). *Embryology of Angiosperms*. Springer, Berlin.

Jürgens, G., Mayer, U., Torrs Ruiz, R.A. Berleth, T. and Misra, S. (1991). Genetic analysis of pattern formation in the *Arabidopsis* embryo. *Development Suppl.*, **1**, 27–38.

Koncz, C., Chua, N.-H. and Schell, J. (eds) (1992). *Methods in Arabidopsis Research*. World Scientific, Singapore.

Koornneef, M. (1990). Linkage map of *Arabidopsis thaliana*. In *Genetic Maps*, 5th Edition (S.J. O'Brien, ed.). Cold Spring Harbor Laboratory Press, pp. 6.94–6.97.

Koornneef, M., Hanhart, C.J. and van der Veen, J.H. (1991). A genetic and physiological analysis of late flowering mutants in *Arabidopsis thaliana*. *Molec. Gen. Genet.*, **229**, 57–66.

Lyndon, R.F. (1990). *Plant Development*. Unwin Hyman, London.

Mansfield, S.G. and Briarty, L.G. (1991). Early embryogenesis in *Arabidopsis thaliana*. II. The developing embryo. *Can. J. Bot.*, **69**, 461–476.

Marton, L. and Browse, J. (1991). Facile transformation of *Arabidopsis*. *Plant Cell Rep.*, **10**, 235–239.

Mayer, U., Torres Ruiz, R.A., Berleth, T., Misra, S., and Jürgens, G. (1991). Mutations affecting body organisation in the *Arabidopsis* embryo. *Nature*, **353**, 402–407.

Meinke, D.W. (1991). Perspectives on genetic analysis of plant embryogenesis. *Plant Cell*, **3**, 857–866.

Meyerowitz, E.M. (1987). *Arabidopsis thaliana*. *Ann. Rev. Genet.*, **21**, 93–111.

Meyerowitz, E.M. (1989), *Arabidopsis* - a useful weed. *Cell*, **56**, 263–269.

Müller, A. (1961). Zur Charakterisierung der Bluten und Infloreszenzen von *Arabidopsis thaliana* (L.) Heynh. *Kulturpflanze*, **9**, 364–393.

Müller, A.J. (1963). Embryonentest zum Nachweis rezessiver Letalfaktoren bei *Arabidopsis thaliana*. *Biol. Zentralbl.*, **82**, 133–163.

Nam, H.-G., Giraudat, J., den Boer, B., Moonan, F., Loos, W.D.B., Hauge, B.M. and Goodman, H.M. (1989). Restriction fragment length polymorphism linkage map of *Arabidopsis thaliana*. *Plant Cell*, **1**, 699–705.

Redei, G.P. (1970). *Arabidopsis thaliana* (L.) Heynh. A review of the genetics and biology. *Bibl. Genet.*, **20**, 1–151.

Sangwan, R.S., Bourgeois, Y. and Sangwan-Norreel, B.S. (1991). Genetic transformation of *Arabidopsis thaliana* zygotic embryos and identification of critical parameters influencing transformation efficiency. *Molec. Gen. Genet.*, **230**, 475–485.

Schiavone, F.M. and Racusen, R.H. (1991). Regeneration of the root pole in surgically transected carrot embryos occurs by position-dependent, proximo-distal replacement of missing tissues. *Development*, **113**, 1305–1313.

Smyth, D.R., Bowman, J.L., and Meyerowitz, E.M. (1990). Early flower development in *Arabidopsis*. *Plant Cell*, **2**, 755–767.

Van Engelen, F.A. and De Vries, S.C. (1992). Extracellular proteins in plant embryogenesis. *Trends Genet.*, **8**, 66–70.

Webb, M.C. and Gunning, B.E.S. (1991). The microtubular cytoskeleton during development of the zygote, proembryo, and free-nuclear endosperm in *Arabidopsis thaliana* (L.) Heynh. *Planta*, **184**, 187–195.

Wu, Y., Haberland, G., Zhou, C. and Koop, H.U. (1992). Somatic embryogenesis, formation of morphogenetic callus and normal development in zygotic embryos of *Arabidopsis thaliana* cultured *in vitro*. *Protoplasma*, **169**, 89–96.

Yanofsky, M.F., Ma, H., Bowman, J.L., Drews, G.N., Feldman, K.A. and Mayerowitz, E.M. (1990). The protein encoded by the *Arabidopsis* homeotic gene agamous resembles transcription factors. *Nature*, **346**, 35–39.

2. *Dictyostelium discoideum*

Robert R. Kay and *Robert H. Insall*

Introduction

The cellular slime mould, *Dictyostelium discoideum*, is a simple eukaryotic micro-organism that lives in the soil. Its amoebae grow and divide as separate individuals as long as their bacterial food is available. When starved, however, they aggregate in their thousands to construct a slug, which goes on to form a fruiting body consisting of a cellular stalk supporting a mass of spores. In *Dictyostelium* development, one sees a multicellular organism constructed not from an egg, but from separate cells. About a third of the amoebae become stalk cells and eventually die, whereas the remainder become spore cells and remain viable. In suitable conditions, the spores hatch out to yield amoebae again (Bonner, 1967; Loomis, 1982).

By the migrating-slug stage of development, the individual amoebae are thoroughly integrated into a unitary organism. The slug is sausage-shaped and secretes an extracellular matrix, the slime sheath, as it moves slowly over the substratum, towards light and along temperature gradients. A slug can be regarded as an embryo, though a motile one. It has an anterior/posterior pattern of prestalk and prespore cells and a dominant, anterior tip which acts as an embryonic organizer. A tip, grafted onto the flank of a host slug, captures some of the cells and leads them off to form a new individual (Raper, 1940); the tip also seems to be the source of an inhibitory gradient that prevents other tips from forming close by (Durston, 1976).

Since development starts from a disorganized mass of amoebae, there can be no input of maternal information to the embryo. Instead, *Dictyostelium* relies on signalling between the cells for self-organization. In consequence, development is remarkably regulative: fruiting bodies can be constructed from between 10 and 100,000 amoebae or from fragments cut from a slug or aggregate.

Dictyostelium as an experimental organism

Dictyostelium development is relatively simple and open to essentially all the experimental techniques that are used in developmental biology (for basic methods, see Spudich, 1987). These range from classic grafting experiments, through cell culture and mutant isolation, to standard biochemical procedures and molecular genetics. *Dictyostelium* cells can be grown on nutrient agar plates, in association with food bacteria with a doubling time of 3hr, or in shaken suspension in rich medium with a doubling time of 8hr (Watts and Ashworth, 1970), or even in defined medium (Franke and Kessin, 1977). It is thus fairly easy to accumulate a sufficient bulk of cells to carry out biochemical work.

Development is initiated by starvation: typically the cells are separated from their food source by differential centrifugation and then plated on non-nutrient agar or filters. Development under these conditions takes about 24hr at 22°C and is reasonably synchronous. Alternatively, cells can be developed in shaken suspension or submerged in buffered salts in tissue-culture dishes. Submerged cells can be induced to differentiate all the way to stalk or spore cells (**2.1–2.3**). This technique is not only convenient for studying their responses to such signal molecules as cAMP (cyclic-AMP), but also allows great control of the cellular environment and has been crucial in the discovery of potential morphogens and factors controlling cell differentiation (Kay, 1987).

Developmental mutants are easy to isolate and maintain. The organism is haploid and mutants can often be recognized by their plaque morphology on bacterial lawns (**2.4**). As development is facultative, mutant strains that are blocked in development can still be maintained as growing amoebae. More recently, molecular genetics has made great progress: *Dictyostelium* can be transformed with vectors that either integrate or remain extra-chromosomal (Leiting and Noegel, 1988; Nellen et al., 1984, 1987) and the expression of genes can be targeted to particular cell types by putting them under the control of a suitable promoter (Haberstroh and Firtel, 1990; Williams et al., 1989). Homologous recombination between inserted and chromosomal DNA is a frequent event (De Lozanne and Spudich, 1987; Manstein et al., 1989) and allows a chromosomal gene to be replaced by a modified or inactive homologue carried in the transforming vector. In this way, null mutants have been created starting only from a cloned gene.

Until recently there has been no routine way of going from mutation to gene in *Dictyostelium*. REMI (Restriction Enzyme Mediated Integration) offers a way out of this situation (Kuspa and Loomis, 1992). Cells are transformed with a plasmid, that has been cut with a restriction enzyme, in the presence of the same restriction enzyme. The enzyme apparently enters the cell and attacks the genomic DNA, creating sites at which the plasmid can integrate. In this way insertional mutants are generated. Once an interesting mutant has been identified, the sequences flanking the insertion can readily be recovered by cutting out the plasmid and transforming into *E. coli*. As a final bonus, the rescued plasmid is now a gene disruption construct:

it can be transformed back into the wild-type and should recreate the original mutation by homologous recombination, thus confirming that the correct gene has been cloned. There are also prospects of cloning genes by complementation of mutants with genomic clone banks, so permitting one to go rapidly from classic mutant to cloned gene (Dynes and Firtel, 1989).

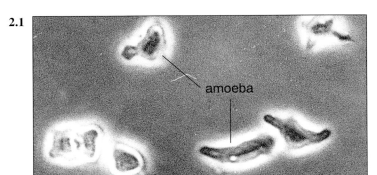

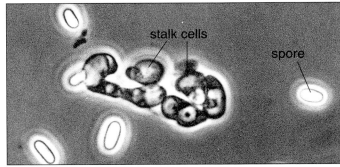

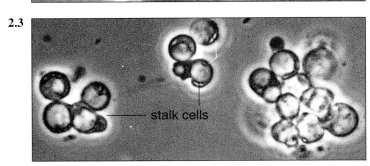

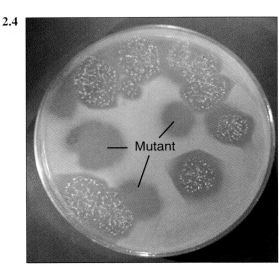

2.1–2.3 Differentiation of amoebae into stalk and spore cells in submerged culture. The phase contrast micrographs show: **2.1**, amoebae, shortly after plating; **2.2**, vacuolated stalk cells and refractile spores formed after 30hr incubation in the presence of 2mM cAMP; **2.3**, pure population of stalk cells formed after 30hr incubation with 2mM cAMP and 15 nM DIF-1. Cells of the sporogenous mutant strain HM29 were incubated in tissue culture dishes at $10^5/cm^2$ in a medium consisting of 10mM MES buffer, 20mM NaCl, 20mM KCl, 1mM $MgCl_2$, 1mM $CaCl_2$ (pH 6.2) plus antibiotics, cAMP and DIF-1, as indicated. (×500; reproduced from Kay (1987), with permission.)

2.4 Plaques of wild-type and mutant cells growing on a lawn of *Klebsiella aerogenes*. The plaques expand by growth of the zones of vegetative cells at their edges. Cells behind the growth zone starve and therefore initiate development. In the wild-type, fruiting bodies form in the centres of the plaques but the mutant (strain HM155) arrests after aggregation so that smooth plaques result. The wild-type-developing strain used (X22) carries a white spore colour marker (spore heads are normally yellow) which is a useful genetic marker. The petri dish is 9cm in diameter.

Normal development

The morphological stages of development are shown diagrammatically in **2.5**, together with the pathways of cell differentiation. After a delay of a few hours, starving amoebae start moving towards signalling centres in response to propagated waves of extracellular cAMP (Tomchik and Devreotes, 1981). Waves are initiated every 4–7min from the centre and propagate out through the field of amoebae at 200–400µm/min (Alcantara and Monk, 1974) with the amoebae moving in towards the centre at about a twentieth of the wave speed. cAMP is detected by a surface receptor having seven transmembrane domains and which is coupled to G-proteins (Klein *et al.*, 1988). Cells respond to cAMP by a transient secretion of cAMP that relays the signal, by switching on various genes and by chemotaxis (Devreotes and Zigmond, 1988; Gerisch, 1987). The rising side of the cAMP wave causes cells to move up the cAMP gradient, towards the signal source; the falling side causes them to stop. In this way, cells are not tempted to pursue the cAMP waves outwards through the aggregation field, but instead move towards the centre in a series of discrete steps.

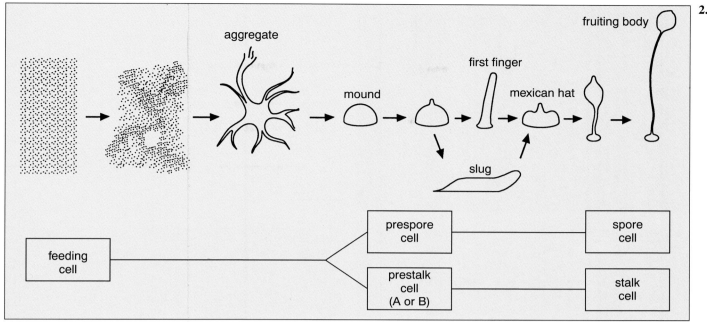

2.5 Outline of development. The top part shows the morphological stages, the bottom the pathways of differentiation followed by individual cells. The first three stages are shown from above, the remainder from the side. Development is initiated by starvation and plating the cells on a suitable surface. In a standard laboratory strain, such as Ax2, fruiting bodies are formed by 24hr. Aggregation takes about 8hr and produces a mound of cells, which forms an apical tip at 12hr and then rapidly elongates upwards to form the first finger. This can fall on its side to form the migrating slug, or fruit on the spot. Amoebae seem to follow a common pathway of differentiation through aggregation but then differentiate into either prestalk or prespore cells in the mound.

Moving and stationary bands of amoebae can be distinguished by dark-field illumination (**2.6**; the moving cells are in the bright, narrow bands). cAMP signals propagate either as concentric waves emanating from a point source, presumably a single cell, or as spiral waves emanating from a loop of cells (both types are shown in **2.6**). Chemotaxis can readily be demonstrated by putting a drop of cAMP next to a group of aggregation-competent cells on agar (**2.7**). After 2hr, cells near the cAMP drop have moved towards it, whereas more remote cells have continued to aggregate and are gathered into small clumps. Aggregating cells soon elongate further and gather into streams (**2.8**), where they are held together in files (**2.9**) by the newly expressed contact-site A glycoprotein (Beug *et al.*, 1973; Noegel *et al.*, 1986).

Aggregation is complete when the cells have gathered together into mounds (**2.10**). Very large aggregates become partitioned into separate entities of up to 10^5 cells, each of which will form a single tip and, generally, a single fruiting body. Little is known of the partitioning mechanism, though it is believed to involve a diffusible inhibitor of tip formation (Kopachik, 1982). Prestalk and prespore cells first differentiate in the mound, before the tip appears, and it is at this stage that the other major signal molecule, a chlorinated alkyl phenone called DIF-1 (Morris *et al.*, 1987) (**2.29**), comes into play.

2.6 Early aggregation fields, viewed at low power with dark field illumination. The light bands represent moving cells, the dark bands stationary cells. cAMP waves are emitted from either a pacemaker loop of cells, which generates spiral waves of regular period, or from a single cell, which produces spasmodic, concentric waves. Where waves meet they annihilate each other, thus defining aggregation territories. Spiral territories usually expand at the expense of neighbouring concentrics, because of their higher signal frequencies. Territories can be up to about 5cm in diameter. Streams are just starting to form in some of the spiral territories (×0.7). (Reproduced with permission from Dr P.C. Newell in *Hormone Receptors and Cellular Interactions in Plants*, (C.M. Chadwick and D.R. Garrod eds), Cambridge University Press, 1986.)

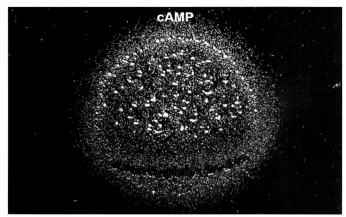

2.7 Demonstration of chemotaxis to cAMP. A 10μl drop of cells was dried into an agar substratum and allowed to become aggregation competent overnight at 7°C. A 3μl drop of 1mM cAMP was then placed next to it. Two hours later, the most adjacent cells had moved towards the cAMP downwards in the figure). Cells in other parts of the drop started to aggregate in this time, making the small bright clusters of cells also visible (×9). Viewed with side illumination.

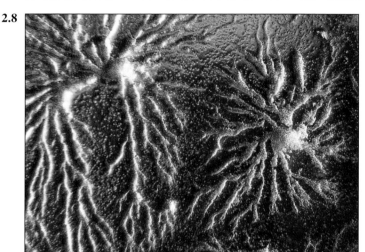
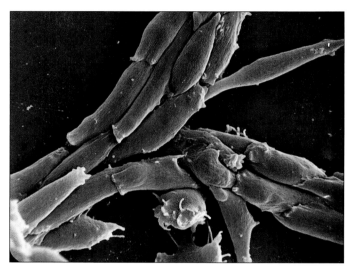

2.8, 2.9 Aggregation streams. In **2.8** (×13) aggregation centres and streams are seen from above; **2.9** (×2160) shows a scanning electron micrograph of elongated cells in an aggregation stream. After perhaps 20 cAMP waves have passed through an aggregation field, the amoebae gather into streams and in this way proceed to the aggregation centre. Streams form because of the relay mechanism for cAMP signals: all cells propagate the cAMP signal and so local inhomogeneities in cell density will produce locally higher cAMP concentrations. These will preferentially attract cells in a self-reinforcing process until streams have formed. Cells in streams are held together by specific contact sites which were first defined by blocking antibodies. Side-by-side cohesion is mediated by contact sites B (**2.9**), which are already present in growing cells and end-to-end cohesion is mediated by contact sites A (**2.8**), which appear during aggregation. (Scanning electron micrograph by Dr G. Gerisch (Martinsreid) and Dr R. Guggenheim (Basel).)

Once the tip has formed, it rapidly elongates upwards, transforming the aggregate into a standing cylinder known as the first finger (**2.11–2.13**). The first finger generally topples onto its side and moves off as a migrating slug which often starts life as long and thin, but usually becomes more compact in shape (**2.14, 2.15**). Migration is for an indeterminate length of time: in some conditions (overhead light, low ammonia concentration) it can be aborted altogether and the first finger is induced to fruit on the spot; in other conditions (unidirectional light, high ammonia concentration), slug migration can be prolonged for days (Newell *et al.*, 1969; Schindler and Sussman, 1977). When culmination is induced, the slug tip stops moving and the rear piles under it so that a mound is again formed. The stalk starts forming in the upper (prestalk) region of the mound and pushes its way down to the substratum, puckering the mound as it does so, to produce the mexican hat stage (**2.16**). Meanwhile, a separate group of prestalk cells, which have sorted to the rear of the prespore zone, form a basal disc of stalk cells, stabilizing the rising fruiting body. The stalk elongates by the addition of cells to the top and carries the prespore cells aloft (**2.17**). Spore cells encapsulate as the stalk is rising and finally a mature fruit is formed which can be up to 5mm high (**2.18**).

2.10–2.13 Development from mound to first finger. **2.10** shows two mounds of cells, just before the tip forms; **2.11**, tipped mound; **2.12**, extending first finger; and **2.13**, a fully extended first finger (All ×70). Prestalk cells first appear scattered in the mound, approximately an hour before the tip forms. By the time the first finger is fully extended (about 1hr after tip formation), the prestalk cells have sorted to form a coherent prestalk zone at the top. This pattern is maintained when the first finger falls on its side to form the migrating slug. The Prestalk/prespore pattern can be visualized with vital stains or specific markers, as shown in **2.19** and **2.22–2.24**. All photographs are of strain NC4 developing on buffered agar and are taken from the side.

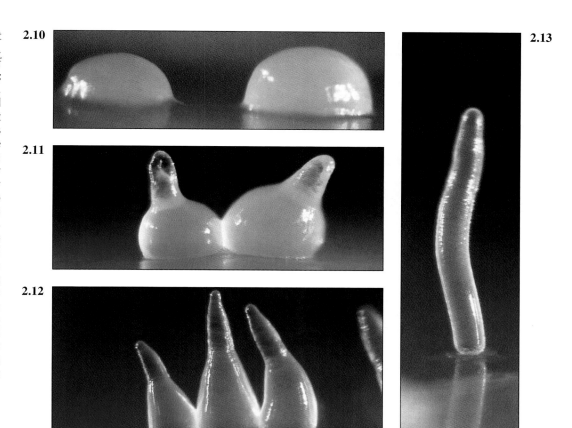

2.14, 2.15 Slugs migrating on an agar substratum. **2.14** shows a young slug (×77) and **2.15**, a mature slug (×38.5). Slugs often start life several millimetres long but then compact to no more than 1–2mm. As they migrate, slugs leave behind a trail – a collapsed tube of slime containing the occasional cell. Phototaxis and thermotaxis of the slug are guided by the tip at the front (to the right in these photographs) and the tip also specifies a developmental axis: a tip grafted onto the flank of a recipient slug leads off part of the tissue to form a separate individual. All photographs are of strain NC4 developing on unbuffered agar and are taken from the side with incident light.

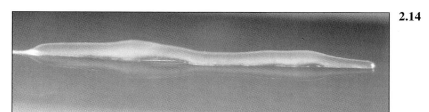

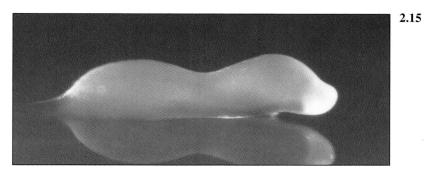

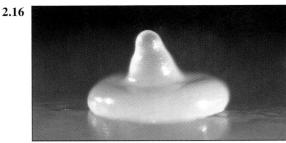

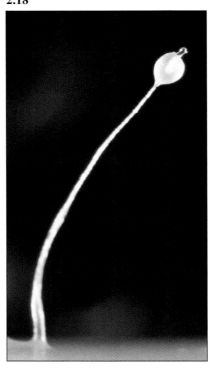

2.16–2.18 Culmination. **2.16** (×64) shows mexican hat; **2.17** (×56), early culminate; and **2.18** (×40), mature fruit. When culmination is induced, a slug stops moving and the prespore tissue piles up under the tip to form a mound. The forming stalk pushes down through the prespore tissue, puckering it to form the mexican hat. The stalk elongates by the addition of cells at the top in a reverse fountain. Spores encapsulate when already aloft. The final fruiting bodies can be several millimetres high. All photographs are of strain NC4, taken from the side with incident illumination.

Markers and cell patterning

Prestalk and prespore cells were first distinguished using vital dyes such as neutral red (Bonner, 1952). These dyes accumulate in acidic compartments, particularly autophagic vacuoles, which are prominent in prestalk cells. This staining is rather variable, but can produce beautiful red-nosed slugs (**2.19**). Prespore cells are also distinguished by unique organelles, the prespore vesicles, which contain precursors of the spore coat and can be detected by electron microscopy (**2.20, 2.21**) (Hohl and Hamamoto, 1969) or by antibody staining (Hayashi and Takeuchi, 1976).

More recently, a number of genes have been cloned whose expression is cell-type specific (Barklis and Lodish, 1983; Jermyn *et al.*, 1987; Mehdy *et al.*, 1983; Williams *et al.*, 1987). Several of the prespore marker mRNAs encode abundant structural proteins of the spore coat, while the definitive prestalk marker mRNAs encode proteins of the extracellular matrix (the ecmA and ecmB proteins). The promoters of various cell-type-specific genes have been hooked up to the lacZ gene and transformed back into cells. Cells expressing lacZ produce β-galactosidase, which produces a blue colour after histochemical staining, thus providing very convenient and sensitive markers for prestalk and prespore cells. This has led to the discovery of sub-types of prestalk cell: prestalk A cells express the ecmA gene and form a cortex in the front of the slug, while prestalk B cells express both the ecmA and ecmB genes and form a central core in the front of the slug (**2.27, 2.28**; Jermyn *et al.*, 1989). Both cell types are also found scattered in the prespore zone, where they were first recognized by neutral red staining and termed anterior-like cells (Sternfeld and David, 1981).

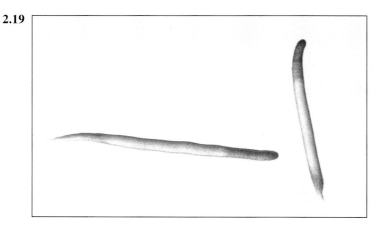

2.19 Slugs stained with neutral red (×23). Neutral red accumulates in acid vesicles, which are prominent in prestalk cells. The anterior of the slugs stains strongest and there are also scattered stained cells in the prespore zone. Sometimes the rear of the slug (rearguard zone) also stains. Taken from above with transmitted light.

2.20, 2.21 Electron micrograph of the junction between the prestalk and prespore zones of a slug. **2.20** shows low power (bar is 5μm; ×2,200); **2.21**, higher power (bar is 2μm; ×6,500). Prespore cells (upper part of both figures) are generally smaller and more electron dense than prestalk cells (lower) and contain a characteristic organelle, the prespore vesicle (**psv**). This contains preformed components of the spore coat. The demarcation between prestalk and prespore cells is quite sharp (dotted line), though prestalk cells are also scattered in the prespore zone. (Reproduced from Kay, *Proc. Natl. Acad. Sci. USA* (1982), **79**, 3228–3231.)

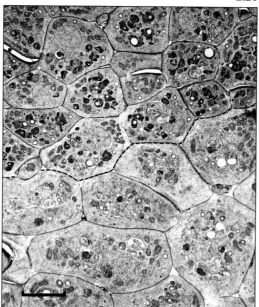

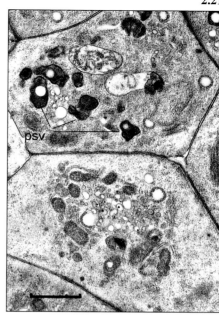

When prestalk A cells first appear in the mound, they are scattered and do not form a coherent mass (**2.22**) (Williams *et al.*, 1989). This surprising result suggests that prestalk cell differentiation is initially induced by a local interaction amongst the cells, rather than by a positional mechanism such as a global morphogen gradient. The scattered prestalk cells in the mound sort to form a coherent prestalk region as the first finger is forming (**2.23, 2.24**). Prestalk A and B cells appear to represent a progression and behave differently during culmination (Jermyn and Williams, 1991; Sternfeld and David, 1982). The original prestalk B core of the slug acts as a rudiment for the stalk of the fruiting body. The stalk then grows by the addition of prestalk A cells to the top. Once the prestalk A cells enter the stalk tube, they rapidly express ecmB, becoming prestalk B cells in the process, and then they vacuolate to form mature stalk cells. The scattered prestalk cells in the prespore zone provide accessory tissues during fruiting. One group sort downward to form the basal disc and another group sort to the bottom and top of the rising prespore mass to form the upper and lower cups (**2.25–2.28**). These cups may have a structural role, perhaps chaperoning the prespore cells up the stalk as they encapsulate and lose mobility.

2.22–2.24 Patterning of prestalk A cells at stages up to first finger formation. **2.22** shows early mound; **2.23**, a mound at about the time of tip formation; and **2.24**, a first finger. Prestalk A cells first appear scattered in the mound, then sort to the top as the tip emerges, forming a coherent block in the first finger and slug. A transformant of Ax2, in which the promoter of the ecmA gene drives β-galactosidase expression was used. (×128; gift from J.G. Williams and K.A. Jermyn.)

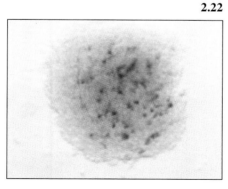

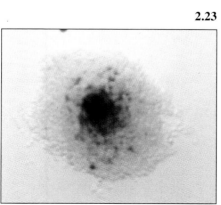

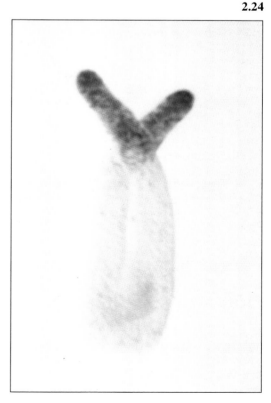

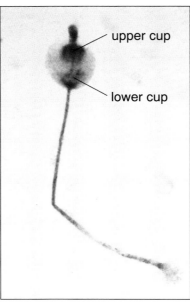

2.25, 2.26 Patterning of prestalk A cells during culmination. **2.25** shows early culminates; **2.26**, late culminate. In culmination, the prestalk A cells from the front of the slug convert into prestalk cells as they enter the stalk tube, but the prestalk A (and prestalk B) cells scattered in the prespore zone form the basal disc and the upper and lower cups of the prespore/spore mass. The lower cup is clearly visible in **2.26**, but the upper cup merges with the rest of the prestalk A zone: it can be distinguished by its expression of ecmB, the prestalk B marker. A transformant of Ax2, in which the promoter of the ecmA gene drives β-galactosidase expression was used. (×70; gift from J.G. Williams and K.A. Jermyn.)

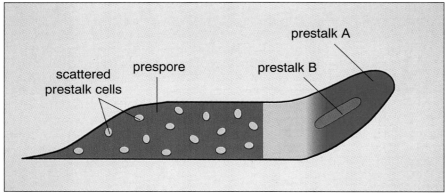
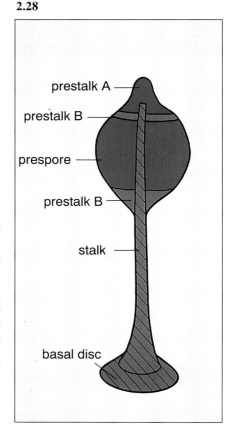

2.27, 2.28 New morphology of: **2.27**, the slug; **2.28**, the culminate. The more complicated morphology of the slug and culminate was revealed using molecular markers for the different cell types. Cell-type specific promoters were linked to reporter genes and transformed back into cells. The positions of the prestalk A (marked by ecmA), prestalk B (marked by ecmB) and prespore (marked by D19) zones in the slug and culminate are shown. In slugs migrating in the dark, the prestalk A marker is very weakly expressed in the rear of the prestalk zone and these cells are called prestalk zero cells. During culmination, the prestalk B zone in the anterior core of the slug forms the stalk primordium. As prestalk A cells are added to it from the the top, they express the ecmB gene becoming prestalk B, then stalk cells. The scattered prestalk A and B cells have a complicated fate. One group, which includes the original rearguard of the slug, forms the basal disc. The rest appear to sort to the top or bottom of the prespore zone to form the upper and lower cups. These cells express ecmB and therefore also become prestalk B cells but they do not go on to become mature stalk cells. (Based on Jermyn et al., 1989; Jermyn and Williams, 1991.)

Diffusible signals and the control of cell differentiation

As well as starvation, the initiation of development appears to require the accumulation of a secreted protein in the medium (Clarke et al., 1988; Jain et al., 1992). This may be important as a way of sensing cell density, so ensuring that development only starts if sufficient cells are present to form a fruiting body. Aggregation depends on secreted cAMP, both to guide cells into aggregates and to induce aggregative gene expression (Gerisch, 1987). cAMP signalling almost certainly continues during later development, with various effects including the induction of expression of all prespore markers tested (Kay, 1979; Williams, 1988).

The choice between prestalk and prespore differentiation is determined by DIF-1 (**2.29**). This molecule induces expression of prestalk markers and suppresses expression of prespore markers. In submerged cell culture, it will redirect the differentiation of potential spore cells into stalk cells (Kay and Jermyn, 1983). Though differentiation of both sub-types of prestalk cell depends on DIF-1, it appears that differences in cAMP concentration direct cells to prestalk A or B differentiation. Ammonia, produced from cell catabolism during development, accumulates to millimolar concentrations in the medium (Schindler and Sussman, 1977) and its best known role is in the control of culmination: slugs tend to carry on migrating in the presence of ammonia and are induced to culminate by its removal. Removal also stimulates stalk cell maturation (Inouye, 1988). Adenosine (Schaap and Wang, 1986; Weijer and Durston, 1985) and SPIF, which is probably methionine (Gibson and Hames, 1988), can modulate prespore cell differentiation, but their roles in development are not yet known.

2.29 The structure of DIF-1: 1-(3,5-dichloro-2,6-dihydroxy-4-methoxyphenyl) hexan-1-one and of DIF-2 and DIF-3. The structure of DIF-1 was determined from less than 100µg of the purified compound, largely by mass spectroscopy (Morris *et al.*, 1987). Two minor activities that also induce stalk cell differentiation have closely related structures: DIF-2 is the pentyl homologue, DIF-3 the mono-chloro analogue. DIF-3 is produced from DIF-1 by a mono-dechlorination, catalysed by DIF-1 dechlorinase (Nayler *et al.*, 1992)

Mutants

Statistical estimates suggest that about 150 genes are essential for development to the end of aggregation, but not for growth, and another 150 genes for the rest of development. Several classes of mutant have been studied extensively (Kessin, 1988). The frigid mutants which include *fdg*A, a G-protein mutant, are unable to respond to cAMP and so cannot aggregate (Kumagai *et al.*, 1989). The streamer mutants make very long aggregation streams and include *stm*F, a mutant in cGMP phosphodiesterase (Coukell and Cameron, 1986). Slugger mutants remain stuck as slugs and only culminate with difficulty. Rapidly developing/sporogenous mutants make spore cells prematurely in development and produce bizarre fruiting bodies (**2.30**, **2.31**); *rde*C is the structural gene for the regulatory subunit of cAMP-dependent protein kinase (Simon *et al.*, 1992). In addition, many mutants are now being made by homologous recombination using cloned genes. Amongst the genes mutated in this way are myosin 2, alpha actinin, gelation factor, various cAMP receptors and G-proteins.

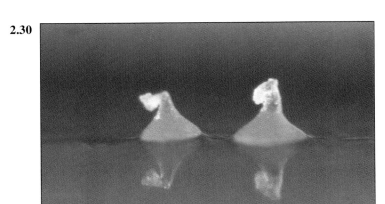 

2.30, 2.31 Final fruiting structures formed by rapidly developing (*rde*) mutants. **2.30** shows *rdeA* (strain HTY507); **2.31**, *rdeC* (strain HTY217). The morphological phenotype can be understood as resulting from premature spore and stalk cell differentiation. At some stage of development, the mound for *rdeC*, terminal differentiation is triggered and further morphological development prevented because the cells are no longer motile. This loss of temporal co-ordination between morphology and cell differentiation can also be thought of as an heterochrony. The *rdeC* locus encodes the regulatory subunit of cAMP-dependent protein kinase (×90).

Other species

The cellular slime moulds of the family *Dictyosteliacae*, comprise about 50 species arranged in two genera: *Dictyostelium* and *Polysphondylium* (Raper, 1984). *Dictyostelids* generally produce fruiting bodies with an unbranched stalk supporting a terminal spore head. *Polysphondylids* produce branched fruiting bodies, with the branches forming whorls along the main stalk (**2.32**). The species most closely related to *D. discoideum* is *D. mucoroides*, the main difference being that *D. muc-*

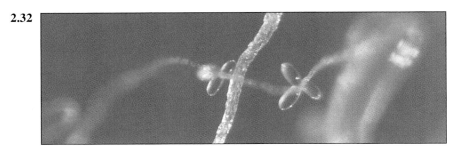

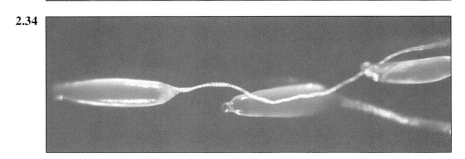

2.32–2.34 Development of other slime mould species. **2.32** shows *P. violaceum* secondary whorls, just forming; **2.33**, *D. mucoroides* slug; and **2.34**, *D. mucoroides* aerial slugs. *P. violaceum* deposits droplets of cells along the stalk, which develop into secondary whorls of fruiting bodies. *D. mucoroides* differs from *D. discoideum* principally in laying down a stalk as it migrates; this allows the slugs to leave the substratum (×90).

oroides slugs lay down a stalk as they migrate (**2.33, 2.34**), whereas *D. discoideum* slugs do not. This allows *D. mucoroides* slugs to lift off the substratum producing characteristic aerial tangles of stalks.

Comparison between species can be useful for filtering out features of development that are universal (amongst slime moulds) from those that are more specific. The chemoattractant during aggregation need not be cAMP: species such as *D. discoideum* and *D. mucoroides* use it, but *Polysphondylium violaceum* uses glorin, a blocked dipeptide (Shimomura, *et al.*, 1982) and *D. lacteum* uses an unidentified pterin. In later development, however, cAMP seems to be used as an extracellular signal in all the species investigated (Schaap and Wang, 1984). DIF-1 has been identified in *D. mucoroides* and is probably present in other species too, while the developmentally regulated production of chlorinated compounds (such as DIF-1 and its metabolites) is present in all species that have been tested (Kay *et al.*, 1992).

The future

Much recent work has concentrated on the cAMP signal transduction pathway, which is now becoming known in some detail (Firtel *et al.*, 1989). A very powerful strategy is the use of homologous recombination to make null mutants that lack particular components of the transduction pathway. Such mutants should allow us to disentangle the roles in chemotaxis and gene expression of the second messengers (cAMP, cGMP and IP3) that are driven off the cAMP receptors.

The present round of molecular work has helped to clarify the questions to be asked about the patterning process in later development. We now know that prestalk cells initially differentiate at scattered positions in the mound and that DIF-1 is their inducer, but we do not know how this is done nor how a small diffusible molecule can affect only a subset of cells. It seems that DIF-1 levels are controlled by negative feedback as DIF-1 induces the DIF-1 dechlorinase enzyme that inactivates DIF-1 (Insall *et al.*, 1992). This provides a way of limiting DIF-1 levels, and hence the number of prestalk cells, in the aggregate. Some sort of threshold mechanism also seems to be required to sharpen up the response of cells to DIF-1. The scattered prestalk cells sort out to form a coherent block by the time the first finger is formed and there are tantalizing hints that sorting is guided by chemotaxix to cAMP. Prestalk cells have a specific cAMP receptor which could be used to detect cAMP signals (Saxe *et al.*, 1993) and over-expression of cAMP phosphodiesterase (expected to attenuate extracellular cAMP signals) delays prestalk cell sorting (Traynor *et al.*, 1992). The extension of the mound into a first finger is an unsolved problem in morphogenesis. Its solution will require an understanding of the motile properties of single cells and of the signalling system that coordinates their behaviour, although it is already known that mutations in certain cytoskeletal proteins can block morphogenesis (De Lozanne and Spudich, 1987; Witke *et al.*, 1992).

Perhaps the most important force shaping work in the immediate future will be the REMI technique for insertional mutagenesis (Kuspa and Loomis, 1992). By this means it should be possible to clone and catalogue most of the 300 genes that are necessary for development, but not for growth. Undoubtably this will throw up a mixture of familiar and novel genes, but in either case the information will provide the foundation from which to try and understand the biological puzzles concealed within the remarkable development of *Dictyostelium*.

References

Alcantara, F. and Monk, M. (1974). Signal propagation during aggregation in the slime mould *Dictyostelium discoideum*. *J. Gen. Microbiol.*, **85**, 321–334.

Barklis, E. and Lodish, H.F. (1983). Regulation of *Dictyostelium discoideum* mRNAs specific for prespore or prestalk cells. *Cell*, **32**, 1139–1148.

Beug, H., Katz, F.E., and Gerisch, G. (1973). Dynamics of antigenic membrane sites relating to cell aggregation in *Dictyostelium discoideum*. *J. Cell Biol.*, **56**, 647–658.

Bonner, J.T. (1952). The pattern of differentiation in amoeboid slime molds. *Am. Nat.*, **86**, 79–89.

Bonner, J.T. (1967). *The Cellular Slime Molds.* Princeton University Press, Princeton, New Jersey.

Clarke, M., Yang, J., and Kayman, S.C. (1988). Analysis of the prestarvation response in growing cells of *Dictyostelium discoideum*. *Dev. Genet.*, **9**, 315–326.

Coukell, M.B. and Cameron, A.M. (1986). Characterization of revertants of stmF mutants of *Dictyostelium discoideum*: evidence that stmF is the structural gene of the cGMP-specific phosphodiesterase. *Dev. Genet.*, **6**, 163–177.

De Lozanne, A. and Spudich, J.A. (1987). Disruption of the *Dictyostelium* myosin heavy chain gene by homologous recombination. *Science*, **236**, 1086–1091.

Devreotes, P.N. and Zigmond, S.H. (1988). Chemotaxis in eukaryotic cells – a focus on leukocytes and *Dictyostelium*. *Annu. Rev. Cell Biol.*, **4**, 649–686.

Durston, A.J. (1976). Tip formation is regulated by an inhibitory gradient in the *Dictyostelium discoideum* slug. *Nature*, **263**, 126–129.

Dynes, J. L. and Firtel, R.A. (1989). Molecular complementation of a genetic marker in *Dictyostelium* using a genomic DNA library. *Proc. Natl. Acad. Sci. USA*, **86**, 7966–7970.

Firtel, R.A., van Haastert, P.J.M., Kimmel, A.R., and Devreotes, P.N. (1989). G-protein linked signal transduction pathways in development: *Dictyostelium* as an experimental system. *Cell*, **58**, 235–239.

Franke, J. and Kessin, R. (1977). A defined minimal medium for axenic strains of *Dictyostelium discoideum*. *Proc. Natl. Acad. Sci. USA*, **74**, 2157–2161.

Gerisch, G. (1987). Cyclic AMP and other signals controlling cell development and differentiation in *Dictyostelium*. *Annu. Rev. Biochem.*, **56**, 853–879.

Gibson, F.P. and Hames, B.D. (1988). Characterization of a spore protein inducing factor from *Dictyostelium discoideum*. *J. Cell. Sci.*, **89**, 387–395.

Haberstroh, L. and Firtel, R.A. (1990). A spatial gradient of expression of a cAMP-regulated prespore cell-type-specific gene in *Dictyostelium*. *Genes Dev.*, **4**, 596–612.

Hayashi, M. and Takeuchi, I. (1976). Quantitative studies on cell differentiation during morphogenesis of the cellular slime mold *Dictyostelium discoideum*. *Dev. Biol.*, **50**, 302–309.

Hohl, H.R. and Hamamoto, S.T. (1969). Ultrastructure of spore differentiation in *Dictyostelium*: the prespore vacuole. *J. Ultrastruct. Res.*, **26**, 442–453.

Inouye, K. (1988). Induction by acid load of the maturation of prestalk cells in *Dictyostelium discoideum*. *Development*, **104**, 669–681.

Insall, R., Nayler, O. and Kay, R.R. (1992) DIF-1 induces its own breakdown in *Dictyostelium*. *EMBO J.*, **11**, 2849-2854.

Jain, R., Yuen, I.S., Taphouse, C.R. and Gomer, R.H. (1992). A density-sensing factor controls development in *Dictyostelium*. *Genes and Develop.*, **6**, 390–400.

Jermyn, K.A., Berks, M., Kay, R.R., and Williams, J.G. (1987). Two distinct classes of prestalk-enriched messenger RNA sequences in *Dictyostelium discoideum*. *Development*, **100**, 745–755.

Jermyn, K.A., Duffy, K.T.I., and Williams, J.G. (1989). A new anatomy of the prestalk zone in *Dictyostelium*. *Nature*, **340**, 144–146.

Jermyn, K.A. and Williams, J.G. (1991). An analysis of culmination in *Dictyostelium* using prestalk and stalk-specific cell autonomous markers. *Development*, **111**, 779–787.

Kay, R.R. (1979). Gene expression in *Dictyostelium discoideum*: Mutually antagonistic roles of cyclic-AMP and ammonia. *J. Embryol. Exp. Morphol.*, **52**, 171–182.

Kay, R.R. (1987). Cell differentiation in monolayers and the investigation of slime mold morphogens. *Methods Cell. Biol.*, **28**, 433–448.

Kay, R.R. and Jermyn, K.A. (1983). A possible morphogen controlling differentiation in *Dictyostelium*. *Nature*, **303**, 242–244.

Kay, R.R., Large, S., Traynor, D. and Nayler, O. (1993). A localized differentiation-inducing-factor sink in the front of the *Dictyostelium* slug. *Proc. Natl. Acad. Sci. USA*, **90**, 487–491.

Kay, R.R., Taylor, G.W., Jermyn, K.A., and Traynor, D. (1992). Chlorine-containing compounds produced during *Dictyostelium* development. Detection by labelling with ^{36}Cl. *Biochem. J.*, **281**, 155–161.

Kessin, R.H. (1988). Genetics of early *Dictyostelium discoideum* development. *Microbiol. Rev.* **52**, 29–49.

Klein, P.S., Sun, T.J., Saxe, C.L., Kimmel, A.R., Johnson, R.L., and Devreotes, P.N. (1988). A chemoattractant receptor controls development in *Dictyostelium discoideum*. *Science*, **241**, 1467–1472.

Kopachik, W. (1982). Size regulation in *Dictyostelium*. *J. Embryol. Exp. Morphol.*, **68**, 23–35.

Kumagai, A., Pupillo, M., Gundersen, R., Miake-Lye, R., Devreotes, P.N., and Firtel, R.A. (1989). Regulation and function of Gα protein subunits in *Dictyostelium*. *Cell*, **57**, 265–275.

Kuspa, A. and Loomis, W.F. (1992) Tagging developmental genes in *Dictyostelium* by restriction enzyme mediated integration of plasmid DNA. *Proc. Natl. Acad. Sci. USA*, **89**, 8803–8807.

Leiting, B. and Noegel, A. (1988). Construction of an extrachromosomally replicating transformation vector for *Dictyostelium discoideum*. *Plasmid*, **20**, 241–248.

Loomis, W.F. (1982). The Development of *Dictyostelium discoideum*. Academic Press, New York.

Manstein, D.J., Titus, M.A., DeLozanne, A., and Spudich, J.A. (1989). Gene replacement in *Dictyostelium* – generation of myosin null mutants. *EMBO J.*, **8**, 923–932.

Mehdy, M.C., Ratner, D., and Firtel, R.A. (1983). Induction and modulation of cell type specific gene expression in *Dictyostelium*. *Cell*, **32**, 763–771.

Morris, H.R., Taylor, G.W., Masento, M.S., Jermyn, K.A., and Kay, R.R. (1987). Chemical structure of the morphogen differentiation inducing factor from *Dictyostelium discoideum*. *Nature*, **328**, 811–814.

Nayler, O., Insall, R. and Kay, R.R. (1992) DIF-1 dechlorinase, a novel cytosolic dechlorinating enzyme from *Dictyostelium discoideum*. *Eur. J. Biochem.*, **208**, 531–536.

Nellen, W., Datta, S., Reymond, C., Sivertsen, A., Mann, S., Crowley, T., and Firtel, R. A. (1987). Molecular biology in *Dictyostelium*: tools and applications. *Methods Cell. Biol.*, **28**, 67–100.

Nellen, W., Silan, C., and Firtel, R. (1984). DNA-mediated transformation in *Dictyostelium discoideum* – regulated expression of an actin gene fusion. *Mol. Cell. Biol.*, **4**, 2890.

Newell, P.C., Telser, A., and Sussman, M. (1969). Alternative developmental pathways determined by environmental conditions in the cellular slime mold *Dictyostelium discoideum*. *J. Bacteriol.*, **100**, 763–768.

Noegel, A., Gerisch, G., Stadler, J., and Westphal, M. (1986). Complete sequence and transcript regulation of a cell-adhesion protein from aggregating *Dictyostelium* cells. *EMBO J.*, **5**, 1473–1476.

Raper, K.B. (1940). Pseudoplasmodium formation and organisation in *Dictyostelium discoideum*. *J. Elisha Mitchell Sci. Soc.*, **56**, 241–282.

Raper, K.B. (1984). *The Dictyostelids*. Princeton University Press, Princeton.

Saxe, C.L. Ginsberg, G.T., Louis, J.M., Johnson, R., Devreotes, P.N. and Kimmel, A.R. (1993) CAR2, a prestalk cAMP receptor required for tip formation in late development of *Dictyostelium*. *Genes and Development*, **7**, 262–272.

Schaap, P. and Wang, M. (1984). The possible involvement of oscillatory cAMP signaling in multicellular morphogenesis of the cellular slime molds. *Dev. Biol.*, **105**, 470–478.

Schaap, P. and Wang, M. (1986). Interactions between adenosine and oscillatory cAMP signaling regulate size and pattern in *Dictyostelium*. *Cell*, **45**, 137–144.

Schindler, J. and Sussman, M. (1977). Ammonia determines the choice of morphogenetic pathways in *Dictyostelium discoideum*. *J. Molec. Biol.*, **116**, 161–169.

Shimomura, O., Suthers, H.L.B., and Bonner, J.T. (1982). Chemical identity of the acrasin of the cellular slime mold *Polysphondylium violaceum*. *Proc. Natl. Acad. Sci. USA*, **79**, 7376–7379.

Simon, M.-N., Pelegrini, O., Veron, M. and Kay, R.R. (1992). Mutation of protein kinase A causes heterochronic development of *Dictyostelium*. *Nature*, **356**, 171–172.

Spudich, J.A. (1987). *Dictyostelium discoideum: molecular approaches to cell biology*. Academic Press, Orlando.

Sternfeld, J. and David, C.N. (1981). Cell sorting during pattern-formation in *Dictyostelium*. *Differentiation*, **20**, 10–21.

Sternfeld, J. and David, C.N. (1982). Fate and regulation of anterior-like cells in *Dictyostelium* slugs. *Dev. Biol.*, **93**, 111–118.

Tomchik, K.J. and Devreotes, P.N. (1981). Cyclic AMP waves in *Dictyostelium discoideum* a demonstration by isotope dilution fluorography. *Science*, **212**, 443–446.

Traynor, D., Kessin, R.H. and Williams, J.G. (1992) Chemotactic sorting to cAMP in the multicellular stages of *Dictyostelium* development. *Proc. Natl. Acad. Sci. USA*, **89**, 8303–8307.

Watts, D.J. and Ashworth, J.M. (1970). Growth of myxamoebae of the cellular slime mould *Dictyostelium discoideum* in axenic culture. *Biochem. J.*, **119**, 171–174.

Weijer, C.J. and Durston, A.J. (1985). Influence of cyclic AMP and hydrolysis products on cell type regulation in *Dictyostelium discoideum*. *J. Embryol. Exp. Morphol.*, **86**, 19–37.

Williams, J.G. (1988). The role of diffusible molecules in regulating the cellular differentiation of *Dictyostelium discoideum*. *Development*, **103**, 1–16.

Williams, J.G., Ceccarelli, A., McRobbie, S., Mahbubani, H., Kay, R.R., Early, A., Berks, M., and Jermyn, K.A. (1987). Direct induction of *Dictyostelium* prestalk gene expression by DIF provides evidence that DIF is a morphogen. *Cell*, **49**, 185–192.

Williams, J.G., Duffy, K.T., Lane, D.P., McRobbie, S.J., Harwood, A.J., Traynor, D., Kay, R.R., and Jermyn, K.A. (1989). Origins of the prestalk-prespore pattern in *Dictyostelium* development. *Cell*, **59**, 1157–1163.

Witke, W., Schleicher, M., and Noegel, A.A. (1992). Redundancy in the microfilament system: abnormal development of *Dictyostelium* cells lacking two F-actin cross-linking proteins. *Cell*, **68**, 53–62.

3. The Sea Urchin

Jeff Hardin

Introduction

The sea urchin embryo has been used for more than a century to study many problems central to developmental biology. During the latter part of the nineteenth century, marine stations in Italy, France, and the United States flourished, and the embryos of marine organisms were found to be favourable material for investigating early embryonic development. The study of echinoderms and, in particular, of sea urchins, that was carried out at these marine stations was influential in the formation of many seminal ideas in developmental biology (for reviews, see the classic texts of Wilson, 1925; Morgan, 1927). Later, in the early part of the twentieth century, the experiments performed on sea urchin embryos using chemical agents and the classic blastomere recombination experiments performed by Hörstadius (1939; 1973) paved the way for ideas about graded distributions of morphogenetic substances in the embryo. The sea urchin also provided useful material for studying many aspects of nucleic acid structure, complexity, and function in the early days of molecular biology (reviewed by Davidson, 1988). The reader interested in the historical role played by sea urchin embryos in the emergence of developmental biology, and the importance and relevance of such experiments today is referred to Wilt (1987), Davidson (1989; 1990) and Livingston and Wilt (1990).

More recently, the sea urchin embryo has been used as a convenient system for studying morphogenetic movements and cell interactions during gastrulation, the changes in gene expression associated with the establishment of tissue territories along the embryonic axes, and phylogenetic variability and associated modifications in early development. In summarizing such work, this chapter provides a brief overview of normal development in the sea urchin embryo and a few case studies illustrating modern uses of this system for studying early developmental events. Where possible, the reader is referred to reviews that treat individual topics in more detail than is possible here.

For methods of maintaining adults, obtaining gametes, and culturing embryos, see Hinegardner (1967; 1975a) and Leahy (1986). General methods for culturing and experimentally manipulating embryos can be found in Harvey (1956) and Hörstadius (1973), while more up-to-date methods are in Schroeder (1986).

Normal development

Sperm, eggs, and fertilization

Sea urchin and sand dollar gametes can be obtained in large numbers by intracoelomic injection of 0.5M KCl or by electrical stimulation; this leads to the shedding of gametes into sea water (in the case of eggs) or 'dry' into a dish (in the case of sperm). Depending on the species, several millilitres of ripe eggs or sperm can be obtained from a single animal and the embryos can be conveniently reared in finger bowls or in stirring cultures. The major stages of early development in the sea urchin are shown in **3.1–3.15**, and each stage will be discussed in turn in the following sections. Mature sea urchin eggs, unlike eggs from many other animals, have completed meiosis and the extrusion of polar bodies in the ovary to produce a haploid gamete (**3.1**). Immediately apposed to the egg plasma membrane is the *vitelline envelope* which contains the glycoproteins essential for species-specific fusion of sperm and egg, while freshly shed eggs are surrounded by a *jelly coat*. Marking the jelly coat with small ink particles reveals the jelly canal, a marker for the animal pole of the egg first described by Boveri (1901) and more recently re-investigated by Schroeder (1980b; see also Maruyama *et al.*, 1985). Another marker of polarity in the unfertilized egg is the subequatorial concentration of orange pigment in some batches of eggs of the Mediterranean sea urchin, *Paracentrotus lividus*, especially those obtained from Villefranche in France (Boveri, 1901; Hörstadius, 1973; Schroeder, 1980a).

Unfertilized eggs possess several other kinds of distinct granules with different distributions within the egg. *Cortical granules* lie immediately beneath the egg surface and are released upon fertilization. These are lamellar structures which contain components necessary for the construction of the *fertilization envelope* (**3.2**) and the *hyaline layer*, an extracellular matrix layer which lies on the outside of the embryo. *Pigment granules* are particularly prominent in species such as *Arbacia punctulata* from America. Other granules, which release their contents following fertilization, but on a much slower time course than the cortical granules, contain extracellular matrix proteins. Some of these granules can be redistributed by centrifuging eggs suspended in sucrose

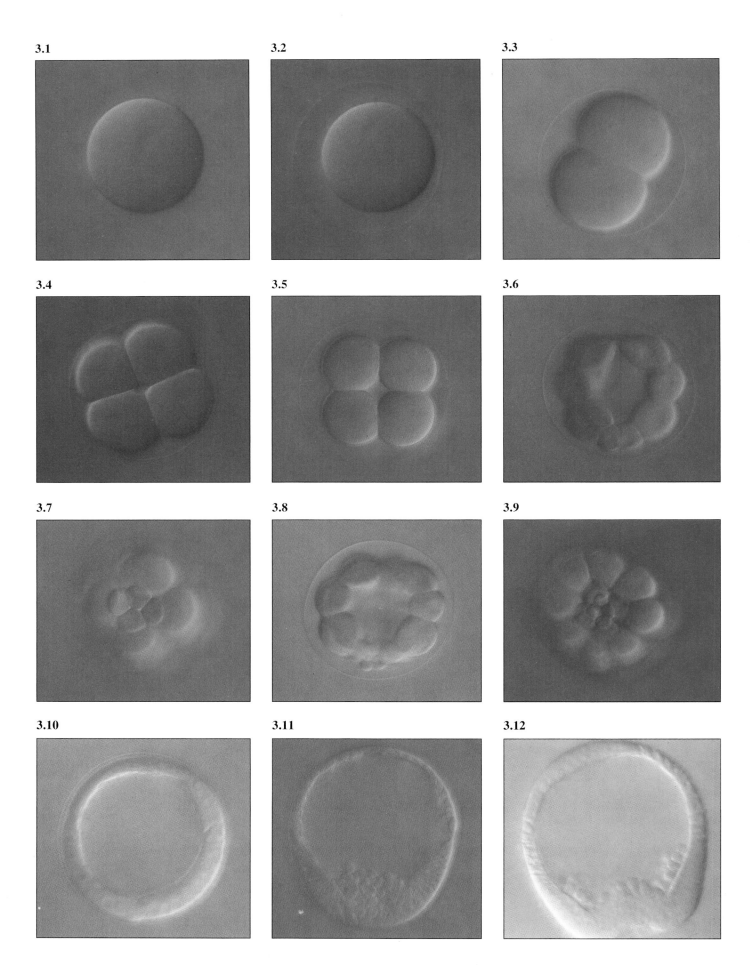

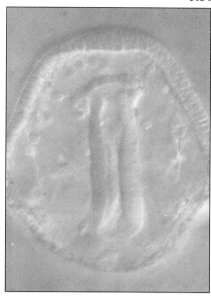
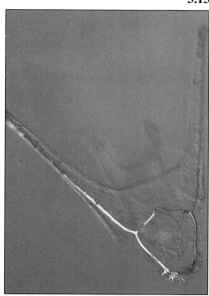

3.1–3.15 Characteristic stages during early development in *Lytechinus variegatus*. The animal pole, when evident, is at the top unless otherwise indicated. **3.1**, unfertilized egg; **3.2**, recently fertilized egg (note the fertilization envelope and the fertilization cone, at the top of the egg); **3.3**, two-cell; **3.4**, 4-cell; **3.5**, 8-cell; **3.6**, 16-cell (the micromeres are at the bottom); **3.7**, 16-cell embryo viewed from the vegetal pole to show the micromeres; **3.8**, 32-cell embryo (the small micromeres are visible at the extreme vegetal pole); **3.9**, vegetal pole view of a 32-cell embryo showing the small micromeres and the larger micromere derivatives immediately above them; **3.10**, blastula stage; **3.11**, mesenchyme blastula; **3.12**, early gastrula (courtesy of C. Ettensohn); **3.13**, late gastrula; **3.14**, prism stage embryo (note the coelomic pouches beginning to form); **3.15**, pluteus (×450).

density gradients with little effect on patterns of development (reviewed in Harvey, 1956; McClay *et al.*, 1990; **3.16**).

Under normal circumstances, sea urchins simply shed gametes directly into the marine environment. Three mechanisms appear to help ensure that interactions between sperm and egg are species-specific. First, in *Arbacia punctulata*, the peptide *resact*, which is contained in the egg jelly, appears to be a species-specific chemo-attractant for sperm (Ward *et al.*, 1985). Second, activation of sperm by egg jelly is also species-specific in some species so that contact between them results in a rapid *acrosome reaction*. The *acrosomal process* which contains proteolytic enzymes, fuses with the sperm plasma membrane and extends dramatically, driven by the rapid polymerization of actin microfilaments. Third, the adhesion of activated sperm to the vitelline layer is mediated by the acrosomal protein bindin, whose binding to the vitelline envelope also appears to be mediated by a species-specific ligand-receptor interaction (Glabe and Vacquier, 1978; Moy and Vacquier, 1979). All regions of the egg surface support sperm attachment and fusion in the sea urchin.

At the site of sperm-egg fusion, localized polymerization of actin filaments in the egg cortex produces the *fertilization cone* (**3.17**). The *fast block to polyspermy*, a rapid depolarization of the egg mediated by an influx of sodium ions, provides an initial barrier to penetration of the egg by more than a single sperm. A slower but more *permanent block to polyspermy* results from *cortical granule exocytosis* and elevation of the *fertilization envelope* (**3.2**) which is triggered by release of calcium ions in a wave that sweeps across the egg. Following its entry into the egg, the sperm nucleus decondenses to form the *male pronucleus*. Microtubules polymerize away from the sperm centriole in the direction of the *female pronucleus*, the two pronuclei migrate towards one another, and ultimately they fuse to form the diploid, zygote nucleus (Bestor and Schatten, 1981). Ultimately, the release of calcium responsible for cortical granule exocytosis results in *activation* of the egg, a complex series of events leading to the initiation of protein and DNA synthesis. Egg activation can be mimicked by a number of treatments which lead to *parthenogenetic activation* and include treatment with butyric acid and hypertonic sea water and calcium ionophores (reviewed by Weidman and Kay, 1986).

Cleavage and the blastula

Sea urchins undergo synchronous, radial, holoblastic cleavages until the blastula stage. The first two cleavages are meridional and perpendicular to one another, passing through the animal and vegetal poles to produce first two and then four cells of equal size (**3.3–3.5** and **3.18**). The third cleavage is equatorial, separating the embryo into an animal and vegetal quartet of cells (**3.5**). At the fourth cleavage, the cells of the animal tier divide equally to produce eight *mesomeres*, while the cells in the vegetal tier divide unequally, producing four large *macromeres*, and four small *micromeres* (**3.6, 3.7** and **3.19**). In

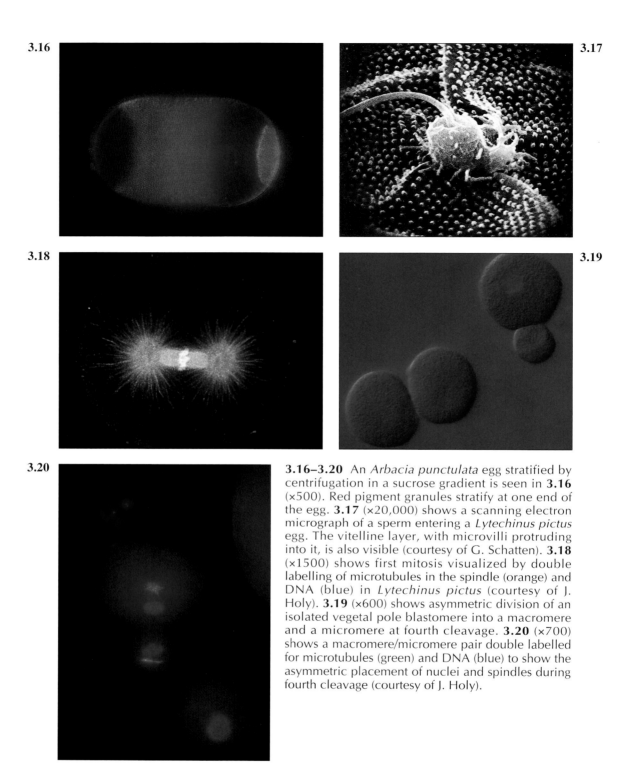

3.16–3.20 An *Arbacia punctulata* egg stratified by centrifugation in a sucrose gradient is seen in **3.16** (×500). Red pigment granules stratify at one end of the egg. **3.17** (×20,000) shows a scanning electron micrograph of a sperm entering a *Lytechinus pictus* egg. The vitelline layer, with microvilli protruding into it, is also visible (courtesy of G. Schatten). **3.18** (×1500) shows first mitosis visualized by double labelling of microtubules in the spindle (orange) and DNA (blue) in *Lytechinus pictus* (courtesy of J. Holy). **3.19** (×600) shows asymmetric division of an isolated vegetal pole blastomere into a macromere and a micromere at fourth cleavage. **3.20** (×700) shows a macromere/micromere pair double labelled for microtubules (green) and DNA (blue) to show the asymmetric placement of nuclei and spindles during fourth cleavage (courtesy of J. Holy).

the vegetal blastomeres of the 8-cell embryo, the nucleus moves to an eccentric, vegetal position, and the mitotic spindle is subsequently assembled eccentrically as well, with the result that the aster is flattened and shortened (**3.20**). The asymmetric fourth cleavage is the first sign that cells distributed along the animal-vegetal axis of the embryo are different and only the micromeres will go on to form primary mesenchyme cells (pmc) that produce the larval skeleton (see below).

At the fifth cleavage, the mesomeres divide equatorially to produce two animal tiers (denoted by Hörstadius as an_1 and an_2), and the macromeres divide meridionally to produce a tier of eight 'half-macromeres'. The micromeres divide asymmetrically to produce a tier of *small micromeres* at the extreme vegetal pole of the embryo, and a tier of larger micromere derivatives immediately above them (**3.8, 3.9**). At the sixth cleavage, all cells divide equatorially to produce a 64-cell embryo with five tiers: the daughters of the an_1 and an_2 cells lie at the animal pole, the **veg**$_1$ and **veg**$_2$ tiers, derived from the macromeres, lie in the vegetal hemisphere, and the micromere descendants lie at the vegetal pole. At the seventh cleavage, all cells divide meridionally to produce a 128-cell blastula. During the blastula stage, cells no longer cleave synchronously: as development proceeds, divisions of local groups of cells remain synchronous, but these regions gradually decrease in size, and eventually the cell cycle lengthens and becomes largely randomized (Dan *et al.*, 1980).

The early blastula is an epithelial monolayer enclosing a central, spherical *blastocoel* (**3.10**) whose cells develop septate junctional contacts between one another (Spiegel and Howard, 1983) and begin to produce the *basal lamina* lining the blastocoel. The exterior, apical ends of the epithelial cells possess numerous microvilli, which are embedded in the hyaline layer, and *apical lamina* (Hall and Vacquier, 1982). The cells of the blastula eventually seal off the internal, embryonic environment from the external environment, with the epithelium becoming impermeable to small sugar molecules by the midblastula stage (Moore, 1940). The forces responsible for formation of the blastocoel remain unknown, although osmotic influx of water into the blastocoel and attachment to the hyaline layer have been suggested as possible factors (Dan, 1960; Dan and Inaba, 1968; Gustafson and Wolpert, 1962). At the midblastula stage, each cell produces a cilium, the embryo begins to rotate within the fertilization envelope, and a *hatching enzyme* is synthesized that is secreted into the space between the embryo and the fertilization envelope. Here it digests the envelope and allows the embryo to hatch and become a free-swimming blastula.

Establishment of the major tissue territories of the embryo

What changes in the shape of the embryo take place during gastrulation, and what consequences do they have for subsequent tissue-specific differentiation? Analyses of cell fate and cell movements both shed light on these questions. By staining individual blastomeres with Nile blue at the 32- and 64-cell stages, Hörstadius (1935) constructed a fate map for the *Paracentrotus lividus* embryo (**3.21**). According to Hörstadius, the **an**$_1$ tier of cells, which is distinct at the 32-cell stage, gives rise to the ectoderm of the animal pole, the **an**$_2$ tier gives rise to equatorial ectoderm, the **veg**$_1$ tier of the 64-cell embryo gives rise to the vegetal pole ectoderm and the **veg**$_2$ tier generates the cells of the archenteron and a group of mesenchymal cells, the *secondary mesenchyme*, which form at the tip of the archenteron, while the large micromere derivatives give rise to the primary mesenchyme which forms the skeletal structures of the larva.

3.21–3.24 Organization and fates of cells in the early sea urchin embryo according to Hörstadius (for *Paracentrotus lividus*; **3.21**) and Cameron and co-workers (*Strongylocentrotus purpuratus*; coloured diagrams; from Davidson, 1988, with permission). The coloured regions refer to the following presumptive tissue territories. (Key: red = skeletogenic cells, yellow = aboral (dorsal) ectoderm, green = oral (ventral) ectoderm, magenta = small micromere derivatives, blue = archenteron and associated structures.)

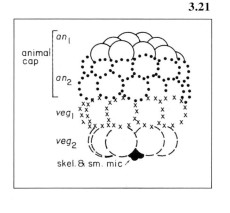

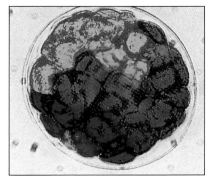

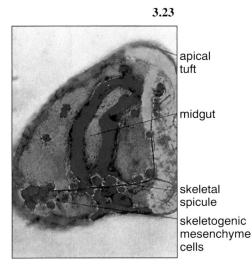

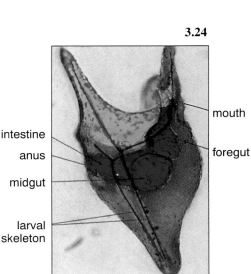

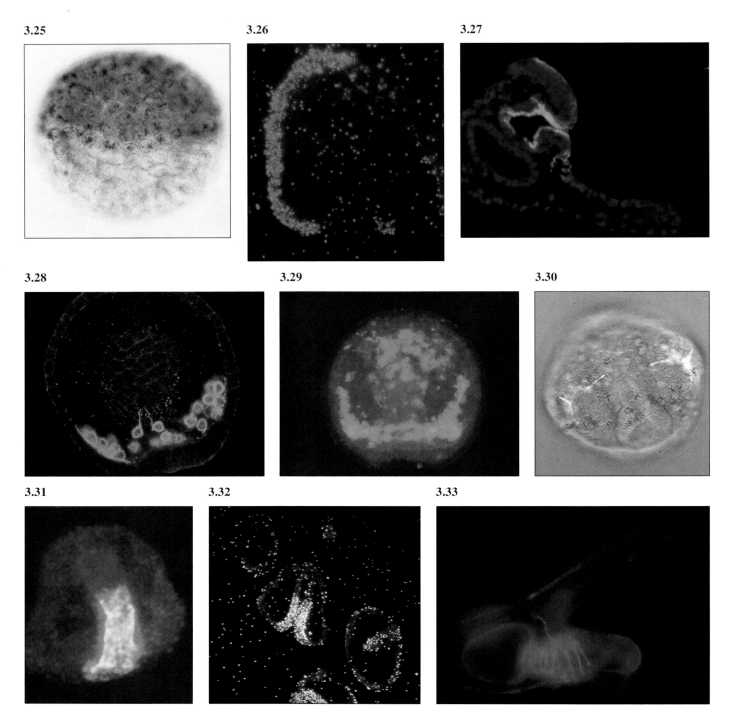

3.25–3.33 Examples of tissue-specific gene expression in the sea urchin embryo. The probes described here are representative, but by no means exhaustive. **3.25** (×450) is a whole mount *in situ* hybridization using an anti-sense BP-10 RNA (courtesy of T. Lepage; Lepage *et al.*, 1992). The boundary of expression sharply demarcates future ectoderm from endoderm and mesoderm. **3.26** (×420) shows an *in situ* hybridization of a *Lytechinus variegatus* embryo using an anti-sense probe for LvS1 (Wessell *et al.*, 1987), a member of the Spec 1 family of genes (Lynn *et al.*, 1983). **3.27** (×350) shows a section of a *Lytechinus variegatus* pluteus immunostained for the Ecto V antigen (V for **v**entral); staining is complementary to aboral markers (courtesy of D. McKlay; Coffman and McClay, 1990). **3.28** (×450) shows whole mount immunostaining for the msp130 homologue in *Lytechinus variegatus*, a cell surface antigen expressed by primary mesenchyme cells and one of a host of probes specific to these cells (e.g., Benson *et al.*, 1987; Drager *et al.*, 1989; Leaf *et al.*, 1987; Wessell and McClay, 1985). **3.29** (×350) shows pigment cells localized to the presumptive arm buds of a *Lytechinus variegatus* prism stage embryo. **3.30** (×350) shows immunostaining for Meso 1, a cell surface epitope present on the surfaces of primary and secondary mesenchyme cells (Wray and McClay, 1988). **3.31** (×400) shows immunostaining for Endo 1, an antigen expressed by the mid- and hindgut of the archenteron (Wessell *et al.*, 1985). **3.32** (×200) shows *in situ* hybridization using an anti-sense probe for LvN1.2, which also localizes to this region (courtesy of G. Wessell; Wessell *et al.*, 1989). **3.33** (×500) shows rhodamine-phalloidin staining for the esophageal muscle bands which surround the foregut (Ishimoda-Takagi *et al.*, 1984; Wessell *et al.*, 1990).

More recently, these lineage studies have been refined and extended to account for distinctions along the dorsoventral axis of the early embryo. Based on unique patterns of gene expression, cell lineages, and one or more characteristic differentiated cell types, sea urchin embryonic cells can be classified into five major tissue territories (**3.22–3.24**; Cameron and Davidson, 1991; Davidson, 1989). These are:

- The *aboral* (or *dorsal*) *ectoderm*, which forms a simple, squamous epithelium.
- The oral (or ventral) ectoderm, which forms the epithelium of the mouth region and the *ciliated band*, a structure that lies at the boundary between oral and aboral ectoderm.
- The *vegetal plate*, which gives rise to the archenteron and its derivatives.
- The *primary mesenchyme cells*, which will produce the larval skeleton.
- The *small micromeres*, which have been reported to contribute to the coelomic pouches (Pehrson and Cohen, 1986).

Each of these tissue territories derives from a specific group of founder cells whose lineages become distinct during cleavage. The lineages of cells lying within these domains can be distinguished by the completion of the sixth cleavage, i.e. when there are approximately 64 cells in the embryo (**3.22–3.24**). The four animal blastomeres of the 8-cell embryo contribute progeny to either oral or aboral ectoderm, and they are termed **Na** and **No** cells (a**n**imal, **o**ral and **a**boral). Some cells from Hörstadius' veg$_1$ tier, derived from the macromeres of the 16-cell embryo, form aboral ectoderm, while others contribute to oral ectoderm. The veg$_2$ tier of Hörstadius gives rise to the structures of the archenteron: the larger (animal) progeny of the micromeres generate skeletogenic mesenchyme, whereas the small (vegetal) progeny of the micromeres contribute to the coelomic pouches.

The clonal boundaries described by Cameron and co-workers are established through invariant cleavages and these boundaries seem to coincide very closely with spatially restricted patterns of gene expression (**3.25–3.33**). That clonal boundaries and patterns of gene expression are co-extensive suggests that the reliability of cleavage may be important in establishing patterns of differentiating tissue within the *unperturbed* embryo. The sea urchin embryo is, however, well known for the ability of its cells to adopt new fates when placed in unusual environments, indicating that the establishment of reliable clonal boundaries does not reflect an underlying 'mosaic' quality of the early cells of the sea urchin embryo, in contrast to some other invertebrate embryos (e.g., nematodes and ascidians).

Gastrulation and post-gastrula development

The various territories of the embryo also differ in their patterns of motility as the embryo is transformed during gastrulation. The study of sea urchin gastrulation has been influential in shaping ideas about general mechanisms of morphogenetic movements, and the work of Gustafson and co-workers in particular demonstrated the power of time-lapse microscopy in elucidating morphogenetic processes (reviewed by Gustafson and Wolpert, 1963; 1967). Just prior to gastrulation, there is a dramatic decrease in the overall rate of cell division, and the embryo comprises about 1000 cells. At the animal pole, a thickened region of epithelium, the *apical plate*, or *acron*, appears with a tuft of cilia that are longer than those found on the rest of the embryo (**3.11**), while the epithelium at the vegetal pole of the embryo flattens and thickens to form the *vegetal plate*.

The onset of gastrulation is marked by the ingression of *primary mesenchyme cells* (pmc) into the blastocoel (**3.11** and **3.34–3.36**), an event accompanied by alterations in cell polarity and loss of the epithelial phenotype (Anstrom and Raff, 1988) and the appearance of new cell-surface determinants and transcripts (see above). When they begin to ingress, pmc become bottle-shaped in profile as the surface area of their apical ends is reduced (Katow and Solursh, 1980), and they eventually detach from the hyaline layer. These cells use bristle-like filopodia to move (reviewed by Solursh, 1986) and require sulphated proteoglycans on their surfaces and/or in the blastocoel for their migration (Lane and Solursh, 1988; Lane and Solursh, 1991; Solursh *et al.*, 1986). Primary mesenchyme cells migrate away from the vegetal plate, but eventually form a ring in the vegetal pole region of the embryo (**3.35**). Ultimately, two clusters of pmc form in the ventrolateral ectoderm and give rise to the *spicule rudiments* of the larva (**3.36**; for further details, see Solursh, 1986, and Decker and Lennarz, 1988).

Following the ingression of pmc, *pigment cells* depart from the vegetal plate (Gibson and Burke, 1985; **3.30**). The vegetal plate then begins to bend inward to form a short, squat cylinder, the *archenteron*. During this initial phase of invagination (*primary invagination*), the archenteron extends $^1/_4$–$^1/_2$ of the way across the blastocoel (**3.12**). A short pause follows primary invagination, after which the archenteron resumes its elongation (*secondary invagination*). At about the time secondary invagination begins, cells at the tip of the archenteron (*secondary mesenchyme cells*) become protrusive, extending long *filopodia* into the blastocoel (**3.13**). Eventually the archenteron elongates across the blastocoel, and its tip attaches to the ventral ectoderm near the animal pole.

By the time the archenteron completes its elongation, pmc have localized into two major clusters in the ventrolateral ectoderm to form spicule rudiments. At the tip of the archenteron, two bilateral outpocketings, the *coelomic pouches*, appear (**3.14**). Ultimately the tip of the archenteron

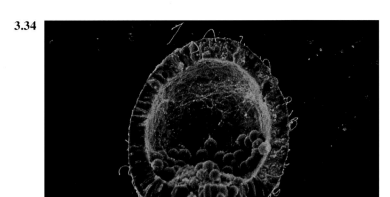
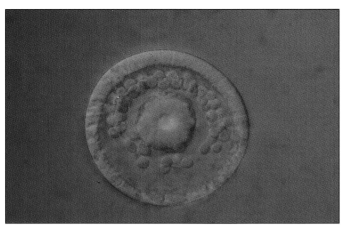
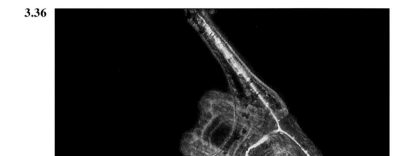

3.34–3.36 Primary mesenchyme cell behaviour during early development. **3.34** (×500) is a scanning electron micrograph of a *Lytechinus variegatus* mesenchyme blastula during ingression of primary mesenchyme cells (courtesy of J. Morrill; magnification. **3.35** (×500) is an *L. variegatus* gastrula viewed from the vegetal pole, showing the aggregation of primary mesenchyme cells into two ventrolateral clusters (magnification. **3.36** (×250) is an *L. variegatus* pluteus viewed with darkfield optics (courtesy of M.A. Alliegro; magnification.

fuses with the ectoderm to form the larval mouth. As the *pluteus* larva develops (**3.15**), the archenteron becomes tripartite, and a host of differentiated tissues appear that include nerve cells in the ectoderm (Bisgrove and Burke, 1987). The left coelomic pouch ultimately forms the hydrocoel which, together with the ectodermal vestibule, gives rise to the *echinus rudiment*, the imaginal structure which generates the juvenile urchin during metamorphosis (Czihak, 1971; Okazaki, 1975a). The eversion of the echinus rudiment is a geometrically complex process and the interested reader is urged to consult the beautiful drawings found in Czihak (1971) and Okazaki (1975a).

The sea urchin as an experimental system

Mutational analysis is not practical in the sea urchin because of the long generation times and the difficulties of rearing embryos through metamorphosis in the laboratory (but see some interesting mutants produced by Hinegardner, 1975b). In contrast to many of the commonly used experimental systems in which genetics is feasible, however, the early embryonic development of the sea urchin can be manipulated directly, thereby allowing epigenetic influences on development to be examined. In particular, the sea urchin system is an excellent system in which to study the role of cell interactions during early development.

Cell interactions in the early embryo

The landmark experiments of Driesch (e.g., 1891) in which he separated early blastomeres of the sea urchin embryo demonstrated that cells derived from the 2- and 4-cell embryo could *regulate* to produce a small embryo possessing most of the normal larval tissues. Since that time, the sea urchin embryo has been used to study how cells become committed to various pathways of differentiation. In particular, some of the most remarkable experiments in the history of embryology were those performed by Hörstadius (1939, 1973) investigating the properties of cells along the animal-vegetal axis of the early sea urchin embryo. By separating unfertilized eggs and early embryos into animal and vegetal halves, separating tiers of cells at the 32- and 64-cell stages, and by juxtaposing various tiers of cells originating at different locations along the animal-vegetal axis, Hörstadius found that a graded influence on differentiation existed along the animal-vegetal axis. These experiments led to the idea that a 'vegetalizing gradient', originating in the unfertilized egg, exerts effects along the embryonic axis. More recently, micromeres have been shown to interact with mesomeres at the level of single pairs of cells, so that eventually the mesomeres form an array of mesodermal and endodermal structures (reviewed by Livingston and Wilt, 1990). The effects of this vegetalizing influence can be mimicked by the *vegetalising agent*, lithium chloride, which confers progressively more vegetal qualities to blastomeres which would not normally possess them (see Hörstadius, 1973). More recent experiments using molecular assays have confirmed that lithium induces alterations in gene expression that include overproduction of endoderm in whole embryos (Nocente-McGrath *et al.*, 1991) and the appearance of mesodermal and endodermal markers in the descendants of single mesomeres (Livingston and Wilt, 1989).

Cell contact also seems to be important in regulating gene expression and differentiation in the early embryo. When mesomeres are isolated as an intact group of eight cells from a 16-cell embryo, they produce a ciliated ball, or *Dauerblastula* (Hörstadius, 1939); when, however, mesomeres are dissociated and recombined, thereby altering cell-cell contacts, they produce spicules and archenterons (Henry *et al.*, 1989). Pervasive alterations in zygotic gene expression also result when cell contacts are continuously prevented from forming by stirring blastomeres in calcium-free sea water (Hurley *et al.*, 1989; Stephens *et al.*, 1989). Cell contact and cell interactions thus seem to be important in specifying cell fate in experimentally treated blastomeres.

In contrast to most cells in the early sea urchin embryo, however, micromeres appear to be committed to a spiculogenic pathway as soon as they appear at the 16-cell stage. This conclusion is largely based on the work of Okazaki (1975b), who developed culture conditions allowing micromeres to differentiate *in vitro*. Cultured micromeres divide, become motile, migrate, and the cells later associate into syncitia and produce spicules (Okazaki, 1975b; see **3.41**). They also undergo several simultaneous adhesion changes at the time that pmc would ordinarily ingress into the blastocoel; these include:

- A loss of an affinity for the hyaline lamina proteins hyalin and echinonectin.
- A loss of an affinity for neighbouring cells.
- An increased affinity for the basal lamina (Fink and McClay, 1985; Burdsal *et al.*, 1991).

That these changes can occur in cultured micromeres suggests that this transformation is entirely autonomous. The fourth cleavage can be equalized by treatment of eggs with low concentrations of sodium dodecyl sulphate and, here, sixteen cells of equal size are produced, and, in some but not all cases, formation of pmc is suppressed (Langelan and Whiteley, 1985).

Cell interactions and the extracellular matrix

In the forming sea urchin embryo, cells interact not only with each other, but with the extracellular environment. The epithelial tissues of the sea urchin embryo are in contact with several extracellular matrix layers: the apical lamina, and hyaline layer, on their apical surfaces, and the basal lamina on their basal surfaces (**3.37, 3.38**). In addition, Dan (1960) has suggested that the hyaline layer is important as a structural support and mechanical integrator of epithelia. Consistent with this idea, treatment of sea urchin embryos with monoclonal antibodies specific for the protein hyalin (a major component of the hyaline layer) results in visible delamination of the hyaline layer from the epithelium, abnormal thickening of the epithelium, and blockage of invagination (Adelson and Humphreys, 1988; **3.39**). If antibody-treated embryos are removed from the antibody, development resumes, and a normal pluteus larva results (Adelson and Humphreys, 1988). These results suggest that the antibody interferes with the mechanical and structural integrity of the epithelium, but also with more general requirements for the initiation of gastrulation. Incubation of *Strongylocentrotus purpuratus* embryos from hatching through to gastrulation in F_{ab} fragments of antibodies which recognize a fibrous apical lamina protein results in disruption of the normal epibolic movements preceding gastrulation, and failure of invagination (Burke *et al.*, 1991), again pointing to an important integrating role for these apical layers during sea urchin embryogenesis.

Treatments affecting the basal lamina also block gastrulation. Incubation of embryos from fertilization onward in β-aminoproprionitrile (BAPN), an inhibitor of lysyl oxidase (an enzyme involved in collagen crosslinking), results in embryos which develop normally to the mesenchyme blastula stage, but fail to progress further. If the drug is removed, even after the embryos have been arrested at the mesenchyme blastula stage for more than 24 hours, the embryos begin to gastrulate and complete development normally (Butler *et al.*, 1987; Wessell and McClay, 1987). In the normal embryo, a fibrillar meshwork is present in the blastocoel along which mesenchyme cells appear to migrate (**3.40**). BAPN treatment results in a poorly constructed basal lamina and poor motility of mesenchyme cells (Butler *et al.*, 1987; Hardin, 1987) and, in addition, antigens normally expressed in the archenteron fail to appear as long as the embryos are incubated with the drug (Wessell and McClay, 1987). The similarity of the effects observed when either extracellular matrix layer is disrupted suggests that a critical period precedes gastrulation during which normal contact with both the basal lamina and the hyaline layer is required in order for gastrulation to begin.

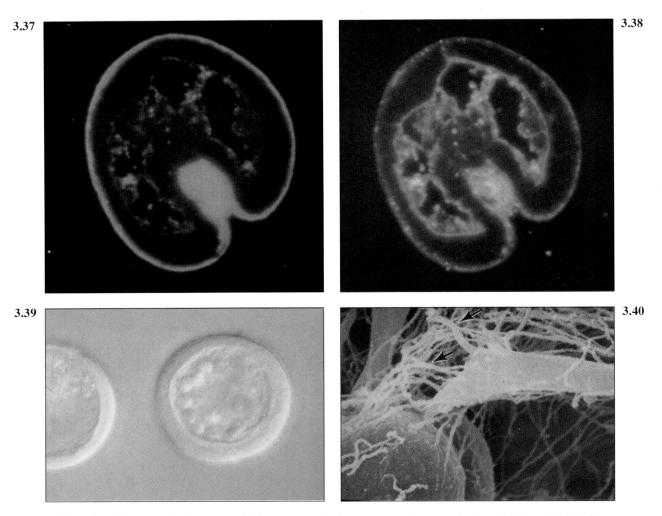

3.37–3.40 The role of the extracellular matrix during sea urchin gastrulation. **3.37** and **3.38** show immunofluorescent localization of the hyaline layer component hyalin and a basal lamina antigen, Ib10 (×550). **3.39** shows the effects of treating a *L. variegatus* embryo with antibodies against hyalin (×400). **3.40** is a scanning electron micrograph of a protrusion from a secondary mesenchyme cell; thin filopodia (arrows) contact fibrils of extracellular matrix (×10,000; courtesy of J. Morrill).

Cell interactions between primary mesenchyme cells and surrounding tissues

Although micromeres are committed early in development to form a pmc (**3.41**), it is also clear that the behaviour of later cells is affected by the embryonic environment. Okazaki *et al.* (1962) observed that pmc localize at sites in the ectoderm where the epithelial cells are thickened, producing an optical effect reminiscent of an oriental fan. When this belt of cells is shifted along the animal-vegetal axis in vegetalized embryos treated with lithium chloride, the pmc localize to the shifted ectoderm (Gustafson and Wolpert, 1961; Okazaki *et al.*, 1962). Likewise, when the normal differentiation of ectodermal tissues is disrupted by treatment of embryos with NiCl$_2$, pmc form spicules, but in a completely radialized pattern (Hardin *et al.*, 1992; **3.42**), even though transplantation experiments suggest that it is only the ectoderm, and not the mesenchyme cells, which are affected (Hardin and Armstrong, 1991). These experiments imply that regionally specific information is contained within the ectoderm which helps specify where pmc localize, but the molecular basis of this localized attachment is still not understood.

The pmc not only receive signals from the environment, but also appear to interact with themselves and other mesenchyme cells. When supernumerary pmc are transplanted into host embryos, the resulting skeleton is indistinguishable from the normal one, even though as many as two to three times the normal number of cells participate in skeleton formation (Ettensohn, 1990b) and it seems that some signal(s) operate which restrict both the size and location of skeletal elements. When all pmc are removed from the embryo, a skeleton arises from a subpopulation of secondary mesenchyme cells which become spiculogenic and produce a normal skeleton (Ettensohn and McClay, 1988; **3.43**), even though secondary mesenchyme cells do not normally participate in spicule production. The

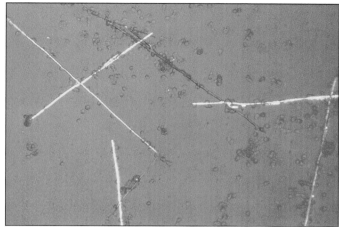
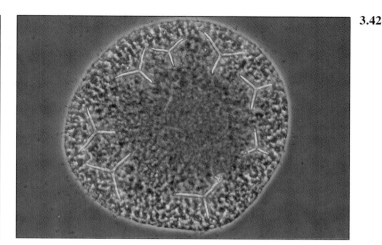
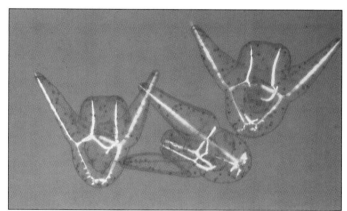

3.41–3.43 Cell autonomy and cell interaction during skeleton formation in the sea urchin embryo. **3.41** shows five picules formed by cultured *Arbacia punctulata* micromeres (×150; courtesy of R. Fink). **3.42** shows radialized skeleton formed in an embryo treated with 0.5 mM NiCl$_2$ from fertilization through the early gastrula stage (×500). **3.43** shows a skeleton produced by 'converted' secondary mesenchyme cells in embryos in which all primary mesenchyme cells were removed (×100; courtesy of C. Ettensohn).

pmc must therefore provide some restrictive signal preventing the secondary mesenchyme cells from differentiating into spicule-producing cells. This interaction displays a remarkably quantitative character for, as one removes more and more pmc from *Lytechinus variegatus*, a progressively greater number of secondary mesenchyme cells convert to compensate for the loss of spiculogenic cells (Ettensohn and McClay, 1988). This restrictive influence must operate over a distance of tens of microns since the two populations of mesenchyme are some distance from one another in the embryo. Based on studies by Ettensohn (1990a), the period during which this interaction occurs ends at about the time the archenteron makes contact with the animal pole (for possible mechanisms, see Ettensohn, 1991).

Cell interactions during archenteron formation

Surprisingly little is known about the forces which promote the inward bending of the archenteron, although many ideas have been put forward. Neither local proliferation of cells nor changes in lateral contact between cells which would ultimately result in bending seem to be involved (reviewed by Hardin, 1990). *Apical constriction* of cells in the vegetal plate has been proposed to account for primary invagination (Odell *et al.*, 1981; reviewed in Ettensohn, 1985b) and there are apically constricted cells in the centre of the vegetal plate, with adjacent cells having expanded apices, so suggesting that they are under tension (Hardin, 1989; **3.44**), but interpretable experimental disruption of this process has been difficult (see Hardin, 1990). *Swelling pressure* generated by secretion of proteoglycans into the lumen of the archenteron has also been suggested as a means by which the archenteron could invaginate (Morrill and Santos, 1985). In support of this idea, a chondroitin sulphate proteoglycan has been localized to the lumen of the archenteron (Lane and Solursh, 1991). Finally, a noticeable amount of *epiboly*, or spreading of the pre-gastrula epithelium, occurs just prior to gastrulation, and this includes tissue immediately adjacent to the vegetal plate (Ettensohn, 1984; Burke *et al.*, 1991). When these movements do not occur properly, invagination fails, so suggesting that they may be important for primary invagination (Burke *et al.*, 1991). Whatever mechanism(s) account for primary invagination, there are no forces outside the immediate vicinity of the vegetal plate required for its invagination, since the vegetal plate can be isolated several hours before primary invagination begins and it will still invaginate on schedule (Moore and Burt, 1939; Ettensohn, 1984).

Considerably more is known about the elongation of the archenteron and here *epithelial cell rearrangement* plays an

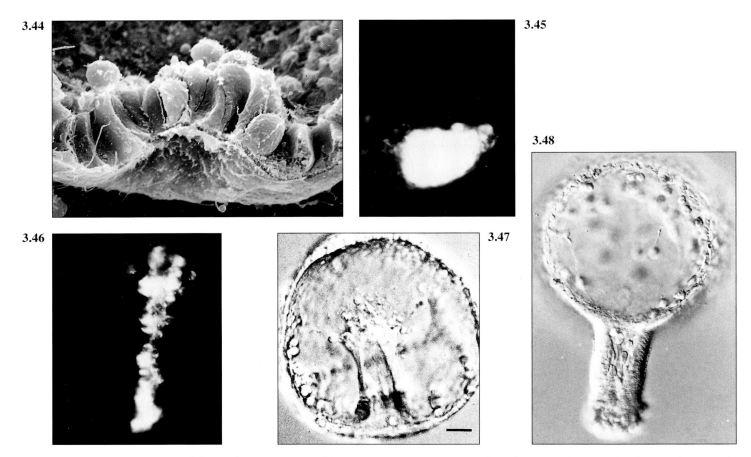

3.44–3.48 Formation of the archenteron *Lytechinus pictus*. **3.44** is a scanning electron micrograph of an early gastrula (×1500). **3.45** and **3.46** (×500) show changes in shape of rhodamine-labelled clones during gastrulation (**3.45**, early gastrula; **3.46**, late gastrula). **3.47** is an embryo is an embryo whose secondary mesenchyme cells have been ablated by a laser microbeam. Elongation of the archenteron ceases at two thirds of its normal final length (×500). **3.48** is an exogastrula produced by treatment with lithium chloride (×500).

important role. Ettensohn (1985a) deduced that, as the archenteron elongates, the number of cells around the circumference of the archenteron decreases, while more direct evidence for cell rearrangement comes from the behaviour of fluorescently labelled patches of cells within the vegetal plate during invagination: patches of such labelled cells gradually extend and narrow as cells interdigitate to lengthen the archenteron (**3.45, 3.46**). Cell rearrangement appears to be the dominant means by which the archenteron elongates, since additional material is not added by mitosis (Stephens *et al.*, 1986) or involution in *L. pictus* (Hardin, 1989). What forces drive this rearrangement? Several observations suggested that the filopodia of secondary mesenchyme cells can exert significant tension and led to the hypothesis that filopodial traction causes the archenteron to elongate (Dan and Okazaki, 1956; Gustafson, 1963; reviewed in Hardin, 1988), but two observations counter this suggestion. When filopodia are ablated with a laser microbeam or when embryos are induced to *exogastrulate*, so producing an evagination rather than an invagination, the archenteron elongates to two-thirds of the normal length, even though secondary mesenchyme cells do not attach and pull in either case (Hardin and Cheng, 1986; Hardin, 1988; **3.47, 3.48**).

The cellular processes which generate autonomous rearrangement are not understood. Direct observation of cell rearrangement in *Eucidaris tribuloides* suggests that cells 'jostle' against one another, and their basal ends display vigorous motility (Hardin, 1989), but how such motility might be translated into directed rearrangement is not known. In any case, completion of archenteron elongation requires the activity of secondary mesenchyme cells. In laser-irradiated embryos in which a few filopodia are left intact, the archenteron will continue to elongate after the two-thirds gastrula stage, but more slowly than normal (Hardin, 1988). At about the time that secondary mesenchyme cells reach the animal pole, there is a transient stretching of cells in the archenteron, apparently in response to filopodial traction (Hardin, 1989).

Cell interactions between secondary mesenchyme cells and the ectoderm

How do secondary mesenchyme cells aid the attachment of the archenteron to a specific site in the ectoderm? When observed closely, the basic behaviour of secondary mesenchyme cells involves continual extension of filopodia (**3.49**); these often remain attached for a time, but eventually detach and collapse, only to be re-extended in a cyclical fashion (Gustafson and Kinnander, 1956; Hardin and McClay, 1990). Analysis of the duration of attachments of filopodia which make contact near the animal pole (**3.50**) indicate that they attach 20–50 times longer than those making attachments at other sites (Hardin and McClay, 1990). Several lines of evidence indicate that secondary mesenchyme cells respond uniquely to this region (Hardin and and McClay, 1990). First, when the animal pole region is pushed towards the tip of the archenteron so that contact is allowed earlier than it would normally occur, the cyclical behaviour of the secondary mesenchyme cells ceases early, and the archenteron stably attaches ahead of schedule (**3.51**). Second, when embryos are squeezed into narrow diameter capillary tubing so that secondary mesenchyme cells cannot attach to the animal pole, they continue their cyclical extension for a longer period of time than in normal embryos. If the embryo is held in such a tube for several hours, some secondary mesenchyme cells eventually detach from the archenteron, migrate to the animal pole, and undergo the change in behaviour seen in normal embryos (**3.52**). If the embryo is released from the tube, it regains a spherical shape, and, as it does so, the tip of the archenteron rapidly attaches to the animal pole. Finally, archenterons attach to the nearest available apical plate region in fused multiple embryos. All of these experiments point to the existence of localized information in the animal pole region which elicits this specific change in the motility and behaviour of secondary mesenchyme cells (Hardin and McClay, 1990).

3.49–3.52 Behaviour of secondary mesenchyme cells during gastrulation in *L. variegatus*. **3.49** (×750) shows a laser scanning confocal image of a midgastrula stained with rhodamine-phalloidin. Filopodia radiate away from the archenteron, producing a 'spray' of protrusions which make contact with the ectoderm. **3.50** (×350) shows a late gastrula; secondary mesenchyme cells cease their protrusive activity when they make contact near the animal pole. **3.51** (×450) shows an embryo imprisoned in nylon cloth; secondary mesenchyme cells attach prematurely at the animal pole. **3.52** (×500) shows an embryo imprisoned in capillary tubing for 2hr. Some secondary mesenchyme cells have detached from the archenteron and congregated at the animal pole (arrow).

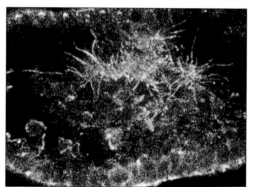

3.49

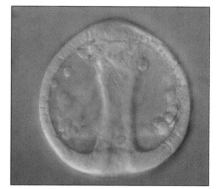

3.50

3.51

3.52

Phylogenetic differences in modes and timing of development

The sea urchin embryo has recently received attention as a system for examining phylogenetic diversity during early development (Raff, 1987; Raff and Wray, 1989; Wray and Raff, 1990). In different sea urchin species, for example, there seem to be several ways in which the archenteron can elongate, all apparently accounted for by differences in embryonic shape and placement of the future oral region (Hardin and McClay, 1990). In addition to variations in embryonic shape, there are also variations in the timing of developmental events with respect to one another when different species are compared (*heterochronies*), with ingression of spiculogenic cells being a good example of this sort of variation. Spicule-producing cells ingress at the mesenchyme blastula stage in the euechinoid sea urchin embryos with a typical larval mode of development, but ingression of spiculogenic cells in *Eucidaris* occurs many hours after invagination of the archenteron has begun, even though these cells are derived from the micromeres (Wray and McClay, 1987; Urben *et al.*, 1988). Even more radical alterations are evident in direct-developing sea urchins where the production of a functioning larval gut does not occur and whose metamorphosis is exceedingly rapid compared to sea urchins which pass through a true larval phase (**3.53–3.56**). Here, features of early development which appear to be devoted to the production of exclusively larval structures are often completely lost (reviewed by Raff, 1987; Raff and Wray, 1989; Wray and Raff, 1990). Other alterations include dramatic differences in cell lineages, mechanisms of cell fate determination along the dorsoventral axis, and mode of gastrulation (reviewed in Raff, 1992)).

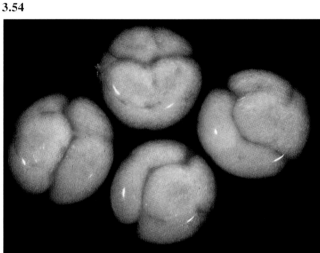

3.53–3.56 Direct development of the Australian sea urchin, *Heliocidaris erythrogramma*. **3.53** shows an unfertilized egg immersed in orange ink to visualize the jelly coat. **3.54** shows the wrinkled gastrula stage. **3.55** shows a four-day-old embryo immediately before metamorphosis, oral view. Tube feet and vestibule are visible. **3.56** shows a juvenile shortly after metamorphosis (×50; courtesy of L. Herlands).

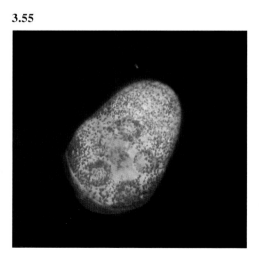

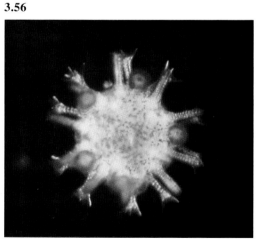

The future

The sea urchin embryo occupies an important place in the history of developmental biology, and continues to be a useful experimental system today. As a model system for uniting experimental embryology and observation with the modern tools of molecular and cell biology, the sea urchin embryo continues to provide unique opportunities for studying early development. Current work is focusing on the roles of cell adhesion during gastrulation, of cell-cell interactions in altering the expression of specific genes, and of DNA binding factors in the regulation of tissue-specific gene expression, a particularly prominent feature of sea urchin development (reviewed by Davidson, 1989). As the tools of modern biology become increasingly refined and powerful, the simplicity of the sea urchin embryo and the ease with which it can be manipulated will provide a rich context in which to study the molecular basis of early development.

References

Adelson, D.L. and Humphreys, T. (1988). Sea urchin morphogenesis and cell-hyalin adhesion are perturbed by a monoclonal antibody specific for hyalin. *Develop.*, **104**, 391–402.

Anstrom, J.A. and Raff, R.A. (1988). Sea urchin primary mesenchyme cells: Relation of cell polarity to the epithelial-mesenchymal transformation. *Dev. Biol.*, **130**, 57–66.

Benson, S., Sucov, H., Stephens, L., Davidson, E. and Wilt, F. (1987). A lineage specific gene encoding a major matrix protein of the sea urchin embryo spicule; I. Authentication of the cloned gene and its developmental expression. *Dev. Biol.*, **120**, 499–506.

Bestor, T.M. and Schatten, G. (1981). Anti-tubulin immunofluorescence microscopy of microtubules present during the pronuclear movements of sea urchin fertilization. *Dev. Biol.*, **88**, 80–91.

Bisgrove, B.W. and Burke, R.D. (1987). Development of the nervous system of the pluteus larva of *Strongylocentrotus drobachiensis*. *Cell Tissue Res.*, **248**, 335–343.

Boveri, T. (1901). Über die Polarität von Ovocyte, Ei und Larve des *Strongylocentrotus lividus*. *Zool. Jahrb. (Abt. Anat.)*, **14**, 630–653.

Burdsal, C.A., Alliegro, M.C. and McClay, D.R. (1991). Tissue-specific temporal changes in cell adhesion to echinonectin in the sea urchin embryo. *Dev. Biol.*, **144**, 327–334.

Burke, R.D., Myers, R.L., Sexton, T.L. and Jackson, C. (1991). Cell movements during the initial phase of gastrulation in the sea urchin embryo. *Dev. Biol.*, **146**, 542–557.

Butler, E., Hardin, J. and Benson, S. (1987). The role of lysyl oxidase and collagen crosslinking during sea urchin development. *Exp. Cell Res.*, **173**, 174–182.

Cameron, R.A. and Davidson, E.H. (1991). Cell type specification during sea urchin development. *Trends Gen.*, **7**(7), 212–218.

Coffman, J.A. and McClay, D.R. (1990). A hyaline layer protein that becomes localized to the oral ectoderm and foregut of sea urchin embryos. *Dev. Biol.*, **140**, 93–104.

Czihak, G. (1971). Echinoids. In *Experimental Embryology of Marine and Freshwater Invertebrates* (G. Reverberi, ed.) pp. 363–506. Amsterdam, North Holland.

Dan, K. (1960). Cyto-embryology of echinoderms and amphibia. *Int. Rev. Cytol.*, **9**, 321–367.

Dan, K. and Inaba, D. (1968). Echinoderma. In *Invertebrate Embryology* (M. Kumé, and K. Dan, eds) pp. 280–332, Belgrade, NOLIT Publishing House.

Dan, K. and Okazaki, K. (1956). Cyto-embryological studies of sea urchins. III. Role of secondary mesenchyme cells in the formation of the primitive gut in sea urchin larvae. *Biol. Bull.*, **110**, 29–42.

Dan, K., Tanaka, S., Yamazaki, K. and Kato, Y. (1980). Cell cycle study up to the time of hatching of the embryos of the sea urchin, *Hemicentrotus pulcherrimus*. *Devel. Growth and Diff.*, **22**, 589–598.

Davidson, E. (1988). *Gene Activity in Early Development*, 3rd Edition. New York, Academic Press.

Davidson, E.H. (1989). Lineage-specific gene expression and the regulative capacities of the sea urchin embryo: a proposed mechanism. *Development*, **105**, 421–445.

Davidson, E.H. (1990). How embryos work: a comparative view of diverse modes of cell fate specification. *Development*, **108**, 365–389.

Decker, G.L. and Lennarz, W.J. (1988). Skeletogenesis in the sea urchin embryo. *Develop.*, **103**, 231–247.

Drager, B.J., Harkey, M.A., Iwata, M. and Whiteley, A.H. (1989). The expression of embryonic primary mesenchyme genes of the sea urchin, *Strongylocentrotus purpuratus*, in the adult skeletogenic tissues of this and other species of echinoderms. *Dev. Biol.*, **133**, 14–233.

Driesch, H. (1892). Entwicklungsmechanische Studien. I. Der Werth der beiden ersten Furchungszellen in der Echinodermenentwicklung. Experimentelle Erzeugen von Theil- und Doppelbildung. *Zeitschr. wiss. Zool.* **53**.

Ettensohn, C.A. (1984). Primary invagination of the vegetal plate during sea urchin gastrulation. *Amer. Zool.*, **24**, 571–588.

Ettensohn, C.A. (1985a). Gastrulation in the sea urchin is accompanied by the rearrangement of invaginating epithelial cells. *Dev. Biol.*, **112**, 383–390.

Ettensohn, C.A. (1985b). Mechanisms of epithelial invagination. *Quart. Rev. Biol.*, **60**, 289–307.

Ettensohn, C.A. (1990a). Cell interactions in the sea urchin embryo studied by fluorescence photoablation. *Science*, **248**, 1115–1118.

Ettensohn, C.A. (1990b). The regulation of primary mesenchyme cell patterning. *Dev. Biol.*, **140**, 261–271.

Ettensohn, C.A. (1991). Mesenchyme cells interactions in the sea urchin embryo. In *Cell-Cell Interactions in Early Development* (J. Gerhart, ed.) pp.175–201, New York, John Wiley and Sons, Inc.

Ettensohn, C.A. and McClay, D.R. (1988). Cell lineage conversion in the sea urchin embryo. *Dev. Biol.*, **125**, 396–409.

Fink, R.D. and McClay, D.R. (1985). Three cell recognition changes accompany the ingression of sea urchin primary mesenchyme cells. *Dev. Biol.*, **107**, 66–74.

Foltz, K., Partin, J.S. and Lennarz, W.J. (1993). Sea urchin egg receptor for sperm: sequence similarity of binding domain to hsp70. *Science*, **259**, 1421–1425.

Gibson, A.W. and Burke, R.D. (1985). The origin of pigment cells in embryos of the sea urchin *Strongylocentrotus purpuratus*. *Dev. Biol.*, **107**, 414–419.

Glabe, C.G. and Vacquier, V.D. (1978). Egg surface glycoprotein receptor for sea urchin sperm bindin. *Proc. Nat. Acad. Sci. USA*, **75**, 881–885.

Gustafson, T. (1963). Cellular mechanisms in the morphogenesis of the sea urchin embryo. Cell contacts within the ectoderm and between mesenchyme and ectoderm cells. *Exp. Cell Res.*, **32**, 570–589.

Gustafson, T. and Kinnander, H. (1956). Microaquaria for time-lapse cinematographic studies of morphogenesis in swimming larvae and observations on sea urchin gastrulation. *Exp. Cell Res.*, **11**, 36–51.

Gustafson, T. and Wolpert, L. (1961). Studies on the cellular basis of morphogenesis in the sea urchin embryo; directed movements of primary mesenchyme cells in normal and vegetalized larvae. *Exp. Cell Res.*, **24**, 64–79.

Gustafson, T. and Wolpert, L. (1962). Cellular mechanisms in the formation of the sea urchin larva. Change in shape of cell sheets. *Exp. Cell Res.*, **27**, 260–279.

Gustafson, T. and Wolpert, L. (1963). The cellular basis of morphogenesis and sea urchin development. *Int. Rev. Cyt.*, **15**, 139–214.

Gustafson, T. and Wolpert, L. (1967). Cellular movement and contact in sea urchin morphogenesis. *Biol. Rev.*, **42**, 442–498.

Hall, G. and Vacquier, V. (1982). The apical lamina of the sea urchin embryo: major glycoproteins associated with the hyaline layer. *Dev. Biol.*, **89**, 160–178.

Hardin, J. (1988). The role of secondary mesenchyme cells during sea urchin gastrulation studied by laser ablation. *Development*, **103**, 317–324.

Hardin, J. (1989). Local shifts in position and polarized motility drive cell rearrangement during sea urchin gastrulation. *Dev. Biol.*, **136**, 430–445.

Hardin, J. (1990). Context-sensitive cell behaviors during gastrulation. *Sem. Dev. Biol.*, **1**, 335–345.

Hardin, J. and Armstrong, N. (1991). Developmental regulation of animal pole targets for mesenchyme cells in the sea urchin embryo. *J. Cell Biol.*, **115**, 464a.

Hardin, J., Coffman, J.A., Black, S.D. and McClay, D.R. (1992). Commitment along the dorsoventral axis of the the sea urchin embryo is altered in response to $NiCl_2$. *Development*, **116**, 671–685.

Hardin, J. and McClay, D.R. (1990). Target recognition by the archenteron during sea urchin gastrulation. *Dev. Biol.*, **142**, 87–105.

Hardin, J.D. (1987). Disruption of collagen crosslinking during sea urchin morphogenesis. In *45th Ann. Proc. Electr. Microsc. Soc. Amer.*, (G.W. Bailey, ed.) pp. 786–787, San Francisco, San Francisco Press.

Hardin, J.D. and Cheng, L.Y. (1986). The mechanisms and mechanics of archenteron elongation during sea urchin gastrulation. *Dev. Biol.*, **115**, 490–501.

Harvey, E.B. (1956). *The American Arbacia and Other Sea Urchins*. Princeton, Princeton University Press.

Henry, J.J., Amemiya, S., Wray, G.A. and Raff, R.A. (1989). Early inductive interactions are involved in restricting cell fates of mesomeres in sea urchin embryos. *Dev. Biol.*, **136**, 140–153.

Hinegardner, R. (1967). Echinoderms. In *Methods in Developmental Biology* (F.H. Wilt, and N.K. Wessels, eds) pp. 130–155, New York, Thomas Y. Crowell.

Hinegardner, R. (1975a). In *The Sea Urchin Embryo* (G. Czihak, ed.). Berlin, Springer-Verlag.

Hinegardner, R.T. (1975b). Morphology and genetics of sea urchin development. *Amer. Zool.*, **15**, 679–689.

Holy, J. and Schatten, G. (1991). Differential behavior of centrosomes in unequally dividing blastomeres during fourth cleavage of sea urchin embryos. *J. Cell Sci.*, **98**, 423–431.

Hörstadius, S. (1935). Über die Determination im Verlaufe der Eiachse bei Seeigeln. *Pubbl. Staz. Zool. Napoli*, **14**, 251–429.

Hörstadius, S. (1939). The mechanics of sea urchin development, studied by operative methods. *Bio. Rev. Cambridge Phil. Soc.*, **14**, 132–179.

Hörstadius, S. (1973). *Experimental Embryology of Echinoderms*. Oxford, Clarendon Press.

Hurley, D.L., Angerer, L.M. and Angerer, R.C. (1989). Altered expression of spatially regulated embryonic genes in the progeny of separated sea urchin blastomeres. *Development*, **106**, 567–579.

Ishimoda-Takagi, T., Chino, I. and Sato, H. (1984). Evidence for the involvement of muscle tropomyosin in the contractile elements of the coelom-esophagus complex in sea urchin embryos. *Dev. Biol.*, **105**, 365-376.

Katow, H. and Solursh, M. (1980). Ultrastructure of primary mesenchyme cell ingression in the sea urchin *Lytechinus pictus*. *J. Exp. Zool.*, **213**, 231–246.

Lane, M.C. and Solursh, M. (1988). Dependence of sea urchin primary cell migration on xyloside- and sulfate-sensitive cell surface associated components. *Dev. Biol.*, **127**, 78–87.

Lane, M.C. and Solursh, M. (1991). Primary mesenchyme cell migration requires a chondroitin sulfate/dermatan sulfate proteoglycan. *Dev. Biol.*, **143**, 389–397.

Langelan, R.E. and Whiteley, A.H. (1985). Unequal cleavage and the differentiation of echinoid primary mesenchyme cells. *Dev. Biol.*, **109**, 464–475.

Leaf, D.S., Anstrom, J.A., Chin, J.E., Harkey, M.A. and Raff, R.A. (1987). A sea urchin primary mesenchyme cell surface protein, msp130, defined by cDNA probes and antibody to fusion protein. *Dev. Biol.*, **121**, 29–40.

Leahy, P.S. (1986). Laboratory culture of *Strongylocentrotus purpuratus* adults, embryo, and larvae. In *Echindoerm Gametes and Embryos* (T. Schroeder, ed.) pp. 1–13, Orlando, Academic Press, Inc.

Lepage, T., Ghiglione, C. and Gache, C. (1992). Spatial and temporal expression pattern during sea urchin embryogenesis of a gene coding for a protease homologous to the human protein BMP-1 and to the product of the *Drosophila* dorsal-ventral gene *tolloid*. *Development*, **114**, 147–164.

Livingston, B.T. and Wilt, F.H. (1989). Lithium evokes expression of vegetal-specific molecules in the animal blastomeres of sea urchin embryos. *Proc. Nat. Acad. Sci. USA*, **86**, 3669–3673.

Livingston, B.T. and Wilt, F.H. (1990). Determination of cell fate in sea urchin embryos. *Bioessays*, **12**, 115–119.

Lynn, D.A., Angerer, L.M., Bruskin, A.M., Klein, W.H. and Angerer, R.C. (1983). Localization of a family of mRNAs in a single cell type and its precursors in sea urchin embryos. *Proc. Nat. Acad. Sci. USA*, **80**, 2656–2660.

Maruyama, Y.K., Nakaseko, Y. and Yagi, S. (1985). Localization of cytoplasmic determinants responsible for primary mesenchyme formation and gastrulation in the unfertilized egg of the sea urchin *Hemicentrotus pulcherrimus*. *J. Exp. Zool.*, **236**, 155–163.

McClay, D.R., Alliegro, M.C. and Black, S.D. (1990). The ontogenetic appearance of extracellular matrix during sea urchin development. In *Recognition and Assembly of Plant and Animal Extracellular Matrix* (R. Mecham and S. Adair, eds) pp. 1–13, New

Moore, A.R. (1940). Osmotic and structural properties of the blastular wall in *Dendraster excentricus*. *J. Exp. Zool.*, **84**, 73–79.

Moore, A.R. and Burt, A.S. (1939). On the locus and nature of the forces causing gastrulation in the embryos of *Dendraster excentricus*. *J. Exp. Zool.*, **82**, 159–171.

Morgan, T.H. (1927). *Experimental Embryology*. Columbia University Press, New York.

Morrill, J.B. and Santos, L.L. (1985). A scanning electron microscopical overview of cellular and extracellular patterns during blastulation and gastrulation in the sea urchin, *Lytechinus variegatus*. In *The Cellular and Molecular Biology of Invertebrate Development* (R.H. Sawyer and R.M. Showman, eds) pp. 3–33, Columbia, S.C., Univ. South Carolina Press.

Moy, G.W. and Vacquier, V.D. (1979). Immunoperoxidase localization of bindin during the adhesion of sperm to sea urchin eggs. *Curr. Top. Dev. Biol.*, **13**, 31–44.

Nocente-McGrath, C., Brenner, C.A. and Ernst, S.G. (1989). Endo16, a lineage-specific protein of the sea urchin embryo, is first expressed just prior to gastrulation. *Dev. Biol.*, **136**, 264–272.

Nocente-McGrath, C., McIsaac, R. and Ernst, S.G. (1991). Altered cell fate in LiCl-treated sea urchin embryos. *Dev. Biol.*, **147**, 445–450.

Odell, G.M., Oster, G., Alberch, P. and Burnside, B. (1981). The mechanical basis of morphogenesis. I. Epithelial folding and invagination. *Dev. Biol.*, **85**, 446–462.

Okazaki, K. (1975a). Normal development to metamorphosis. In *The Sea Urchin Embryo* (G. Czihak, ed.) pp. 177–232, Berlin, Springer-Verlag.

Okazaki, K. (1975b). Spicule formation by isolated micromeres of the sea urchin embryo. *Amer. Zool.*, **15**, 567–581.

Okazaki, K., Fukushi, T. and Dan, K. (1962). Cyto-embryological studies of sea urchins. IV. Correlation between the shape of the ectodermal cells and the arrangement of the primary mesenchyme cells in sea urchin larvae. *Acta Embryol. Morphol. Exp.*, **5**, 17–31.

Pehrson, J.R. and Cohen, L.H. (1986). The fate of the small micromeres in sea urchin development. *Dev. Biol.*, **113**, 522–526.

Raff, R.A. (1987). Constraint, flexibility, and phylogenetic history in the evolution of direct development in sea urchins. *Dev. Biol.*, **119**, 6–19.

Raff, R.A. (1992). Evolution of developmental decisions and morphogenesis: the view from two camps. *Development,* 1992, Supplement, 15–22.

Raff, R.A. and Wray, G.A. (1989). Heterochrony: developmental mechanisms and evolutionary results. *J. Evol. Biol.*, **2**, 409–434.

Schroeder, T.E. (1980a). Expressions of the prefertilization axis in sea urchin eggs. *Dev. Biol.*, **79**, 428–443.

Schroeder, T.E. (1980b). The jelly canal marker of polarity for sea urchin oocytes, eggs, and embryos. *Exp. Cell Res.*, **128**, 490–494.

Schroeder, T.E. (1986). *Echinoderm Gametes and Embryos*. Orlando, Academic Press, Inc.

Solursh, M. (1986). Migration of sea urchin primary mesenchyme cells. In *The Cellular Basis of Morphogenesis* (L. Browder, ed.) pp. 391–431, New York, Plenum Press.

Solursh, M., Mitchell, S.L. and Katow, H. (1986). Inhibition of cell migration in sea urchin embryos by β-D-xyloside. *Dev. Biol.*, **118**, 325–332.

Spiegel, E. and Howard, L. (1983). Development of cell junctions in sea urchin embryos. *J. Cell Sci.*, **62**, 27–48.

Stephens, L., Hardin, J., Keller, R. and Wilt, F. (1986). The effects of aphidicolin on morphogenesis and differentiation in the sea urchin embryo. *Dev. Biol.*, **118**, 64–69.

Stephens, L., Kitajima, T. and Wilt, F. (1989). Autonomous expression of tissue-specific genes in dissociated sea urchin embryos. *Development*, **107**, 299–307.

Urben, S., Nislow, C. and Spiegel, M. (1988). The origin of skeleton forming cells in the sea urchin embryo. *Roux's Arch. Dev. Biol.*, **197**, 447–456.

Ward, G.E., Brokaw, C.J., Garbers, D.L. and Vacquier, V.D. (1985). Chemotaxis of *Arbacia punctulata* spermatozoa to resact, a peptide from the egg jelly layer. *J. Cell Biol.*, **101**, 2324–2329.

Weidman, P.J. and Kay, E.S. (1986). Egg and extracellular coats: isolation and purification. In *Echinoderm Gametes and Embryos* (T.E. Schroeder, ed.) pp. 113–138, Orlando, Academic Press, Inc.

Wessell, G.M., Goldberg, L., Lennarz, W.J. and Klein, W.H. (1989). Gastrulation in the sea urchin embryo is accompanied by the accumulation of an endoderm-specific mRNA. *Dev. Biol.*, **136**, 526–536.

Wessell, G.M. and McClay, D.R. (1985). Sequential expression of germ-layer specific molecules in the sea urchin embryo. *Dev. Biol.*, **111**, 451–463.

Wessell, G.M. and McClay, D.R. (1987). Gastrulation in the sea urchin embryo requires the deposition of crosslinked collagen within the extracellular matrix. *Dev. Biol.*, **121**, 149–165.

Wessell, G.M., Zhang, W. and Klein, W.H. (1990). Myosin heavy chain accumulates in dissimilar cell types of the macromere lineage in the sea urchin embryo. *Dev. Biol.*, **140**, 447–454.

Wilson, E.B. (1925). *The Cell in Development and Heredity,* 3rd Edition, New York, The Macmillan Co.

Wilt, F.H. (1987). Determination and morphogenesis in the sea urchin embryo. *Develop.*, **100**, 559–575.

Wray, G.A. and McClay, D.R. (1988). The origin of spicule-forming cells in a primitive sea urchin (*Eucidaris tribuloides*) which appears to lack primary mesenchyme. *Develop.*, **103**, 305–315.

Wray, G.A. and McClay, D.R. (1989). Molecular heterochronies and heterotopies in early echinoid development. *Evolution*, **43**, 803–813.

Wray, G.A. and Raff, R.A. (1990a). Pattern and process heterochronies in the early development of sea urchins. *Sem. Dev. Biol.*, **1**, 245–251.

Wray, G.A. and Raff, R.A. (1990b). Novel origins of lineage founder cells in the direct-developing sea urchin *Heliocidaris erythrogramma*. *Dev. Biol.*, **141**, 41–54.

4. *Caenorhabditis elegans*, the Nematode Worm

Ian A. Hope

Introduction

The current interest in the nematode *Caenorhabditis elegans* began approximately 25 years ago when Sidney Brenner selected this species as the most suitable for studies of metazoan development and nervous systems (Brenner, 1974). The basis of this selection rested on the anatomical simplicity of nematodes, which nevertheless possess the major differentiated cell types of higher animals, and the tractability of *C. elegans* to the genetic approach. Over the past two decades or so, progress has been impressive: the cell lineage from egg to adult (Sulston and Horvitz, 1977; Sulston *et al.*, 1983) and the anatomy of the nervous system have been completely described (White *et al.*, 1986), genetic investigations of numerous developmental problems are co-ordinated within a universally-agreed, systematic nomenclature (Edgley and Riddle, 1989), a physical map of the *C. elegans* genome is nearing completion (Coulson *et al.*, 1988) and a project to sequence the entire genome is underway (Lewin, 1990; Sulston *et al.*, 1992). Furthermore, the number of laboratories seeking to understand the mechanisms controlling animal development through genetic and molecular investigations of *C. elegans* is rising rapidly as the advantages of this organism become increasingly appreciated[1].

Background biology

C. elegans is a free-living nematode worm. The nematode, round worms should not be confused with the platyhelminth flatworms or the segmented, annelid worms. Nematodes are pseudocoelomates, an evolutionarily ancient group often placed at the base of the protostome branch of the metazoan phylogenetic tree (**4.1**).

The natural environment of *C. elegans* is in the soil where the organism feeds mainly on bacteria. In the laboratory the organism is conveniently grown at 20°C, on agar plates (**4.2**) or

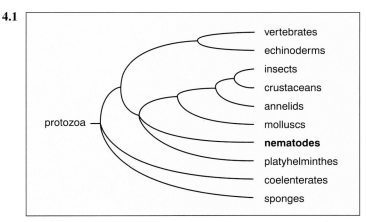

4.1 A phylogenetic tree to indicate the probable evolutionary relationship of nematodes to the other major animal groups. (This figure was compiled from information in Raff and Kaufman (1983), Poinar (1983), Gilbert (1991) and Alberts *et al.* (1989).)

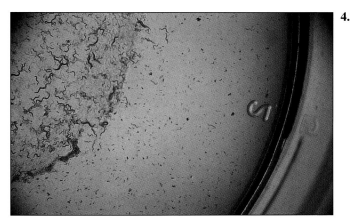

4.2 *C. elegans* as grown in the laboratory. A light micrograph of a *C. elegans* culture on the agar surface of a 50mm diameter petri dish. Adult worms are 1mm long (x3).

---

[1] Techniques for handling *C. elegans* are given in Sulston and Hodgkin (1988) while mutant strains can be obtained directly from the laboratories in which they were generated or, for strains containing canonical mutations, from Bob Herman, Caenorhabditis Genetics Centre, University of Minnesota, 1445, Gortner Avenue, St. Paul, MN 55108-1095, USA.

in liquid culture. *Escherichia coli* as the food source is grown separately and provided before addition of C. elegans. The worms are small, up to 1mm in length and 80μm in diameter, and may be moved in bulk in liquid or individually using a dissecting microscope. The life cycle is short (3.5 days) and the brood size is large (>300), so that, if a single adult *C. elegans* hermaphrodite is placed on a 9cm plate seeded with bacteria, some 500,000 individuals will have been generated in a little over a week.

General anatomy

C. elegans has two sexes, a self-fertilizing hermaphrodite and a male, with both sexes having a very similar general anatomy (**4.3–4.6**; White, 1988). Two concentric tubes of tissue run the length of each individual and are separated by the pseudocoelom. The outer tube is composed of cuticle, hypodermis, muscle and nerve cells. The inner tube is the gut which opens anteriorly at the tip of the head and ventrally, close to the tail.

Electron microscopy has revealed multiple layers in the extracellular cuticle, each having a characteristic fibre organization (Bird, 1980). Several components of the cuticle have been identified but little is known of the mechanism of assembly. The hypodermis, which secretes the cuticle, is one cell deep over the entire worm and most of the cells are multinucleate. Beneath the hypodermis, the 95 mononucleate, body-wall muscle cells are arranged in four longitudinal bands, each two cells wide, running the length of an individual, two bands sub-dorsal and two sub-ventral (Waterston, 1988). The co-ordinated contraction of ventral versus dorsal quadrants causes sinusoidal waves to pass down the individual's body, propelling the worm along the agar surface on its side. Most of the nerve cells are located in the nerve ring surrounding the pharynx, in the ventral nerve cord or in the tail ganglia (White *et al.*, 1986).

The gut consists of a muscular pharynx and an intestine. Apart from the 20 large muscle cells, the pharynx also contains 20 neurons, 9 epithelial cells and 9 marginal cells (Albertson and Thomson, 1976). Bacteria are taken in and then concentrated and crushed by the pharynx before being passed on to the intestine. The intestine is composed of 20 cells arranged as 9 rings, each one cell wide.

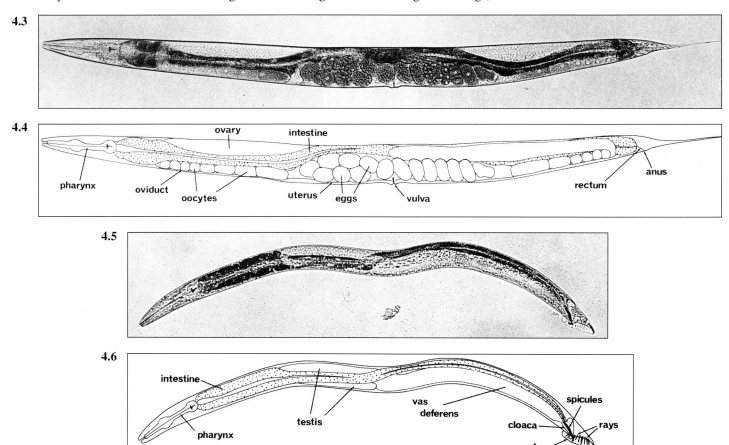

4.3–4.6 Light micrographs and line drawings showing the major anatomical features of *C. elegans*. **4.3** and **4.4** show an adult hermaphrodite and **4.5** and **4.6** show an adult male, all in lateral views. (**4.3** and **4.5** ×100. Reprinted with permission from Sulston and Horvitz, 1977.)

The tubular gonad is located in the pseudocoelomic space (Kimble and Hirsh, 1979; Hodgkin, 1988). In the hermaphrodite, the gonad is bilobed and symmetrical about the single mid-ventral opening at the vulva, with each arm of the gonad extending either anteriorly or posteriorly from the uterus (**4.7**) through a spermatheca and oviduct before reflexing back towards the centre of the animal through the ovary to the distal tip. The germ line matures with passage down the gonad (Kimble and Ward, 1988) and proliferates by mitosis at the distal end of the gonad, with meiosis occurring towards its bend. The first 150 germ cells produced by each arm differentiate as sperm for self-fertilization, while subsequent germ cells differentiate as oocytes which bud off from the syncytium prior to fertilization in the spermatheca. Fertilized eggs are retained in the uterus for several hours before they are laid through the vulva.

The single-lobed male gonad opens at the cloaca near the tail. The testis is at the distal tip and runs anteriorly from a mid-ventral position before reflexing back posteriorly through the seminal vesicle and vas deferens to reach the cloaca. The germ line shows a pattern of maturation along the gonad similar to that in the hermaphrodite, except that only sperm are produced. After mating, male sperm out-compete hermaphrodite sperm for fertilization of the hermaphrodite oocytes.

The sexes are most easily distinguished under the dissect-

4.7 Symmetry of the hermaphrodite gonad about the vulva. A transformant containing a fusion of a *C. elegans* promoter to the *E. coli* lacZ gene has been stained for β-galactosidase activity, so revealing expression in the uterus (×100; Hope, 1991).

ing microscope by their tail morphologies (**4.3–4.6** and **4.68** and **4.72**). The hermaphrodite tail is a simple long spike, while that of the male is an elaborate structure consisting of an extended cuticular fan and associated nerve and muscle cells specialized for mating with the hermaphrodite.

Normal development

C. elegans is transparent throughout its life cycle and development can therefore be followed by direct observation using differential interference contrast (DIC) or Nomarski optics. Using this technique to follow the fate of individual cells in live embryos, the entire *C. elegans* cell lineage has been determined (Sulston and Horvitz, 1977; Sulston *et al.*, 1983).

Embryonic development

Embryogenesis lasts 14hr and can be divided into two phases (**4.8**; Wood, 1988). For the first 350min, cell division proceeds rapidly to generate an ovoid of 550 cells showing little overt differentiation. (All times quoted are from first cleavage, for development at 20°C). The second phase involves differentiation and morphogenesis to generate the elongated worm during which there are very few additional cell divisions.

Development of the early embryo is described in terms of the invariant cell lineage (**4.9–4.11**; Sulston *et al.*, 1983). The initial divisions are unequal, producing daughters of different size to generate the founder cells, AB, C, D, E, MS and P_4. The divisions of each founder cell and its progeny are equal and occur at a particular, characteristic rate. The synchrony of the divisions of descendants of each founder cell does eventually break down, but is a useful marker for cell identification. Some of the founder cells produce a single tissue (E gives rise to all of and only the intestine), while other founder cells produce many different tissue types (C produces muscle, neuronal and hypodermal cells).

Gastrulation starts 100min after first cleavage when the embryo consists of 28 cells. The two daughters of E, lying ventrally and posteriorly, are the first to sink inwards. These are followed by P_4 and the eight descendants of MS. Later, descendants of C and D and some cells from the AB lineage also enter through the ventral cleft before it closes after 230–290min of development.

The extent of elongation to the comma, $1\frac{1}{2}$-, 2- and 3-fold embryo, provides an obvious marker of progress during the second phase of embryogenesis. Embryonic elongation is achieved by change in cell shape and not through directional cell division (Priess and Hirsh, 1986). Upon completion of elongation, cuticle is laid down (at 660min) to maintain the worm's shape. Twitching, which follows the differentiation of muscle cells, begins at 400min as the contractile proteins within muscle cells become properly organized and the elongating worm then twists continuously within the egg until hatching. Neuronal processes also grow out during this second phase of embryogenesis.

Studies on *C. elegans* embryogenesis have used several markers of cell type and differentiation. The germline can be identified throughout development using monoclonal anti-

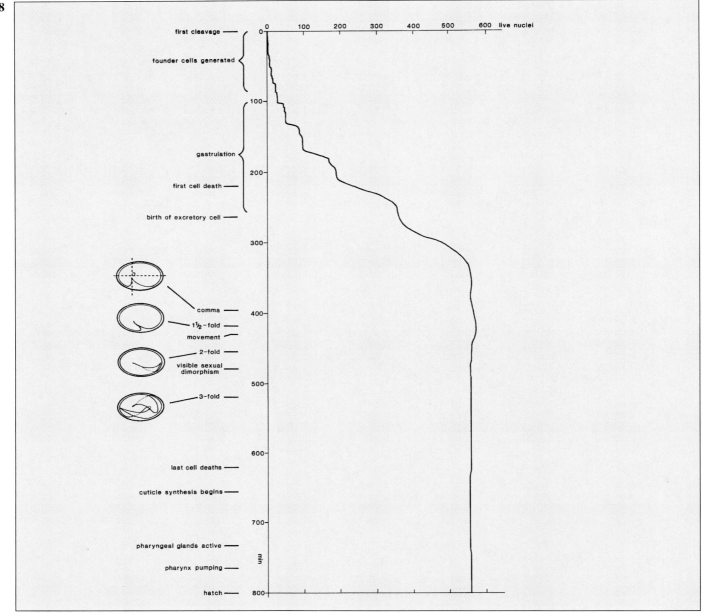

4.8 Graph of cell number against time during embryogenesis. The times of key events during embryogenesis are also indicated. (Reprinted with permission from Sulston et al., 1983.)

bodies which recognize cytoplasmic P granules (**4.12–4.19**; Strome and Wood, 1983). Two markers have been used for gut-cell differentiation: rhabditin granules can be detected in the E-cell lineage of live embryos with polarization or fluorescence optics (**4.20–4.23**; Babu, 1974), while a gut-specific carboxylesterase can be detected earlier in fixed embryo preparations (Edgar and McGhee, 1986). A monoclonal antibody recognizing the belt desmosomes of hypodermal cells has been used in studies on embryonic elongation (**4.24–4.29**; Priess and Hirsh, 1986). Monoclonal antibodies which recognize various muscle cell components (Francis and Waterston, 1985) have been used as markers of muscle cell differentiation. Finally, β-galactosidase is now being used as a reporter of gene expression patterns (Fire et al., 1990) both in general screens (Hope, 1991) and in investigations of specific genes (e.g. Krause et al., 1990). These studies promise to yield many cytogenetic markers of differentiation which should be useful in other studies of C. elegans development.

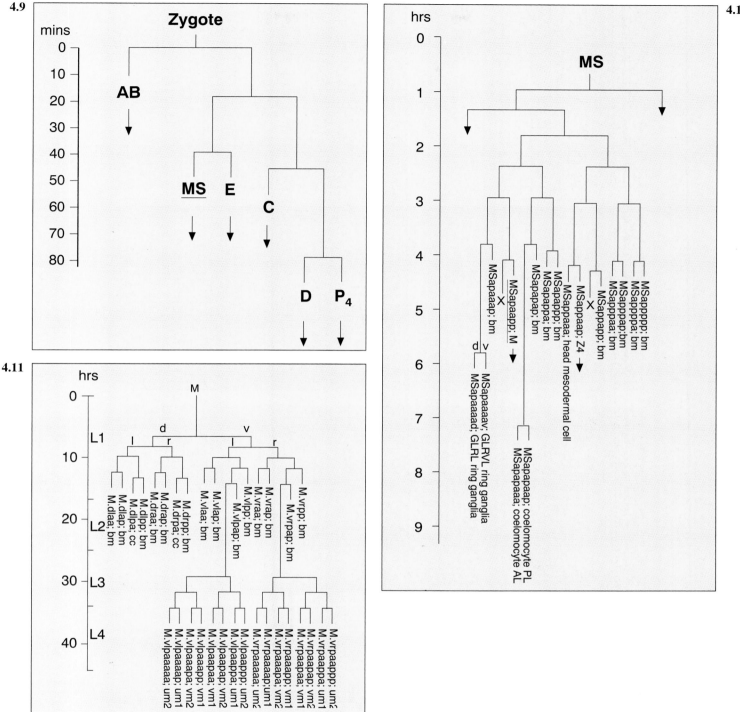

4.9–4.11 The entire cell lineage is known for *C. elegans*, from zygote to adult (Sulston *et al.*, 1983; Sulston and Horvitz, 1977). The lineage has been represented in a graphic form with time (at 20°C) down the vertical axis and each horizontal line being a cell division. All cell divisions with an anterior/posterior component are arranged with the anterior daughter to the left. Where no anterior/posterior component is present, the direction of cell division is indicated by **d** and **v** for dorsal and ventral daughter or **l** and **r** for left and right daughter. Further cell division is indicated with an arrow. **4.9** shows the asymmetric divisions occurring during the first 80min of development which give rise to the founder cells. **4.10** shows the embryonic cell divisions for one of the four granddaughters of the founder cell MS (underlined in **4.9**). This represents about 2.5% of the embryonic cell divisions. Two of the progeny of this set of cell divisions, M and Z4 go on to divide postembryonically. **4.11** shows the postembryonic cell divisions of the blast cell M (underlined in **4.10**). This represents about 8% of the postembryonic cell divisions. Each cell has a unique name, dictated by the pattern of cell divisions by which it is generated. The final fate of each cell after differentiation is also indicated: **bm**, **um** or **vm** for body wall, uterine or vulval muscle cell, **cc** for coelomocytes. (This figure was compiled from information in Sulston *et al.* (1983), Sulston and Horvitz (1977).)

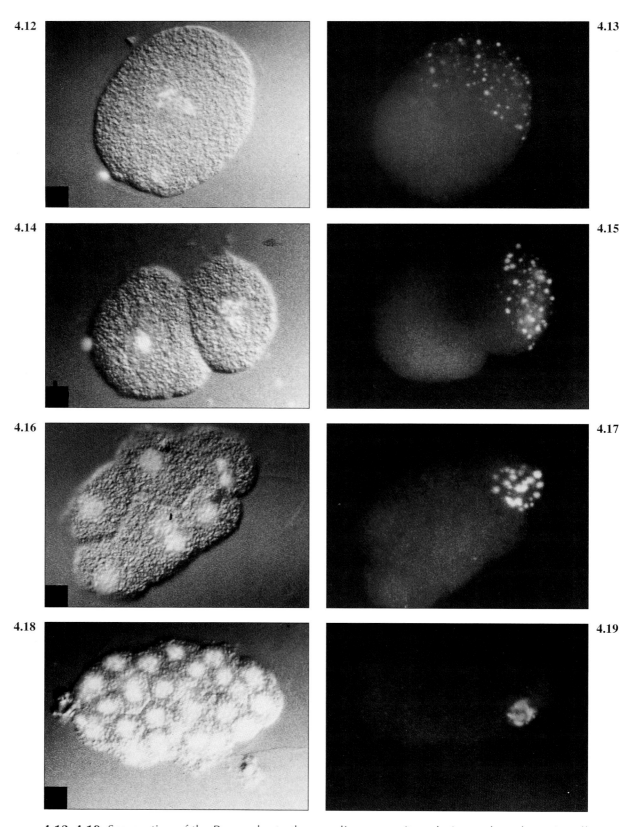

4.12–4.19 Segregation of the P-granules to the germline progenitors during early embryonic cell divisions of *C. elegans*. The figures on the left (**4.12, 4.14, 4.16** and **4.18**) show DAPI staining of the nuclei and those on the right (**4.13, 4.15, 4.17** and **4.19**) show immunofluorescent staining of the cytoplasmic P-granules. **4.12** and **4.13**, **4.14** and **4.15**, **4.16** and **4.17**, and **4.18** and **4.19** show 1-, 2-, 8- and 28- cell embryos, respectively. (×1000; photograph courtesy of S. Strome.)

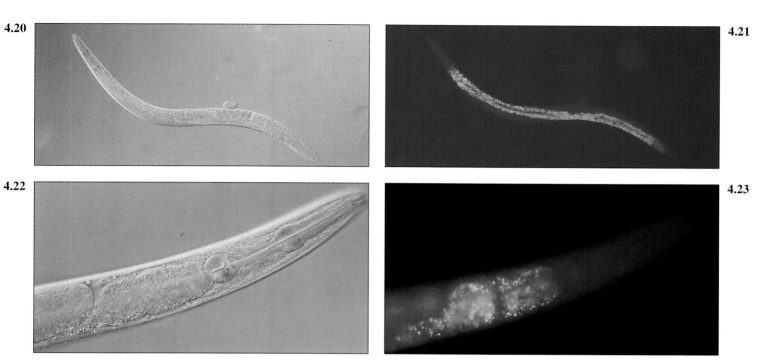

4.20–4.23 Autofluorescent gut granules. Anaesthetized adult hermaphrodites photographed under differential interference contrast optics (**4.20, 4.22**) or upon indirect UV illumination (**4.21, 4.23**). (**4.20** and **4.21** ×100; **4.22** and **4.23** ×400.)

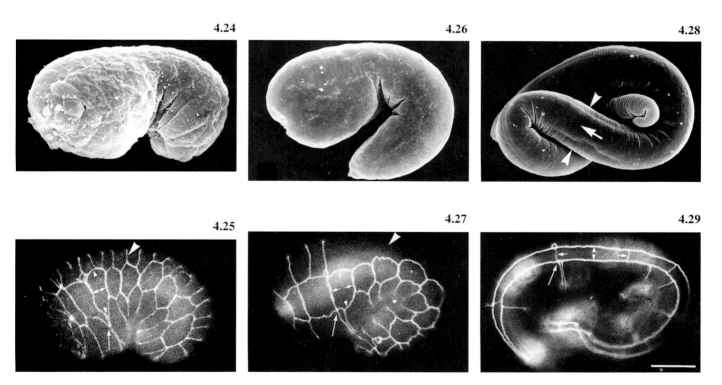

4.24–4.29 Embryonic elongation. *C. elegans* embryos before (**4.24, 4.25**), in the early stages of (**4.26, 4.27**) and in later stages of (**4.28, 4.29**) elongation. The upper panels are scanning electron micrographs. The lower panels are indirect immunofluorescence micrographs obtained using a monoclonal antibody which recognizes desmosomes of hypodermal cells. Note the change in cell shape which accompanies elongation. (×100. Reprinted with permission from Priess and Hirsh, 1986.)

Postembryonic development

Embryogenesis generates the first larval stage, L1, which has a very similar general anatomy to the other three larval stages, L2, L3 and L4, and to the adult (**4.30–4.37**). Growth during postembryonic development is regular, with little perturbation accompanying the shedding of cuticle with moulting at the end of each larval stage (Byerly *et al.*, 1976). If the L1 stage is starved and present at a sufficiently high population density, an alternative developmental pathway is followed to generate a dauer larva instead of an L3 (Riddle, 1988). The dauer is relatively thin and lethargic, with an increased chemical

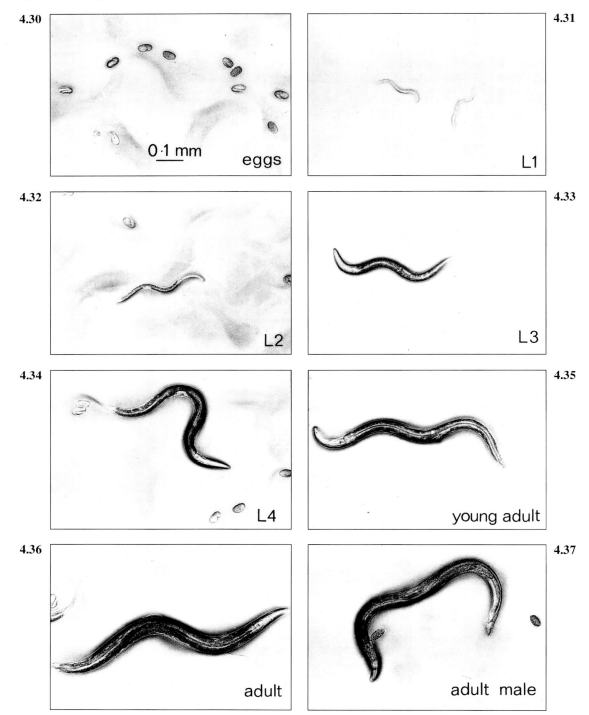

4.30–4.37 Bright-field photomicrographs of egg (**4.30**), larval (**4.31–4.34**) and adult forms (**4.35–4.37**) of *C. elegans*. The L1 hatches from the egg and the L1/L2, L2/L3, L3/L4 and L4/adult molts occur at 14, 29, 37, 46 and 59 hours respectively, after fertilization, at 20°C. At each of these stages of development the individuals are approximately 250μm, 370μm, 480μm, 640μm and 850μm in length, respectively. (Photograph courtesy of J.Sulston.)

resistance, a modified sensory anatomy and an absence of pharyngeal pumping. The dauer form is thus specialized for dispersal and long-term survival. Once food is restored, the dauer larva re-initiates development and moults to produce an L4.

Postembryonic development is more than simply growth. The number of somatic nuclei, approximately doubles to 959 in the hermaphrodite and 1031 in the male (**4.38**). The somatic cells generated postembryonically contribute mainly to the sexually dimorphic structures such as the gonads, the vulva of the hermaphrodite and the male tail. At hatching the gonad primordium of both sexes consists of just two germ-line and two somatic cells, and is located mid-ventrally (Kimble and Hirsh, 1979; **4.39, 4.40**). The distinct gonads of the adult forms are generated through different patterns of cell division and gonadal elongation.

As for embryonic development, the postembryonic cell-lineage and cell/nuclear migrations have been completely described (Sulston and Horvitz, 1977, Sulston, 1988). The reproducibility of the lineage in pattern and timing means that postembryonic development of relevant structures can be followed precisely (e.g. cell division in the lateral hypodermis occurs regularly throughout the larval stages). The limited, natural variation in the cell-lineage between individuals is restricted to postembryonic development. This variation consists of mutually exclusive, binary decisions by eleven pairs of ancestrally and positionally related cells. Very few cells undergo long range migrations during *C. elegans* development; most cells are born where they are required (Sulston and Horvitz, 1977). Despite the continuation of growth, somatic cell divisions have not been observed in the adult.

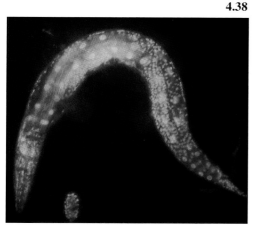
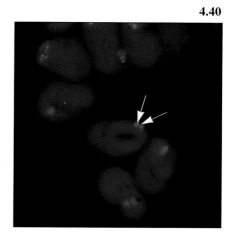

4.38 The nuclei of an adult hermaphrodite viewed by indirect UV illumination after incubation with the fluorescent nuclear stain, DAPI. The large nuclei of the polyploid intestinal cells are particularly prominent. The large numbers of nuclei in the gonadal syncytium can also be seen (×100).

4.39, 4.40 The four cells of the gonad primordium in a mature embryo. A *C. elegans* strain (Hope, 1991) transformed with a lacZ fusion giving β-galactosidase expression in the two somatic gonad precursor cells (arrows, **4.39**). It was also stained with a monoclonal antibody to the P granules of germ-line cells (arrows, **4.40**; ×400).

Experimental manipulations

C. elegans is not particularly suitable for direct surgical manipulations as the embryos are small and their integrity appears important for development. Furthermore, the larval and adult forms are maintained under hydrostatic pressure. Nevertheless, several very informative manipulation experiments have been described.

Axis perturbation

In normal development, the first cleavage is unequal and along the anterior-posterior axis to give a larger anterior AB cell and a smaller posterior P_1 cell. The germline progenitors at the posterior pole go on to divide unequally, whereas the other founder cells divide equally. If a hole is generated at either the posterior or anterior end of a zygote using a laser microbeam, cytoplasm with or without the nucleus can be extruded (Schierenberg, 1989). The segment containing the nucleus, inside or outside of the egg, will go through several rounds of division which may be equal or unequal (**4.41–4.46**). The

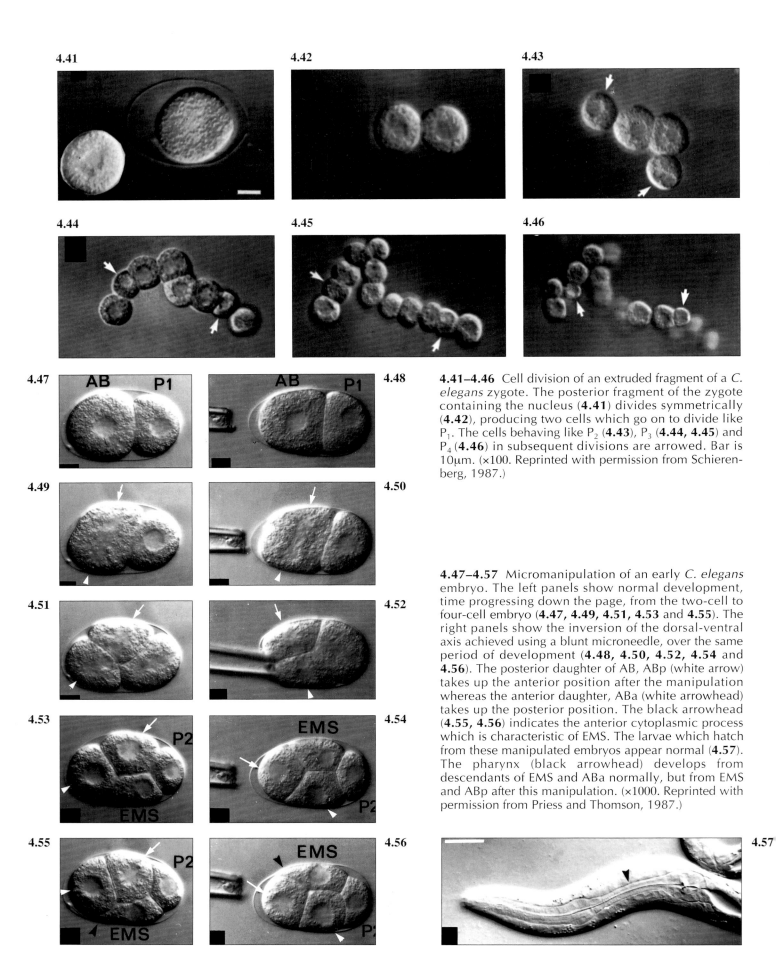

4.41–4.46 Cell division of an extruded fragment of a *C. elegans* zygote. The posterior fragment of the zygote containing the nucleus (**4.41**) divides symmetrically (**4.42**), producing two cells which go on to divide like P_1. The cells behaving like P_2 (**4.43**), P_3 (**4.44, 4.45**) and P_4 (**4.46**) in subsequent divisions are arrowed. Bar is 10μm. (×100. Reprinted with permission from Schierenberg, 1987.)

4.47–4.57 Micromanipulation of an early *C. elegans* embryo. The left panels show normal development, time progressing down the page, from the two-cell to four-cell embryo (**4.47, 4.49, 4.51, 4.53** and **4.55**). The right panels show the inversion of the dorsal-ventral axis achieved using a blunt microneedle, over the same period of development (**4.48, 4.50, 4.52, 4.54** and **4.56**). The posterior daughter of AB, ABp (white arrow) takes up the anterior position after the manipulation whereas the anterior daughter, ABa (white arrowhead) takes up the posterior position. The black arrowhead (**4.55, 4.56**) indicates the anterior cytoplasmic process which is characteristic of EMS. The larvae which hatch from these manipulated embryos appear normal (**4.57**). The pharynx (black arrowhead) develops from descendants of EMS and ABa normally, but from EMS and ABp after this manipulation. (×1000. Reprinted with permission from Priess and Thomson, 1987.)

capacity for unequal cleavage, known as the germline-like cleavage potential, appears to be autonomously determined by an activity associated with the posterior cytoplasm.

At the second cleavage, the ventral-dorsal axis and, by implication, the left-right axis are apparent with ABp, the posterior daughter of AB, occupying the dorsal surface. Using a blunt microneedle to push against the surface of the egg, the positions of the daughters of AB can be reversed so that ABp takes up an anterior position and ABa occupies what should have been the ventral surface (**4.47–4.57**; Priess and Thomson, 1987). Division of P_1 then gives P_2 at the posterior pole as normal, but EMS occupies what should have been the dorsal surface. Such operated embryos go on to develop normally showing that the dorsal-ventral axis is not determined prior to the 2-cell stage. Furthermore, the individuals which develop from the operated embryos have normal left-right asymmetries, a result implying that the left-right axis is likewise not determined prior to the 2-cell stage. Subsequently, similar manipulations but of older embryos did generate individuals with left-right reversals showing that the left-right axis is not determined prior to the 6-cell stage when the left-right asymmetry is first manifested (Wood, 1991). These experiments suggest that extensive cell-cell interactions direct early embryogenesis and so contradict the classic, mosaic view of nematode development which predicts that determination is mainly cell-autonomous.

Laser microsurgery

Individual cells in a living embryo or anaesthetized larva can be destroyed using a laser microbeam (Sulston and White, 1980). A pulsed-dye laser is pre-aligned on to a selected cell by epi-illumination. The cell is killed using repeated applications of sub-lethal laser pulses which seems to leave neighbouring cells unaffected after the operation (**4.58–4.61**); the effect on subsequent development can be evaluated. The procedure is rapid and several cells may be killed in an individual to investigate the effects of multiple ablations.

Many of the cell ablation experiments are consistent with the mosaic view of nematode development. After the operation, the embryo continues to develop, but the remaining cells fail to compensate for the lack of structures which would have derived from the killed cell. Ablation of muscle precursors from the C or D lineages thus leads to larvae with gaps in the longitudinal bands of muscle (Sulston *et al.*, 1983), a result to be expected if development was cell-autonomous.

The *C. elegans* cell lineage does, however, show some natural variation that involves mutually exclusive fates of lineally related cells (see above, postembryonic development). When either cell of such a pair is ablated, the remaining cell always adopts the same fate (called the primary fate). The normal fate of the other cell of the pair is called the secondary fate. In the developing gonad, for example, Z1ppp and Z4aaa are a cell pair for which the primary fate is to be an anchor cell and the secondary fate is to be a ventral uterine precursor cell

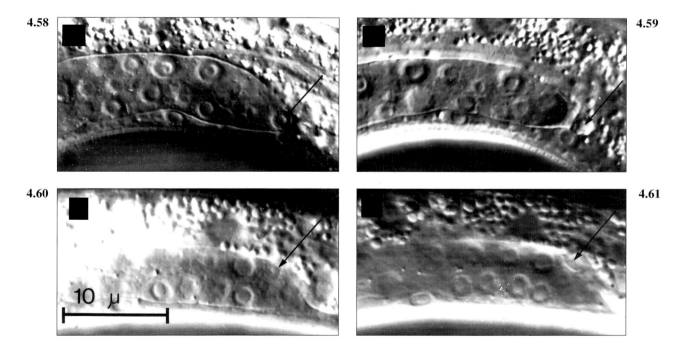

4.58–4.61 Laser microsurgery. Nomarski micrographs of distal tip cells (arrows) in an hermaphrodite (**4.58, 4.59**) and a male (**4.60, 4.61**) before (**4.58, 4.60**) and after (**4.59, 4.61**) laser ablation. These experiments revealed the influence of the distal tip cell on gonad development. (Reprinted with permission from Kimble and White, 1981.)

(Kimble, 1981). Ablation experiments have revealed that the two cells seem to communicate in competition for the primary fate, the 'loser' taking the secondary fate, so achieving the mutual exclusion of cell fates that is observed. The two cells of such a pair have equivalent developmental potential and are said to belong to an 'equivalence group' (Sulston and White, 1980).

In addition to the Z1ppp/Z4aaa pair, laser ablation experiments have also revealed five other equivalence groups which do not show natural variation during normal *C. elegans* development. Equivalent developmental potential was only revealed when the cell which normally adopts the primary fate was destroyed. The cell which would have taken the secondary fate then takes the primary fate instead. A more complex example is that of the equivalence group generating the vulva, as this involves six cells and a tertiary fate (Sulston and White, 1980).

Despite the limitations on direct manipulation of *C. elegans*, the two techniques described in this section, micromanipulation and cell ablation, have yielded very important results. Although some work supports the traditional view of cell-autonomous development, other experiments provide many examples of non-autonomous development and we do not know the full extent of intercellular communication involved in *C. elegans* development may have.

In vitro development

Shearing and other mechanical manipulations have been used to strip the egg shell and vitelline membrane from early embryos. Under appropriate culture conditions, such embryos can continue developing for several hours and some even complete embryogenesis normally to hatch as L1s. These permeabilized embryos have been treated with drugs such as cytochalasin B to block cytokinesis (Cowan and McIntosh, 1985) or aphidicolin to block DNA synthesis (Edgar and McGhee, 1988). Alternatively, individual early blastomeres have been physically isolated from the exposed embryos (Schierenberg, 1989; Laufer *et al.*, 1980). The subsequent capacity of drug-treated embryos or isolated blastomeres to differentiate particular cell types has been monitored using markers for hypodermal cells, muscle cells, gut cells and germline cells.

These studies suggested that cell-autonomous determinants are important, at least for development of the germline (Schierenberg, 1989) and intestine (Laufer *et al.*, 1980). As each cell in cleavage-arrested embryos expressed only one of three markers tested (Cowan and McIntosh, 1985), support was also obtained for the law of exclusivity of differentiation, which proposes that a single cell will only follow one pathway of differentiation to the exclusion of others. And, as a final example, experiments with a DNA synthesis inhibitor have suggested that expression of several differentiation markers depend on a critical round of DNA replication and, by implication, a critical cell division (Edgar and McGhee, 1988).

Developmental mutants

C. elegans was originally selected for investigation of animal development because of its suitability for genetic analysis (Brenner, 1974). *C. elegans* can be maintained easily in the laboratory and in very large numbers. The life-cycle of 3.5 days at 20°C is very fast for a metazoan and means that a series of genetic crosses can be performed rapidly. *C. elegans* grows well from 15–25°C (Byerly *et al.*, 1976) and temperature-sensitive mutants can be obtained readily. Self-fertilization in the hermaphrodite has several useful implications. Heterozygotes automatically produce homozygotes amongst their self-progeny with no need for back-crossing and strains naturally tend towards homozygosity with repeated passage. Also, even hermaphrodites severely debilitated by mutation can produce progeny by self-fertilization. The existence of males was vital to permit the combining mutations.

When the genetic investigation of *C. elegans* began, a deliberate attempt was made to make the nomenclature completely systematic and universal (**4.62–4.65**). Each gene is represented by three letters referring to the mutant phenotype within broad categories, followed by a distinguishing number (e.g. *unc-1* for *unc*o-ordinated). Furthermore, every mutation is given a unique allele number which includes a letter(s) identifying the laboratory in which the mutation was obtained (e.g. *unc-1(e719)*). *C. elegans* strains can be stored in a frozen state which means that all characterized mutations may be retained in case they are of future use. The Caenorhabditis Genetics Centre has co-ordinated the genetic mapping and maintains canonical mutant stocks for each gene (Edgley and Riddle, 1989). There is a single reference wild-type strain called Bristol or N2 (Brenner, 1974) and stocks, which were frozen when the strain was originally established, are still being thawed out. Samples of this original strain can thus be obtained which are only a few generations old.

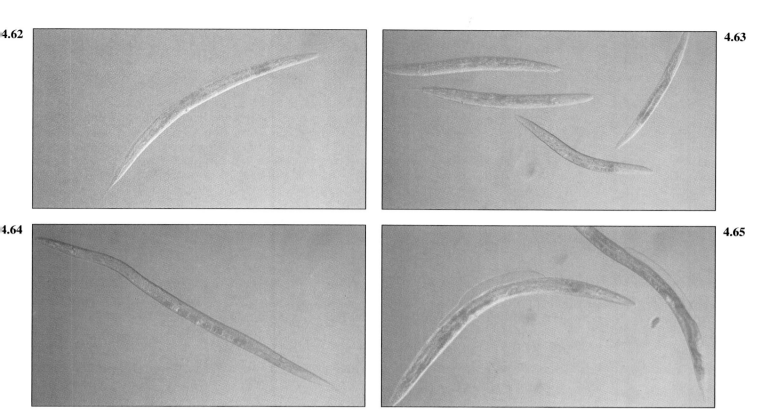

4.62–4.65 A few mutant phenotypes. Nomarski micrographs of *C. elegans* adults: wild-type (**4.62**), *dpy-2(e8)* which is shorter than wild-type or dumpy (**4.63**), *lon-2(e678)* which is longer than wild-type (**4.64**), and *bli-4(e937)* with large blisters of the cuticle (**4.65**; all ×100).

Cellular development

Over 50 genes have been identified which affect the *C. elegans* cell lineage in a specific way (Horvitz, 1988, 1990). Mutations in these genes were often isolated because they perturbed development of a particular, morphological structure such as the vulva (**4.66, 4.67**; Horvitz and Sulston, 1980), and the specific affect on the cell lineage was only revealed upon closer examination of the development of these mutants. But other cell lineage mutants have been isolated in direct screens by simply examining living individuals from a mutagenized parent using Nomarski optics (Hedgecock and Thomson, 1982).

Most of the cell lineage mutations have been referred to as homeotic (Sternberg and Horvitz, 1984), by analogy to the homeotic mutations of *Drosophila*. The phenotypes of the mutations involve exchange of one cell's fate for that of another, often of a related cell, and many cells usually undergo

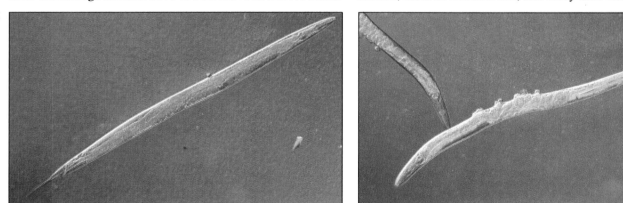

4.66, 4.67 Cell lineage mutants. Nomarski micrographs of a wild-type *C. elegans* (**4.66**) with smooth outline and a *lin-1(e1777)* mutant (**4.67**) with multiple, protruding, vulval-like structures (×100).

the same type of transformation. In some cases, particular cells may retain the same fate as their mother cell and go through a re-iterative pattern of divisions (e.g. *unc-86*; Chalfie *et al.*, 1981). In other mutants particular cells may adopt the normal fates of their sisters (e.g. *unc-17*; Sternberg and Horvitz, 1984). Alternatively, a cell may change its fate to that of an analogous cell (e.g. *lin-22*; Sternberg and Horvitz, 1984). Analogous cells (Sulston, 1988) have a clearly related, but not necessarily identical, fate even though they may not be ancestrally related in the cell lineage. The distinction between equivalent cells (i.e., cells with equivalent developmental potential) may be affected (e.g. *lin-12*; Greenwald, 1989). Finally, there are heterochronic mutants, such as *lin-14*, in which particular cells appear to take the fate of a lineally related cell from a different time in development (Ambros and Horvitz, 1984).

The last two examples, *lin-12* and *lin-14*, are genes likely to be directly involved in developmental decisions. For both of these genes, mutations exist with opposite phenotypic effects and recessive, loss-of-function mutations have the inverse effect of dominant, gain-of-function mutations. In other words, the absence or presence of active gene product appears a critical parameter in the developmental decision. *lin-12* has been particularly well characterized and appears to be the receptor for the intercellular communication that occurs between particular equivalent cells (Greenwald, 1989).

Sex determination

The primary signal determining the sexual phenotype in *C. elegans* is the X chromosome to autosome ratio (X/A; Hodgkin, 1988). Both sexes have a diploid complement of the 5 autosomes (A; I, II, III, IV and V), but hermaphrodites have two X chromosomes and males have only one. The hermaphrodite chromosomal complement is thus XXAA and the X/A ratio is 1. The male chromosomal complement is XOAA and the X/A ratio is 0.5.

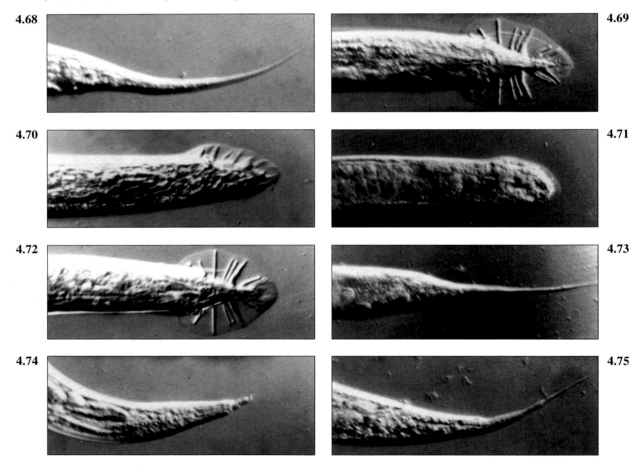

4.68–4.75 Mutations causing sexual transformations. Nomarski micrographs of tail phenotypes for XX individuals (**4.68–4.70**) and an XXX individual (**4.71**) which would normally be hermaphrodite with a tail spike as for the wild-type (**4.68**) and XO individuals (**4.72–4.75**) which would normally be male with a highly specialized tail as for the wild type (**4.72**). The figure shows a *tra-1* (null), XX male (**4.69**), *tra-2* (null), XX, incomplete male (**4.70**), a *tra-1* (null), XXX, abnormal male (**4.71**), an *her-1* (null), XO hermaphrodite (**4.73**), a *tra-1* (dominant)/+, XO female (**4.74**) and a *tra-1* (dominant), XO female. (×400. Reprinted with permission from Hodgkin, 1988.)

The mechanism by which the X/A ratio directs development of sexual phenotype has been extensively investigated by genetic and, more recently, by molecular approaches (Kuwabara and Kimble, 1992). The X/A ratio controls three processes: somatic sexual differentiation, germ-line sexual differentiation, and X-chromosome-dosage compensation which is necessary to equalize the expression of genes on the X chromosome for the two sexes. The X/A ratio is interpreted for all three processes by four master regulator genes (Meyer, 1988). There are four genes known to affect dosage compensation specifically (Plenefisch *et al.*, 1989) and whose products appear important for reducing the levels of X-linked transcripts in hermaphrodites. Somatic and germ-line sexual differentiation are controlled through similar regulatory pathways for which seven components have been identified. Mutations in these genes can cause XO individuals to develop as perfectly normal hermaphrodites or cause XX individuals to develop as perfectly normal males (**4.68–4.75**).

Some mutations result in development as females, rather than hermaphrodites, with all the germ-line cells becoming oocytes. Interactions between mutations in these different genes have led to a proposed regulatory cascade (Hodgkin, 1987) with slight differences for specification of the soma and germ line. Most of the genes involved in *C. elegans* sex determination have been cloned (Hodgkin, 1990) to investigate the molecular mechanisms involved.

Many developmental problems have been addressed in *C. elegans* by the genetic approach and areas not discussed here include the development of muscle (Anderson, 1989), neurons (e.g. Desai *et al.*, 1988), vulva (Horvitz and Sternberg, 1991), and early embryogenesis (Kemphues, 1989) dauer formation (Riddle, 1988) and programmed cell death (**4.76–4.81**; Ellis and Horvitz, 1986; Driscoll and Chalfie, 1992). The mechanisms behind most of these developmental processes are now being pursued at a molecular level through cloning of the relevant genes.

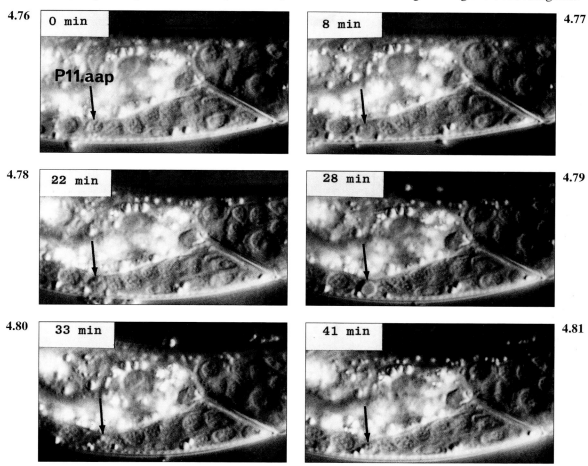

4.76–4.81 Programmed cell death. Nomarski micrographs in temporal series showing the death of P11aap (arrow) in an L1 hermaphrodite. Lateral views. Bar is 20μm. (Reprinted with permission from Sulston and Horvitz, 1977.)

Case studies
Molecular topics

Transformation of *C. elegans* with recombinant DNA was achieved five years ago (Fire, 1986; Stinchcomb *et al.*, 1985) and the efficiency of the technique is still being improved (Mello *et al.*, 1991). DNA is micro-injected into either the oocytes or the syncytial ovary of adult hermaphrodites (**4.82**– **4.84**) and several *C. elegans* genes have been used as genetic markers to identify transformants at subsequent generations. Most transformed lines contain the exogenous DNA as a large extrachromosomal array which is not transmitted to all members of a brood, but it is possible to obtain completely stable, integrative transformants, with the exogenous DNA integrated into a chromosome. Transformation of a mutant strain with concommitant rescue of the mutant phenotype has been used as evidence for identifying genes from within the physical genome map.

The *C. elegans* physical genome map is an ordered library of genomic DNA clones (Coulson *et al.*, 1986, 1988, 1991). Generation of this library was facilitated by the small size of the *C. elegans* genome (approximately 10^8 base pairs) and involved the assembly of contigs, contiguous regions of the genome represented by overlapping clones (**4.85**). At the time of writing the library was down to fewer than 40 contigs (i.e., there are now less than 40 gaps in the library). This ordered and nearly complete library of genomic DNA clones provides the ideal starting point for a project to sequence the entire genome and *C. elegans* appears to have become the model metazoan for the human genome sequencing project (Lewin, 1990; Sulston *et al.*, 1992). As genes and other genetic markers have been cloned, links have been established that tie together the colinear genetic and physical maps. The physical

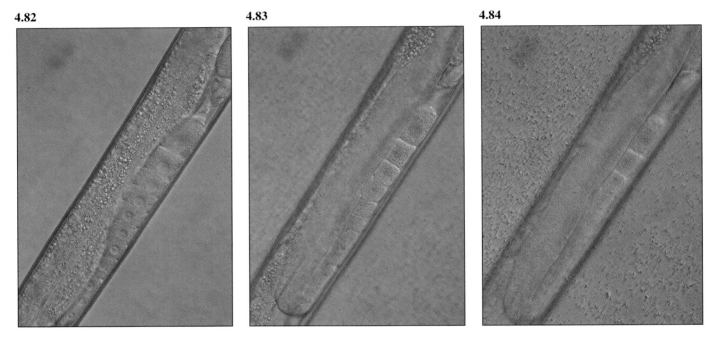

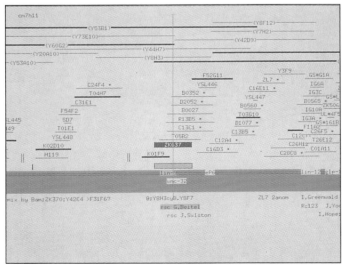

4.82–4.84 Sites of micro-injection used for *C. elegans* transformation. Nomarski micrographs of an hermaphrodite focussing at different depths to show oocytes in the proximal arm (**4.82**), the turn of the gonad (**4.83**) and the nuclei in the syncytial distal arm (**4.84**; all ×400).

4.85 The *C. elegans* physical genome map. The bulk of the map consists of overlapping genomic DNA, cosmid clones depicted by short horizontal lines (middle of figure). The longer horizontal lines (top) are genomic DNA clones in yeast artificial chromosome vectors. Positions of genes, e.g. *lin-9* (light blue), are also indicated. ZK637 (dark blue) was the first cosmid sequenced as part of the genome sequencing project. (This image, representing less than 0.5% of the physical map, was taken from ACEDB, a *C. elegans* computer database under construction by J. Thierry-Mieg and R. Durbin.)

map thus consists of a catalogued stock of genomic DNA clones and a computer database detailing the relationships among the clones and between the clones and the genetic markers. Information and clones in the physical map are freely distributed.

The physical genome map has provided a focal point of communication between scientists studying *C. elegans*. Genes identified by mutation have been rapidly cloned by extrapolation between physical and genetic maps and time-consuming chromosome walking can thus be avoided. Conversely, genes identified using molecular approaches can be linked to previously characterized mutations. The physical map can also lead different laboratories to realize that they have independently come to be studying the same gene.

The nervous system

The nervous system of the adult *C. elegans* hermaphrodite has been described completely at the cellular level (White *et al.*, 1986, Chalfie and White, 1988). Electron micrographs of serial sections through the entire worm were used to follow the processes of every individual neuron and to determine neuronal connectivity (**4.86, 4.87**). There are 302 neurons in an adult *C. elegans* hermaphrodite and 381 in a male and these can be classified into 118 different types on the basis of morphology or connectivity. These properties in combination with positional differences mean that each neuron is unique and so could be assigned a unique label (**4.88**). Of course, the invariant cell lineage means that each nerve cell also has a distinct developmental history.

Evidence on the use of the neurotransmitters acetylcholine, GABA, serotonin, dopamine and octopamine within the *C. ele-*

4.86, 4.87 The *C. elegans* nervous system was reconstructed from electron micrographs of serial sections. An electron micrograph (**4.86**; ×4000) of a transverse section through the ventral cord is shown here with a line drawing (**4.87**) illustrating the identities of the nerve processes. The cord, consisting of nerve processes and a hypodermal ridge, is bounded by a basal lamina (**BL**). The body wall muscle cells send processes (muscle arms, **MA**) to the nerve cord for synapse through the basal lamina. Within muscle cells can be seen the regular arrays of contractile fibres and dense bodies (**DB**). Apart from M for muscle cell and HDC for hypodermal cell all other labels are neuronal identities. (Reprinted with permission from White *et al.*, 1986.)

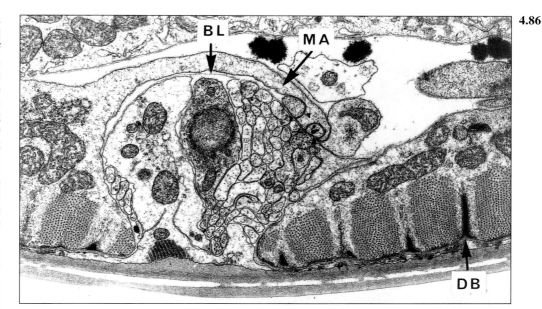

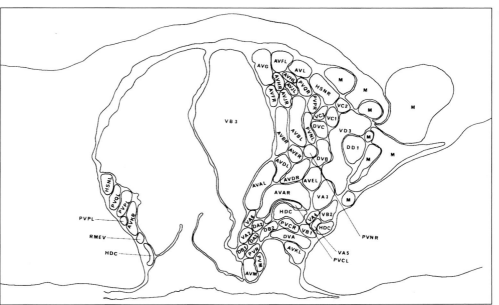

gans nervous system is largely indirect. More direct evidence is available for other nematodes such as *Ascaris* which are large enough for electrophysiological experiments and current information suggests that these nematodes' nervous systems have very similar circuitry.

C. elegans was originally selected for genetic investigation of the nervous system and some mutations have been obtained which have a direct effect on nerve cell development. Examples include mutations in *unc-5*, *unc-6* and *unc-40* which affect dorsalward and/or ventralward circumferential cell migration and axon outgrowth (Hedgecock *et al.* 1990). *unc-5* and *unc-6* have been cloned and sequenced. *unc-6* appears to encode a component of the extracellular matrix (Ishi *et al.*, 1992). *unc-5* appears to encode a transmembrane protein which could interact with the extracellular matrix (Leung-Hagesteijn et al., 1992). Cellular and molecular techniques, along with the thorough cellular and connectivity descriptions, should elucidate the mechanisms behind the assembly and activity of this organism's entire nervous system.

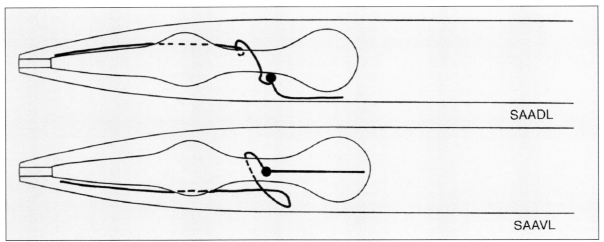

4.88 The path of each neuronal process in *C. elegans* has been described. The paths takens by the interneurons SAADL and SAAVL are depicted here. For each, processes from the cell body extend both posteriorly and anteriorly, the anterior projections initially running in the circumpharyngeal nerve ring. (Reprinted with permission from White *et al.*, 1986.)

Muscle cell structure

C. elegans muscle cell components have been identified using monoclonal antibodies and through genetic analysis (Waterston, 1988). A preparation of the relatively insoluble muscle cell proteins has been used as antigen for monoclonal antibody production (Francis and Waterston, 1985). The antibodies not only recognized some well-known muscle cell proteins (e.g. myosins, paramyosins, actins), but also identified previously uncharacterized components. The positions of these structural proteins in the muscle cell and aspects of their assembly during development have been revealed by indirect immunofluorescence microscopy (**4.89, 4.90**). The antibodies have also had a role in evaluating phenotypes of mutations affecting the musculature. In addition, muscle cell structure has been examined by direct visualization using polarized light (**4.91**; Sulston and White, 1980).

The genetic analysis has been facilitated by the fertility of *C. elegans* mutants with severely defective body wall muscle (Anderson, 1989). Large numbers of mutations affecting muscle

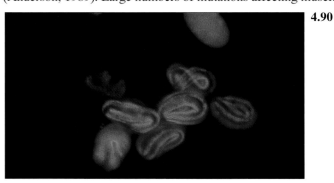

4.89, 4.90 Embryos at various stages of development are shown in **4.89**. **4.90** shows the same field under indirect immunofluorescence; the embryos are stained with a monoclonal antibody (Epstein *et al.*, 1982) which recognizes body wall muscle myosin (×400).

cell function have been obtained identifying minor muscle cell proteins and yielding information on the roles of the different proteins and of domains within proteins. The description of the molecular anatomy of *C. elegans* body wall muscle cells appears close to completion and an outline of the molecular mechanisms of muscle protein assembly is emerging.

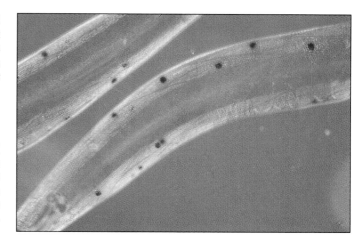

4.91 Polarized light reveals the rhomboid, body wall and muscle cells in an adult *C. elegans* hermaphrodite. Each muscle cell quadrant is two cells wide and two quadrants can be seen in this focal plane. The micrograph is of a transformant which expresses β-galactosidase in the single nucleus (blue) of each body wall muscle cell (×400; Hope, 1991).

The future

As this chapter has shown, there has been impressive progress in our understanding of *C. elegans* development over the past few years. With the cloning of *C. elegans* developmental genes, areas of research previously analysed genetically have become the subject of molecular analysis and the current challenge is to determine the activities of the gene products.

There are, however, other areas of work that are attracting increasing attention, one of which is the genetic investigation of the early stages of *C. elegans* development (Kemphues, 1989). In the past progress here has been restricted by difficulty in interpreting mutant phenotypes, as mutations that affect early embryogenesis often cause embryos to arrest with approximately the complete number of cells, but with apparently little differentiation and no elongation. Although the nature of these developmental defects may not readily be determined from direct examination of mutant embryos, these observations do suggest that either the cells of the early embryo show considerable interdependence and/or gene products are used repeatedly in different embryonic processes. We can however expect progress in this area because genes expressed early in embryogenesis are now being cloned and these should provide many cytogenetic reporters (**4.92**) additional to the few markers of early differentiation presently available.

In a more general context, molecular studies have begun to reveal common control mechanisms for development across the animal kingdom. Many homeobox genes, for example those which are important in control of development in other species, are present in *C. elegans* (Burglin *et al.*, 1989). Moreover, a homeobox cluster may also exist whose arangement matches those in insects and vertebrates, with the position of function along the anterior-posterior axis reflecting the order of the homeoboxes along the chromosome (Kenyon and Wang, 1991). A second example is given by a *C. elegans* equivalent of MyoD which was identified on the basis of sequence homology (Krause *et al.*, 1990). In vertebrates, MyoD appears important in specification of muscle-cell type and expression in muscle cell precursors suggests that the *C. elegans* MyoD may have a similar role, although the nature of this role has recently become less clear (Chen *et al.*, 1992).

The biggest impact on *C. elegans* research may, however, come from the genome sequencing project (Lewin, 1990; Sulston *et al.*, 1992). At the time of writing, sequence data from this project has just begun to be released and the rate of sequence generation is expected to increase rapidly, reaching 3Mbp per year by 1993. Computer analysis of the sequence data will identify many new genes for characterization and the quantity of information to be obtained suggests that this approach will help elucidate many areas of *C. elegans* develop-

4.92 Differentiation markers may facilitate investigation of early *C. elegans* development. This Nomarski micrograph shows mixed stage embryos of a transformant strain which expresses nuclear localized, β-galactosidase (blue) in a subset of cells, early in development (Hope, 1991; ×400).

ment. Here, computer databases will be vital to make best use of this information and two are being established with different aims. ACEDB, **a** **C. e**legans **d**ata**b**ase, (Durbin and Thierry Mieg, personal communication) will handle genetic and molecular data related to the genome project, while WCS, the **W**orm **C**ommunity **S**ystem (Schatz, personal communication) is intended to be an 'interactive computer environment' containing 'all' available information on *C. elegans*.

Although development from egg to adult may well have been described more completely for *C. elegans* than for any other metazoan, we still understand but a small proportion of the mechanistic detail behind this complex process which allows life to be propagated. A greater understanding of development in *C. elegans*, as elsewhere, is going to require a combination of the genetic and molecular approaches with the higher levels of cellular and organismal biology.

Acknowledgements

I would like to thank J. Young, A. Lynch and J. White for critical reading of the manuscript, and S. Strome, J. Sulston, J. Hodgkin, J. White, J. Priess, J. Kimble and E. Schierenberg for generously providing figures used in this chapter.

References

Alberts, B., Bray, D., Lewis, J., Raff, M., Roberts, K. and Watson, J.D. (1989). *Molecular Biology of the Cell*. Garland Publishing, New York.

Albertson, D.G. and Thomson, J.N. (1976). The pharynx of *Caenorhabditis elegans*. *Phil. Trans. Roy. Soc. London*, **275**, 299–325.

Ambros, V. and Horvitz, H.R. (1984). Heterochronic mutants of the nematode *Caenorhabditis elegans*. *Science*, **226**, 409–16.

Anderson, P. (1989). Molecular genetics of nematode muscle. *Annu. Rev. Genet.*, **23**, 507–525.

Babu, P. (1974). Biochemical genetics of *Caenorhabditis elegans*. *Molec. Gen. Genet.*, **135**, 39–44.

Bird, A.F. (1980). The nematode cuticle and its surface. In *Nematodes as Biological Models*, Vol. **2**. (B.M. Zuckerman ed.) pp. 213–236, B.M. Academic Press, New York.

Brenner, S. (1974). The genetics of *Caenorhabditis elegans*. *Genetics*, **77**, 71–94.

Burglin, T.R., Finney, M., Coulson, A. and Ruvkun, G. (1989). *Caenorhabditis elegans* has scores of homeobox-containing genes. *Nature*, **341**, 239–243.

Byerly, R. C., Cassada, R.C. and Russel, R.L. (1976). The life-cycle of the nematode *Caenorhabditis elegans*. *Dev. Biol.*, **51**, 23–33.

Chalfie, M., Horvitz, H.R. and Sulston, J.E. (1981). Mutations that lead to reiterations in the cell lineages of *C. elegans*. *Cell*, **24**, 59–69.

Chalfie, M. and White, J. (1988). The nervous system. In *The Nematode Caenorhabditis Elegans*, (W.B. Wood, ed.) pp. 337–392, Cold Spring Harbor Laboratory, New York.

Chen, L., Krause, M., Draper, B., Weintraub, H. and Fire, A. (1992) Body-wall muscle formation in *Caenorhalsditis elegans* embryos that lack the MyoD homolog *hlh–1*. *Science*, **256**, 240–243.

Coulson, A., Sulston, J., Brenner, S. and Karn, J. (1986). Towards a physical map of the genome of the nematode *Caenorhabditis elegans*. *Proc. Natl. Acad. Sci. USA*, **83**, 7821–7825.

Coulson, A., Waterston, R., Kiff, J., Sulston, J. and Kohara, Y. (1988). Genome linking with yeast artificial chromosomes. *Nature*, **335**, 184–186.

Coulson, A., Kozono, Y., Lutterbach, B., Shownkeen, R., Sulston, J. and Waterston, R. (1991). YACS and the *C. elegans* genome. *Bioessays*, **13**, 413–417.

Cowan, A.E., and McIntosh, J.R. (1985). Mapping the distribution of differentiation potential for intestine, muscle and hypodermis during early development in *Caenorhabditis elegans*. *Cell*, **41**, 923–932.

Desai, C., Garriga, G., McIntire, S.L. and Horvitz, H.R. (1988). A genetic pathway for the development of the *Caenorhabditis elegans* HSN motor neurons. *Nature*, **336**, 638–646.

Driscoll, M. and Chalfie, M. (1992) Developmental and abnormal cell death in *C. elegans*. *Trends in Neurosciences*, **15**, 15–19.

Edgar, L.G. and McGhee, J.D. (1986). Embryonic expression of a gut-specific esterase in *Caenorhabditis elegans*. *Dev. Biol.*, **114**, 109–118.

Edgar, L.G. and McGhee, J.D. (1988). DNA synthesis and the control of embryonic gene expression in *C. elegans*. *Cell*, **53**, 589–99.

Edgley, M.L. and Riddle, D.L. (1989). The nematode *Caenorhabditis elegans*. *Genetic Maps*, **5**, 111–133.

Ellis, H.M., and Horvitz, H.R. (1986). Genetic control of programmed cell death in the nematode *C. elegans*. *Cell*, **44**, 817–829.

Epstein, H.F., Miller, D.M., Gossett, L.A. and Hecht, R.M. (1982). Immunological studies of myosin isoforms in nematode embryos. In *Muscle Development. Molecular and Cellular Control*, (M.L. Pearson, and H.F. Epstein, eds) pp. 7–14, Cold Spring Harbor Lab., New York.

Fire, A. (1986). Integrative transformation of *Caenorhabditis elegans*. *EMBO J.*, **5**, 2673–2680.

Fire, A., Harrison, S.W. and Dixon,D. (1990). A modular set of lacZ fusion vectors for studying gene expression in *Caenorhabditis elegans*. *Gene*, **93**, 189–198.

Francis, G.R. and Waterston, R.H. (1985). Muscle organization in *Caenorhabditis elegans*: Localization of proteins implicated in thin filament attachment and I-band organization. *J. Cell Biol.*, **101**, 1532–1549.

Gilbert, S.F. (1991). *Developmental Biology*. Sinauer Associates Inc., Sunderland, Mass.

Greenwald, I. (1989). Cell-cell interactions that specify certain cell fates in *C. elegans* development. *Trends in Genetics*, **5**, 237–241.

Hedgecock, E.M. and Thomson, J.N. (1982). A gene required for nuclear and mitochondrial attachment in the nematode *Caenorhabditis elegans*. *Cell*, **30**, 321–330.

Hedgecock, E.M., Culotti, J.G. and Hall, D.H. (1990). The *unc-5*, *unc-6* and *unc40* genes guide cicumferential migrations of pioneer axons and mesodermal cells on the epidermis in *C. elegans*. *Neuron*, **2**, 61–85.

Hodgkin, J. (1987). Sex determination and dosage compensation in *Caenorhabditis elegans*. *Ann. Rev. Genet.*, **21**, 133–154.

Hodgkin J. (1988) Sexual dimorphism and sex determination. In *The Nematode Caenorhabditis Elegans*, (W.B. Wood, ed.) pp. 243–280, Cold Spring Harbor Laboratory, New York.

Hodgkin, J. (1990). Sex determination compared in Drosophila and Caenorhabditis. *Nature*, **344**, 721–728.

Hope, I.A. (1991). Promoter trapping in *Caenorhabditis elegans*. *Dev.*, **113**, 399–408.

Horvitz, H.R. (1988) Genetics of cell lineage. In *The Nematode Caenorhabditis Elegans*, (W.B. Wood, ed.) pp.157–190, Cold Spring Harbor Laboratory, New York.

Horvitz, H. R. (1990). Genetic control of *Caenorhabditis elegans* cell lineage. *Harvey Lectures*, **84**, 65–77.

Horvitz, H.R., and Sulston, J.E. (1980). Isolation and genetic characterization of cell-lineage mutants of the nematode *Caenorhabditis elegans*. *Genetics*, **96**, 435–454.

Horvitz, H.R. and Sternberg, P.W. (1991). Multiple intercellular signalling systems control the development of the *Caenorhabditis elegans* vulva. *Nature*, **351**, 535–541.

Ishii, N., Wadsworth, W.G., Stern, B.D., Culotti, J.G. and Hedgecock, E.M. (1992) *UNC–6*, a laminin-related protein, guides cell and pioneer axon migrations in *C. elegans*. *Neuron*, **9**, 873–881.

Kemphues, K.J. (1989). Caenorhabditis. In *Genes and Embryos* (D.M. Glover, and B.D. Hames, eds.) pp. 95–126, IRL Press, Oxford.

Kenyon, C. and Wang, B. (1991). A cluster of Antennapedia-class homeobox genes in a non-segmented animal. *Science*, **253**, 516–517.

Kimble, J. (1981). Alterations in cell lineage following laser ablation of cells in the somatic gonad of *Caenorhabditis elegans*. *Dev. Biol.*, **87**, 286–300.

Kimble, J., and Hirsh, D. (1979). The post-embryonic cell lineages of the hermaphrodite and male gonads in *Caenorhabditis elegans*. *Dev. Biol.*, **70**, 396–417.

Kimble, J. and Ward, S. (1988) Germ-line development and fertilization. In *The Nematode Caenorhabditis Elegans*, (W.B. Wood, ed.) pp.191–214, Cold Spring Harbor Laboratory, New York.

Kimble, J.E. and White, J.G. (1981). On the control of germ cell development in *Caenorhabditis elegans*. *Dev. Biol.* **81**, 208–219.

Krause, M., Fire, A., Harrison, S.W., Priess, J. and Weintraub, H. (1990). CeMyoD accumulation defines the body wall muscle cell fate during *C. elegans* embryogenesis. *Cell*, **63**, 907–919.

Kuwabara, P.E. and Kimble, J. (1992) Molecular genetics of sex determination in *C. elegans*. *Trends in Genetics*, **8**, 164–168.

Laufer, J.S., Bazzicalupo, P. and Wood, W.B. (1980). Segregation of developmental potential in early embryos of *Caenorhabditis elegans*. *Cell*, **19**, 569–577.

Leung–Hagesteijn, C., Spence, A.M., Stern, B.D., Zhou, Y., Su, M.–W., Hedgecock, E.M. and Culotti, J.G. (1992) UNC–5, a transmembrane protein with immuno globulin and thrombospondin type 1 domains, guides cell and pioneer axon migrations in *C. elegans*. *Cell*, **71**, 289–299.

Lewin, R. (1990). A worm at the heart of the genome project. *New Scientist*, **127**, 38–42.

Mello, C.C., Kramer, J.M., Stinchcomb, D. and Ambros, V. (1991). Efficient gene transfer in *C. elegans*: extrachromosomal maintenance and integration of transforming sequences. *EMBO J.*, **10**, 3959–3970.

Meyer, B. (1988). Primary events in *C. elegans* sex determination and dosage compensation. *Trends in Genetics*, **4**, 337–342.

Plenefisch, J.D., Delong, L. and Meyer, B.J. (1989). Genes that implement the hermaphrodite mode of dosage compensation in *Caenorhabditis elegans*. *Genetics*, **121**, 57–76.

Poinar, G.O. (1983). *Natural History of Nematodes*. Prentice Hall Inc., Englewood Cliffs, New Jersey.

Priess, J.R., and Hirsh, D.I. (1986). *Caenorhabditis elegans* morphogenesis: The role of the cytoskeleton in elongation of the embryo. *Dev. Biol.*, **117**, 156–173.

Priess, J. R., and Thomson, J.N. (1987). Cellular interactions in early *C. elegans* embryos. *Cell*, **48**, 241–250.

Raff, R.A. and Kaufman, T.C. (1983). *Embryos, Genes and Evolution*. Macmillan Publishing Co. Inc., New York.

Riddle, D.L. (1988) The dauer larva. In *The Nematode Caenorhabditis Elegans*, (W.B. Wood, ed.) pp.393–412, Cold Spring Harbor Laboratory, New York.

Schierenberg, E. (1987). Reversal of cellular polarity and early cell-cell interaction in the embryo of *Caenorhabditis elegans*. *Dev. Biol.*, **122**, 452–463.

Schierenberg, E. (1989). Cytoplasmic determination and distribution of developmental potential in the embryo of *Caenorhabditis elegans*. *Bioessays*, **10**, 99–104.

Sternberg, P.W. and Horvitz, H.R. (1984). The genetic control of cell lineage during nematode development. *Ann. Rev. Genet.*, **18**, 489–524.

Stinchcomb, D.T., Shaw, J.E., Carr, S.H. and Hirsh, D. (1985). Extrachromosomal DNA transformation of *Caenorhabditis elegans*. *Mol. Cell. Biol.*, **5**, 3484–3496.

Strome, S. and Wood, W.B. (1983). Generation of asymmetry and segregation of germ-line granules in early *C. elegans* embryos. *Cell*, **35**, 15–25.

Strome, S. (1989). Generation of cell diversity during early embryogenesis in the nematode *Caenorhabditis elegans*. *Int. Rev. Cytol.*, **114**, 81–123.

Sulston, J. (1988) Cell lineage. In *The Nematode Caenorhabditis Elegans*, (W.B. Wood, ed.) pp. 123–156, Cold Spring Harbor Laboratory, New York.

Sulston, J., Du, Z., Thomas, K., Wilson, R., Hillier, L., Staden, R., Halloran, N., Green, P., Thierry–Mieg, J., Qiu, L., Dear, S., Coulson, A., Craxton, M., Durbin, R., Berks, M., Metzstein, M., Hawkins, T., Ainscough, R. and Waterston, R. (1992). The *C. elegans* genome sequencing project: a beginning. *Nature*, **356**, 37–41.

Sulston, J. and Hodgkin, J. (1988) Methods. In *The Nematode Caenorhabditis Elegans*, (W.B. Wood, ed.) pp. 587–606 Cold Spring Harbor Laboratory, New York.

Sulston, J.E. and Horvitz, H.R. (1977). Post-embryonic cell lineages of the nematode, *Caenorhabditis elegans*. *Dev. Biol.*, **56**, 110–156.

Sulston, J. E. and White, J.G. (1980). Regulation and cell autonomy during postembryonic development of *Caenorhabditis elegans*. *Dev. Biol.*, **78**, 577–597.

Sulston, J.E., Scherienberg, E., White, J.G. and Thomson, J.N. (1983). The embryonic cell lineage of the nematode *Caenorhabditis elegans*. *Dev. Biol.*, **100**, 64–119.

Waterston, R.H. (1988). Muscle. In *The Nematode Caenorhabditis Elegans*, (W.B. Wood, ed.) pp. 281–336, Cold Spring Harbor Laboratory, New York.

White, J. (1988). The anatomy. In *The Nematode Caenorhabditis Elegans*, (W.B. Wood, ed.) pp. 81–122, Cold Spring Harbor Laboratory, New York.

White, J.G., Southgate, E., Thomson, J.N. and Brenner,S. (1986). The structure of the nervous system of the nematode *Caenorhabditis elegans*. *Phil. Trans. Roy. Soc. London.*, **314**, 1–340.

Wood, W.B. (1988). Embryology. In *The Nematode Caenorhabditis Elegans*, (W.B. Wood, ed.) pp. 215–242, Cold Spring Harbor Laboratory, New York.

Wood, W.B. (1991). Evidence from reversal of handedness in *C. elegans* embryos for early cell interactions determining cell fates. *Nature*, **349**, 536–538.

5. Molluscs

Jo A.M. van den Biggelaar, Wim J.A.G. Dictus and *Florenci Serras*

Introduction

Limpets and top shells, sea slugs and sea hares, ormers, whelks and winkles, snails and oysters, mussels, clams and gapers, squids and octopods all belong to the phylum of the soft-bodied, often hard-shelled animals: the Mollusca. Aside from the arthropods, the more than 50,000 living species and 35,000 fossil forms together constitute the largest group of animals without backbones. Despite the enormous variety of molluscs, they all illustrate a common design and this is because their morphogenesis is similar.

In all species except the cephalopods, the early development starts with spiral cleavage (either left-handed, *sinistral,* or right-handed, *dextral*) which leads to the formation of the blastula. Within a single species, the cell pattern is invariant, while there are only minor differences among species, and this makes it possible to trace the origin of various parts of the body back to the founder cells. The blastula forms a gastrula which then turns into a larva (called a *trochophore*) which has an apical tuft and a central ciliated belt, the *prototroch,* which may later expand and form lobes. The trochophore will develop into a second larval form, the *veliger,* or directly into the adult form. Four species of molluscs that are often used as model systems for developmental biology are the snails *Lymnaea* (**5.1**), *Ilyanassa* and *Nassarius* (**5.2**), and the common limpet, *Patella* (**5.3**).

The diversity of the molluscs arises from the way in which the trochophores and veligers differentiate, but, in this chapter, we will focus on the early events in molluscan embryogenesis. We will pay particular attention to the nature of the spiral cleavage patterns and the determination of cell fates which are responsible for so much of the later organization of the animals and briefly discuss evidence on the mechanisms responsible for molluscan morphogenesis. The chapter ends with a discussion of the advantages of molluscan embryogenesis as a model system and the particular molecular problems meriting investigation.

5.1 The pond snail *Lymnaea stagnalis* during the deposition of an egg mass.

5.2 The marine mud snail *Nassarius reticulatus*. Some snails burrow in the sand so that only their siphons are visible.

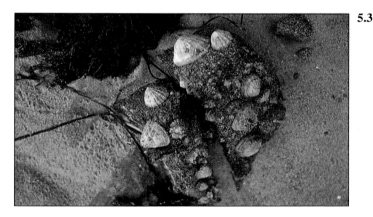

5.3 The marine limpet *Patella vulgata* in its natural environment.

The molluscan phylum

Before discussing their development in any detail, it will be useful to give a short survey of the different groups which form part of the phylum Mollusca (Haszprunar, 1988). Representative genera are mentioned in parenthesis.

Class 1: Aplacophora

The *aplacophorans* are bilateral-symmetrical, worm-shaped hermaphroditic molluscs with a poorly developed head; characteristic molluscan structures like foot, shell and mantle are missing, and the body does not show the torsion displayed by other molluscs. (*Neomenia.*)

Class 2: Polyplacophora

These are molluscs with an ovoid, bilaterally symmetrical dorso-ventrally flattened body, covered with eight overlapping shell plates; the head is indistinct, whereas foot and mantle are well developed. Males and females are generally distinct (dioecious). (*Chiton.*)

Class 3: Monoplacophora

These are dioecious molluscs with a single symmetrical shell, and serially repeating structures (e.g., six pairs of nephridia, and five to six pairs of gills). (*Neopilinia.*)

Class 4: Gastropoda

Subclass 1:
Steptoneura (**5.4**). Gastropods with a nervous system with three ganglia (no parietal ganglion). The subclass includes two orders: Archaeogastropods – gastropods with simple tentacle nerves; and Apogastropoda – gastropods with adjacent cerebral and pleural ganglia, a bifurcated tentacle nerve, a parapedal commissure, and internal fertilization primarily by use of spermatophores.

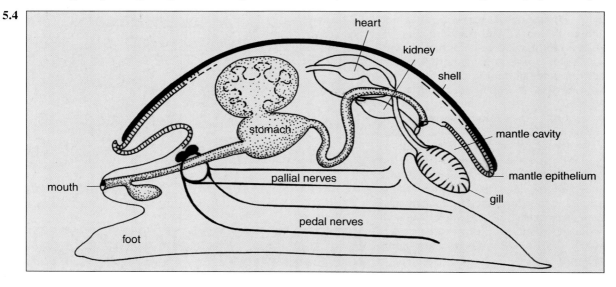

5.4 Diagram of a primitive gastropod.

Subclass 2:
Euthyneura. Gastropods with a pentaganglionate nervous system. There are three orders: the first is Architectibranchia (bubble shells) – shell generally present but reduced, or in some even absent (e.g., *Haminoea, Philine*); the second is Opisthobranchia – these exclusively marine species are simultaneous or protandric hermaphrodites, they often lack a mantle cavity and shell, or if the shell is reduced it is covered by mantle or pedal folds (e.g., *Aplysia* (sea hare), *Berthella, Doris, Archidoris* and *Aeolidia* (sea slugs)); Pulmonata – these species have their mantle cavity transformed into a pulmonary cavity, they are usually fresh water or terrestrial animals and are simultaneous hermaphrodites (e.g., *Lymnaea* (pond snail), *Physa, Bithynia* and *Limax* (land slug)).

Subclass 3:
Pulmonata. These species have their mantle cavity transformed into a pulmonary cavity. They are usually fresh water or terrestrial animals and are simultaneous hermaphrodites. Examples are *Lymnaea* (pond snail), *Physa, Bithynia* (land slug) and *Limax*.

Class 5: Scaphopoda

These are exclusively marine, dioecious species, with a tusk-like shell that is open at both ends. (*Dentalium* (tusk shell).)

Class 6: Bivalvia

These generally dioecious molluscs are untorted and bilaterally symmetrical; they are enclosed in a bivalve shell with a hinge at the dorsal side. Examples include *Phola* (piddock), *Spisula* (trough shell), *Mytilus* (common mussel) and *Ostrea* (oyster).

Class 7: Cephalopoda

These dioecious molluscs have a bilaterally symmetrical body, a well-developed head with eyes analogous to those of vertebrates, and a head encircled by tentacles or arms homologous to the anterior part of the molluscan foot. (*Nautilus, Sepia, Loligo, Octopus.*)

Oocyte maturation

In the ovary, oocytes do not develop beyond the germinal vesicle stage and the cell cycle is arrested at the meiotic prophase. During oogenesis, the oocytes are in contact with the ovarian wall and with follicle cells (**5.5**) and the point where the egg contacts the ovarian wall invariably becomes the vegetal pole. In species with reduced contact area, a micropyle in the egg envelope may indicate the original place of contact. In eggs of *Patella*, the layer of follicle cells begins to loosen from the egg, starting at the original basal, but now vegetal pole, and is finally stripped off at the animal pole (**5.6**). As the animal-vegetal axis is related to the anterior-posterior axis of the embryo, the longitudinal axis of the embryo is influenced, but not necessarily irreversibly determined, by the polarity of the egg in the ovary (van den Biggelaar and Guerrier, 1983).

After ovulation and elimination of the follicle cells, the meiotic arrest is released, the germinal vesicle breaks down and maturation is initiated, but, at the first meiotic metaphase,

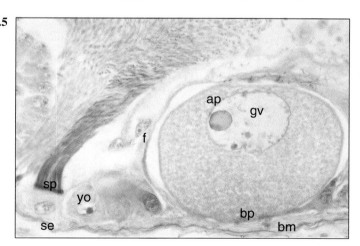

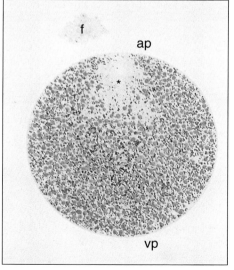

5.5 Eggs and sperm in ovotestis. An oocyte in the ovotestis of *Lymnaea stagnalis* (methyl green pyronin stained). The oocyte has a large germinal vesicle facing the apical pole, while its basal pole rests on the wall of the ovotestis. In the lower left corner, a group of spermatozoa can be distinguished. (×440. Key: **ap**, apical pole; **bm**, basement membrane of the ovotestis; **bp**, basal pole; **f**, follicle cells; **gv**, germinal vesicle; **se**, Sertoli cell bearing spermatozoa; **sp**, spermatozoa; **yo**, young oocyte.)

5.6 An oocyte of *Patella vulgata*. The clear cytoplasm marks the animal pole. Slightly left of the animal pole, the group of follicle cells that have been eliminated from the oocyte can be seen. (×280. Key: **ap**, apical pole; **f**, follicle cells; **vp**, vegetal pole; * marks the position of the maturation spindle.)

further progress is again blocked. These natural discontinuities in the maturation divisions allow molluscs such as *Spisula, Patella,* and *Mytilus* to be used as models for the molecular analysis of the cell cycle (Swenson *et al.,* 1986; Draetta *et al.,* 1989; Guerrier *et al.,* 1990; Dubé and Dufresne, 1990; Dubé *et al.,* 1991; van Loon *et al.,* 1991).

Gametes can be released in three ways (for review, see Tompa *et al.,* 1984). Eggs and sperm may be shed freely into the water or the spawn may be deposited as an egg mass, composed of a number of capsules with one or a few eggs and surrounded by a gelatinous or mucus layer. The third way of spawning is associated with brood care in cloacal pouches, pallial grooves, suprabranchial cavities, gills or under the foot.

Fertilization and ooplasmic segregation

Prior to fertilization, the visible apical-basal polarity of the egg in the ovary is the first indication of a heterogeneity in the distribution of the maternal ooplasmic components. Fertilization triggers a further increase in the heterogeneity of these inclusions. Raven (1945) described a remarkable example of such ooplasmic segregation in the newly laid egg of the pond snail *Lymnaea stagnalis.* At the metaphase of the first maturation division, a distinct vegetal pole plasm can be seen (**5.7**) and, during the first meiotic anaphase, this plasm spreads beneath all the cortex except that of the animal region (**5.8**). Another example of ooplasmic segregation is demonstrated by the regional difference in colour in vitally stained eggs of *Aplysia* (Ries and Gersch, 1936; **5.9, 5.10**).

The organization of the ovarian egg and the zygote is causally related to the organization of the body plan, and the presence of distinct ooplasmic components in specific domains of the zygote provides them with distinct morphogenetic capacities. A famous example found in a number of molluscs is the sequestration of the vegetal ooplasm in the form of a polar lobe during the first divisions. During cytokinesis, the lobe remains attached to one of the two daughter cells, into which it is absorbed after cleavage (**5.11–5.13** and **5.14–5.16**).

Experimentation clearly shows that a re-organization of the egg structure is initiated following fertilization. If, for instance, the vegetal fragment of an oocyte of *Dentalium* is fertilized, a polar lobe is formed, with the size of this lobe being proportional to the size of the fragment. In contrast, the vegetal fragment of a fertilized egg forms a normally sized polar lobe, independent of the volume of the fragment (Verdonk *et al.* 1971). This result indicates that the domain which will form the polar lobe has not been determined before fertilization; after it, size regulation is no longer possible. Other types of experiments also demonstrate that the organization of the egg becomes more rigid after fertilization. Freeman and Lundelius (1982) found that, up to the end of the second maturation division but not later, the chirality of the cleavage pattern in the sinistral eggs (see below) of *Lymnaea peregra* could be reversed by injection of ooplasm from a dextral dividing egg. Guerrier (1968) observed that, up to the end of the second maturation division, it was possible to change the animal-vegetal polarity of the zygote in *Limax maximus*, but that, after extrusion of the second polar body, this was no longer possible.

The morphogenetic significance of the polar lobe has been demonstrated by deletion experiments. After removal of the polar lobe or the vegetal part of the egg prior to polar lobe formation, the embryo fails to develop a dorso-ventral body axis. In *Bithynia*, the morphogenetic significance of the polar lobe appears to be associated with a special structure, the so-called vegetal body which is located in the polar lobe during first cleavage (**5.11–5.13**; Dohmen and Verdouk, 1974). In *Nassarius, Bithynia* and *Crepidula* and a number of other species which will form polar lobes, the surface of the vegetal part of the egg has a series of ridges (**5.14–5.16,** and **5.17**), but, during lobe formation, these patterns are restricted to the surface of the polar lobe (Dohmen and van der Mey, 1977). The role of the polar lobe in the regulation of gene expression has been analysed most extensively in *Ilyanassa* (Collier, 1983, 1989) and *Nassarius* (van Dongen *et al.,* 1985).

Cleavage

The pattern of successive cleavages is extremely regular in all members of the phylum so far studied, apart from the cephalopods. Within a species, the cleavage pattern is invariant and there are only minor variations among species. It is thus possible to trace the lineage of the individual blastomeres without the use of cell lineage markers.

This regularity of the cell pattern is associated with the spiral character of cleavage. Spiral cleavage is a modification of radial cleavage and is found in a great number of protostome phyla and subphyla (Plathyhelminthes, Gastrotricha, Nemertina, Rotifera, Sipuncula, Mollusca, Annelida, Crustacea, Echiura), whereas the radial type is more typical for the deuterostome phyla (Echinodermata, Hemichordata, Chordata).

5.7, 5.8 Ooplasmic segregation in *Lymnaea*. **5.7** shows the vegetal pole plasm (blue) in an oocyte at the transition from the first meiotic metaphase into anaphase (adapted from Raven, 1945; key: **ap**, apical pole; **vp**, vegetal pole; **vpp**, vegetal pole plasm). In **5.8** arrows indicate the direction of the spreading of the vegetal pole plasm under the surface of an oocyte at late anaphase of the first maturation division (×450; adapted from Raven, 1945).

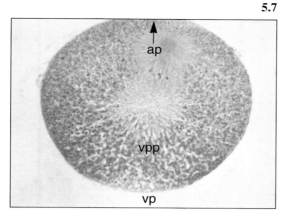
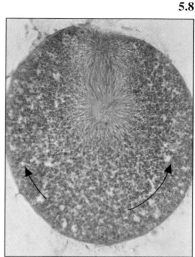

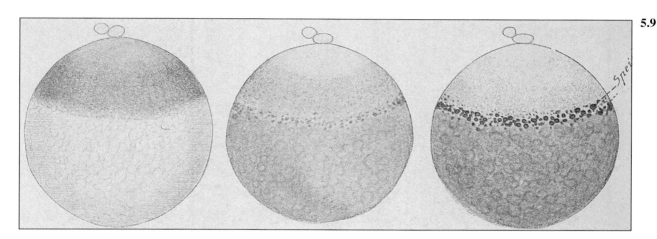

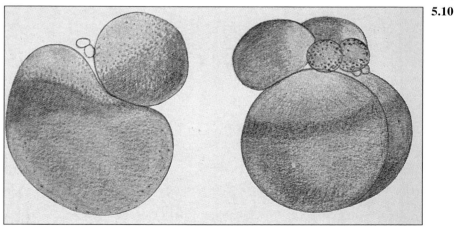

5.9, 5.10 Ooplasmic segregation in *Aplysia*. **5.9** shows the differences in the rate of the oxidation of leucomethylene blue in a mature egg of *Aplysia*. Left: most rapid oxidation in the animal hemisphere, 1min after the application of the dye. Centre: differential distribution of the reduced methylene blue 5–10min later. Right: accumulation of methylene blue in the supra-equatorial ring of *Speichergranula*. (From Ries and Gersch, 1936, with permission Stazione Zoologica di Napoli). In the left of **5.10** a 2-cell stage of *Aplysia* after staining with vital blue shows strong colour differences between the different zones of the blastomeres. The yolky, vegetal hemisphere has a natural yellow green colour. On the right of this figure is a 6-cell stage of *Aplysia* after staining with toluidine blue. The vegetal hemisphere has a greenish colour because of the natural yellowish yolk granules. The two small micromeres are rich in the Speichergranula (×625).

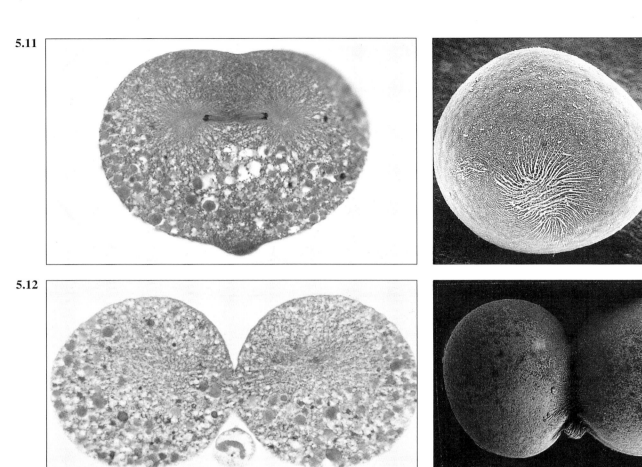

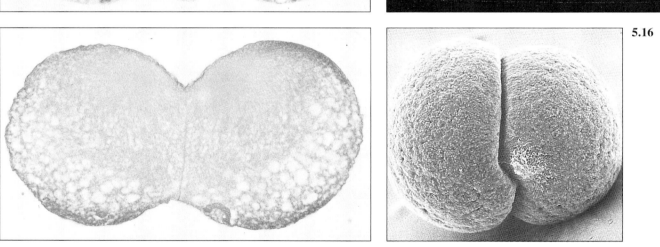

5.11–5.13 Polar lobe formation in *Bithynia*. Longitudinal sections through eggs of the fresh water snail *Bithynia tentaculata* during the formation and fusion of the first polar lobe (courtesy of Dr M.R. Dohmen). **5.11** (×350) shows the formation of the polar lobe at late anaphase of the first cleavage. The protruding polar lobe region contains the dark stained vegetal body (haematoxylin-eosin stain). **5.12** (×400) shows the maximal constriction of the polar lobe (including the vegetal body) at the end of cytokinesis. The cup-shaped vegetal body is placed upside down in the polar lobe (haematoxylin-eosin stain). **5.13** (×400) shows the 2-cell stage, the right blastomere has been fused with the lobe, and includes the vegetal body (methyl green and pyronin stain).

5.14–5.16 Surface views during polar lobe formation. **5.14** (×400) shows an SEM picture of an uncleaved egg of the marine slipper limpet, *Crepidula fornicata* after the first meiotic division illustrating the vegetal domain with folds in the presumptive polar lobe region (Academic Press, with permission). **5.15** (×350) shows a lateral view of an egg of *Bithynia* during the first cleavage, illustrating the formation of the first polar lobe (Cambridge University Press, with permission). **5.16** (×400) shows a latero-vegetal view of a 2-cell stage of *Bithynia* after the fusion of the polar lobe with the blastomere showing the folds characterizing the polar lobe region. (Courtesy of Dr M.R. Dohmen.)

A. Spiral cleavage

In the deuterostomes, the eggs show a radially cleaving pattern with the spindles placed alternately at right angles or parallel to the animal-vegetal egg axis. From the 4-cell stage onwards, the odd cleavages thus double the number of tiers of cells along the animal-vegetal axis, whereas the even cleavages double the number of cells within a tier. This means that, in the absence of dorso-ventral organization of the zygote and with the animal-vegetal segregation of the ooplasm as the only source of heterogeneity, each even cleavage doubles the number of cells with the same constitution, and hence with equivalent developmental capacities. This large number of equivalent cells may help explain the regulative character of embryogenesis in deuterostomes.

In contrast to the spindles in radial cleavage, those arising during spiral cleavage take up a position oblique to the egg axis, although this is hard to see during the first cleavage. A faint indication of the spiral character does, however, become evident at the second division. The cells AB and CD (**5.17–5.22**) each divide to give slightly more animal and slightly more vegetal daughter cells and, as a consequence, the four A–D cells do not lie in a single plane. The two cells, A and C, touch one another at the animal pole, where they form the animal cross-furrow (**5.18**), and B and D touch at the vegetal pole to form the vegetal cross-furrow (**5.17**). At third cleavage, the spiral character is obvious. Seen from the animal pole, the four spindles are not parallel to the egg axis and, according to the species, the upper part of each spindle points slightly clockwise or counterclockwise relative to the deeper part. In each quadrant, therefore, the animal cell is rotated clockwise or counterclockwise with respect to its vegetal sister cell (**5.19, 5.20**). Each new plane of division tends to intersect the preceding one at right angles, and clockwise and counterclockwise cleavages thus regularly alternate with each other. If the third cleavage has been dextral (**5.19, 5.20**), the fourth will be sinistral, etc. (**5.21, 5.22**).

A species is said to be dextral or dexiotropic if the odd cleavages are clockwise and the even cleavages counterclockwise. In contrast, a species is sinistral or laeotropic if the odd cleavages are counterclockwise. Eventually, this initial pattern of cell division leads to the corresponding chirality in the adult shell, with both being determined by the maternal genome (Boycott and Diver, 1923; Freeman and Lundelius, 1982). The mechanism of spiral cleavage may be related to the organization of cortical microfilaments in the contractile ring during cytokinesis (Meshcheryakov, 1990; Meshcheryakov and Beloussov, 1975).

As the cleavages are not parallel to the animal-vegetal egg axis, and as molluscs show a clear heterogeneity along that axis, each division must double the number of qualitatively different cells. In the absence of dorso-ventral organization, and with the assumption that the first two cleavages are parallel to the egg axis, all blastomeres within a single quadrant are different, and each only has one equivalent cell in each of the other three quadrants. This small number of equivalent cells appears to be related to the nonregulative character of spirally cleaving embryos.

B. Blastomere nomenclature

The regular arrangement of the cells during the successive cleavage stages makes it possible to identify the individual blastomeres and to follow their genealogy. For this purpose, a notation system is used in which the cell name reflects its lineage (Wilson, 1892; Conklin 1897; Robert, 1902) and this system is shown in **5.23** and **5.24**.

The first division splits the zygote into blastomeres AB and CD. At second cleavage, AB divides into A and B, and CD into C and D. At third cleavage, each quadrant divides into a smaller animal micromere and a larger vegetal macromere. The first quartet of micromeres and macromeres are called 1a–1d and 1A–1D, respectively. During the fourth, fifth and sixth cleavages, each macromere similarly divides into a micromere (2a–2d, 3a–3d and 4a–4d) and a macromere (2A–2D, 3A–3D and 4A–4D). Four quartets of micromeres usually form, but in some species one or more additional micromere quartets are generated. The macromeres are always the most vegetal cells, and are denoted by capitals and a coefficient which indicates the number of micromere quartets that they have formed.

Micromere derivatives are named according to the number of the quartet and the quadrant from which they derive, together with an exponent number. Because of the spiral cleavage pattern, each cell divides into a slightly more animal cell (given the exponent 1) and a slightly more vegetal cell (exponent 2). After the first division of the first quartet cells the animal tier is thus called $1a^1$–$1d^1$ and the vegetal tier $1a^2$–$1d^2$. During the next division, the cells $1a^1$–$1d^1$ divide into the animal tier $1a^{11}$–$1d^{11}$ and the vegetal tier $1a^{12}$–$1d^{12}$; $1a^{12}$–$1d^{12}$ then divides into the upper tier $1a^{121}$–$1d^{121}$ and the lower tier $1a^{122}$–$1d^{122}$, etc. The same system is applied to name the derivatives of the other quartets. The genealogy of each blastomere can thus be derived from its name. The series of exponents indicated in the superscript represent the number of divisions by which that particular cell has been formed. For instance, the cell $2a^{1212}$ stems from the micromere 2a via a series of four divisions: the division of 2a into $2a^1$ and $2a^2$, of $2a^1$ into $2a^{11}$ and $2a^{12}$; of the latter into $2a^{121}$ and $2a^{122}$, and finally of $2a^{121}$ into $2a^{1211}$ and $2a^{1212}$.

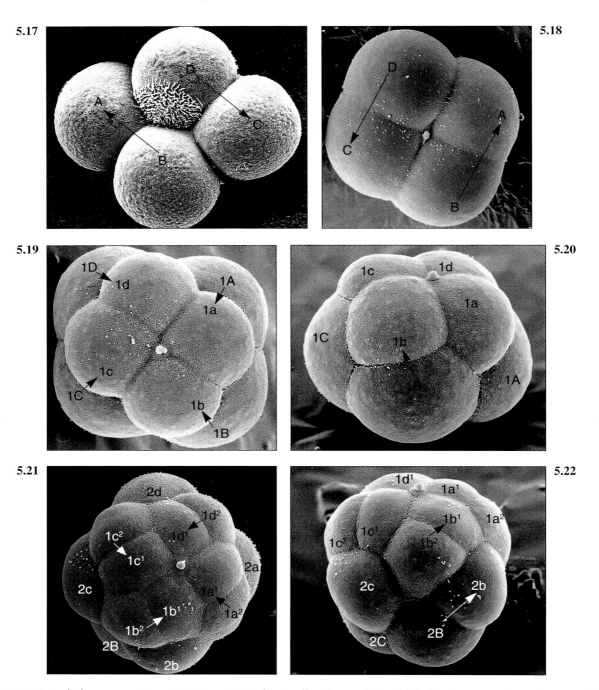

5.17–5.22 Spiral cleavage pattern. **5.17** is an SEM of a 4-cell embryo of *Crepidula fornicata* seen from the vegetal pole. As the D quadrant is provided with the folds of the polar lobe domain, the quadrants can be identified with certainty. The dorsal D and the ventral B quadrants are in touch at the vegetal cross-furrow and put between the two lateral quadrants A and C. **5.18–5.22** are scanning micrographs of embryos of *Patella vulgata*. In the absence of a polar lobe, the naming of the presumptive lateral and median quadrants is arbitrary; neither A can be distinguished from C, nor B from D. The arrows indicate the position of the mitotic spindles during the preceding cleavage. **5.18** is an animal view of a 4-cell stage. The small animal cross-furrow between the two presumptive lateral quadrants is scarcely visible. **5.19** is an animal view of an 8-cell embryo showing the clockwise rotation of the first quartet of micromeres. **5.20** is a lateral view of an 8-cell embryo. **5.21** is an animal view of a 16-cell embryo showing the position of the cells $1a^1$–$1d^1$ around the animal pole anticlockwise from the sister cells $1a^2$–$1d^2$. The large cells at the periphery of the graph are the second quartet cells located anticlockwise from the corresponding first quartet cells. **5.22** is a lateral view of a 16-cell embryo, showing the mutual position of the blastomeres in a separate quadrant (**5.17** ×260, **5.18–5.22** ×250).

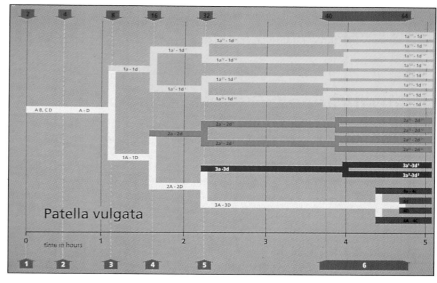

5.23 Cell lineage in *Patella vulgata*. First quartet cells in yellow, second quartet in light blue, third quartet in dark blue. Yellow and blue represent the presumptive ectoderm. Fourth quartet and macromeres in green, except the mesentoblast 4d which is in red. Green is the presumptive endoderm. Red is the mesentoblast, stem cell of the mesodermal bands and part of the endoderm. The numbers in the arrows at the lower line show the numbers of the cleavages. Notice that the sixth arrow indicates the asynchronous sixth cleavage. The numbers in the arrows at the upper line show the onset of the successive cleavage stages.

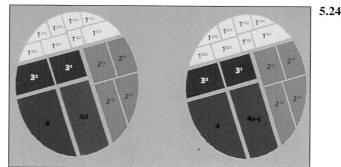

5.24 Fate map of molluscan embryo. The position and the developmental fates of the blastomeres in the four quadrants are shown. The odd quartets are placed clockwise and the even quartets are placed anticlockwise in each of the quadrants. The colours correspond with those used in **5.23**. On the left is the D quadrant and on the right the A, B and C quadrants.

Specification of the quadrants

The first two cleavages divide the molluscan egg into four cells with distinct lineages. The blastomeres of the four quadrants each contribute in a specific way to the development of the dorsal, ventral, right and left parts of the embryo. B and D are the median quadrants, A and C are the lateral quadrants. The dorsal part of the embryo mainly derives from the dorsal or D quadrant. The B quadrant contributes to the formation of the ventral part, whereas A and C contribute to the left and right part of the embryo, respectively. In some species, the four quadrants are specified by unequal division; in others, the specification is determined by later cellular interactions. The two types will be described separately.

A. Specification by unequal division

In a number of molluscs, the four quadrants are different from the early onset of development as the first two cleavages are unequal. Here, the egg is sometimes subdivided into three smaller quadrants A, B and C, and a larger D quadrant, while, in species like *Aplysia*, A and B are larger than C and D. The formation of a larger D quadrant may occur in two different ways. The first happens when the early cleavages are associated with the formation of a large (*Dentalium* and *Nassarius*) or small (*Bithynia* and *Crepidula*) polar lobe (**5.11–5.13, 5.14–5.16**). After completion of the first cleavage, the lobe fuses with the blastomere to which it is attached (cell CD). At the second cleavage, AB divides equally, whereas CD forms a second polar lobe which subsequently fuses with one of the two daughter cells (D).

The second way in which a larger D quadrant can form may also be the result of an unequal cleavage, but without a polar lobe and this occurs in most lamellibranchs (e.g. *Pholas, Spisula, Anodonta, Unio,* and *Dreissensia*). During the first mitosis, the spindle is initially symmetrically placed at right angles to the plane through the egg axis and the sperm entrance point. Prior to cleavage, the spindle moves off centre to the right or left in a horizontal plane normal to the plane through the sperm entrance point and the egg axis. The first division is consequently unequal, with AB being smaller than CD. At second cleavage, AB divides equally; as the spindle in CD is displaced in the direction of the sperm entrance point, CD divides into a small C and a large D cell. At the end of the second cleavage the embryo consist of a large D quadrant and the three equally sized quadrants A, B and C. Because of the random movement of the first mitotic spindle the four quadrants may be arranged clockwise or counterclockwise, without any significance for later development (Guerrier, 1970).

B. Specification by cellular interaction

In a number of species, the first two cleavages are more or less equal and do not lead to the development of different quadrants (e.g. *Patella vulgata*; van den Biggelaar, 1977). Although the quadrants are initially equivalent, the future plane of bilateral symmetry is biased by the cell pattern at the 4-cell stage. The two blastomeres which form the animal cross furrow appear to become the two lateral quadrants, while those which form the vegetal cross furrow become the two presumptive median quadrants. The dorso-ventral axis is therefore defined, but its polarity remains to be specified.

This specification only occurs during the interval between the formation of the third and fourth quartet of micromeres as the result of an inductive influence of the animal cells on one of the macromeres (van den Biggelaar and Guerrier, 1979). From the lineage in the embryo of *Patella* shown in **5.23**, it will be evident that all blastomeres divide synchronously. At the sixth cleavage, however, synchrony between the tiers of cells along the animal-vegetal axis is lost, although it is maintained within the individual tiers of micromeres (**5.25**). The macromeres do not divide simultaneously as 3D divides later than 3A, 3B and 3C (**5.28**) and it alone divides unequally, with 4D being much smaller than 4d (**5.26**). The cleavage delay and the pronounced unequal division of 3D reliably indicate the dorso-ventral polarity of the embryo.

The specification of the dorsal quadrant appears to be preceded by an asymmetery in the position of macromere 3D during the interval between the fifth and the sixth cleavage. After the fifth cleavage, the micromeres of the second quartet move away from the centre of the embryo to leave a free blastocoelic cleft between the animal micromeres and the vegetal macromeres 3A–3D. All four macromeres extend in the animal direction and make contact with the animal micromeres, but these contacts are only maintained with one of the two cross-furrow macromeres (**5.29**). The central position (**5.30**) and the exclusive contacts with the animal micromeres appear to lead to the division delay and the unequal division of the central macromere 3D into the smaller macromere 4D and the larger 4d cell (van den Biggelaar, 1977).

The macromeres 3A, 3B and 3C contribute to the formation of the endoderm alone, but the complex history of 3D causes it to have a different developmental fate. Its daughter cells will form endoderm and mesoderm, with macromere 4D being an endodermal precursor cell and micromere 4d a mesentoblast which contributes to the endoderm and forms the stem cells of the adult mesoderm. Serras and Speksnijder (1990) have shown that the induction of the D quadrant in *Patella* is associated with the appearance of a higher amount of F actin in 3D than in any other macromere or in any of the micromeres, as shown by labelling the embryo with fluorescent phalloidin, a marker for F actin. After division of 3D, only the presumptive endodermal 4D macromere retains the higher density of F actin (**5.27**; Serras and Speksnijder, 1990). This result shows that the identity of the four quadrants of the *Patella* embryo is lost between the fifth and the sixth cleavages (van den Biggelaar and Guerrier, 1979).

The causal relation between the interaction of the animal micromeres and the central macromere and the determination

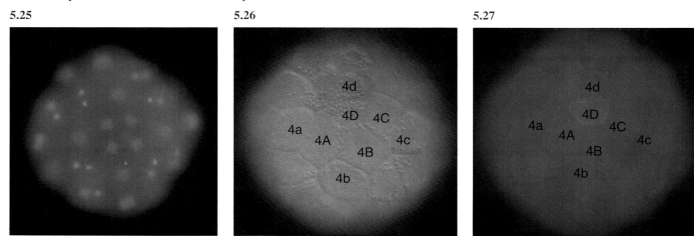

5.25–5.27 *Patella* embryos at sixth cleavage. **5.25** shows an embryo of *Patella vulgata* at the transition from the 32- into the 40-cell stage seen from the animal pole. The four groups of primary trochoblasts are in late telophase and form 16 cells (compare **5.31**), whereas the eight cells around the animal pole are still in interphase. **5.26** and **5.27** are micrographs of a 64-cell embryo of *Patella* seen from the vegetal pole. Only the position of the fourth quartet of micromeres and the macromeres has been indicated. The greater part of the mesentoblast 4d protrudes into the centre of the embryo, and cannot, therefore, be seen in a surface view. The position of the micromeres 4a–4d and of the macromeres has been indicated. **5.26** is a Nomarski picture showing the unequal division of 3D. **5.27** is a TRITC-phalloidin labelling of the same embryo showing the higher labelling density in the presumptive endodermal cell 4D. (×300. Adapted from Serras and Speksnijder, 1990.)

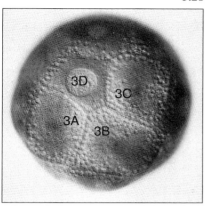 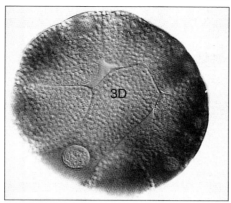 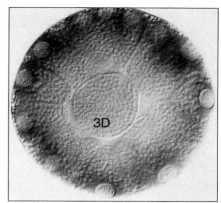

5.28–5.30 Landmarks of the D-quadrant in the embryo of *Patella* at sixth cleavage. **5.28** is a Nomarski picture of an embryo at the transition of the 60 into the 63-cell stage by the division of the macromeres 3A, 3B and 3C. The macromere 3D is still in interphase. **5.29** is a lateral view of a 32-cell embryo showing the central position of the apex of the macromere 3D underneath the micromeres at the animal pole. **5.30** is an optical cross-section through the centre of a 32-cell embryo showing the central position of 3D (×300).

of the mesodermal stem cells has been demonstrated in experiments where contact has been inhibited or changed. Inhibition of this contact prevents the division asynchronies within the macromeres and other tiers of cells; the difference in cleavage pattern of the macromeres consequently fails to appear and the embryos thus remain radially symmetric (van den Biggelaar and Guerrier, 1979; Martindale *et al.*, 1985; Kühtreiber *et al.*, 1988). Removal of a presumptive lateral micromere at the 8-cell stage reliably ensures that the quadrant anticlockwise to it develops as the dorsal quadrant (Arnolds *et al.*, 1983). It is not yet clear how the animal micromeres induce the central macromere 3D.

The maintenance of equipotentiality of the quadrants until the period between the formation of the third and fourth quartet of micromeres appears to be a common aspect of the early development of species with an equal 4-cell stage, examples being *Patella vulgata* (van den Biggelaar and Guerrier, 1979), *Lymnaea stagnalis* (Arnolds *et al.*, 1983) *Lymnaea palustris* (Martindale *et al.*, 1985) and *Haminoea callidegenita* (Boring, 1989).

The later stages of development
The blastula

After about six cleavages, the embryo forms a blastula, with fresh water species usually developing a blastocoel, whereas marine species do not. The early blastula stage appears to be a transition point after which cell specification begins to be influenced by cellular interactions, cell differentiation starts and differences occur in cell proliferation, while intercellular communication by gap junctions is initialized.

The 2–4 primary trochoblasts in each quadrant (depending on the species) are the first cells that stop dividing and start to differentiate (**5.31**). At around the time of the previous cell division, their capacity for the intercellular exchange of small molecular weight substances through the gap junctions is diminished or lost (Serras and van den Biggelaar, 1987; Serras *et al.*, 1990). The four clones of uncoupled primary trochoblasts demarcate a cross-like pattern in the animal hemisphere known as the 'molluscan cross' and the cells of this cross remain coupled by gap junctions (**5.32**; Serras and van den Biggelaar, 1987). This finding may indicate an incompatibility between the maintenance of an exchange of small molecular weight substances and the cessation of further divisions with the onset of differentiation into ciliated cells.

The gastrula

The blastula is transformed into a gastrula when the archenteron forms and blastulae with and without a blastocoel gastrulate differently. In the presence of a blastocoel, gastrulation begins with the reduction of the external surface of the macromeres and a flattening of the embryo (**5.34**). By the depression thus formed at the vegetal pole and by the subsequent invagination of the endodermal micromeres, the archenteron is deepened (**5.35**) and the descendants of the mesentoblast 4d attain a position between the endo- and ectodermal cells (**5.33**). In the absence of a blastocoel, the blastula is flattened in the animal-vegetal direction and, just before gastrulation, a depression forms at the vegetal pole. This is followed by an invagination of the endodermal cells and an epibolic extension of the presumptive ectodermal cells over the endodermal blastomeres.

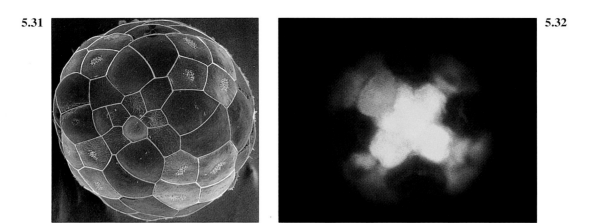

5.31, 5.32 The Molluscan Cross. **5.31** is an SEM micrograph of the animal hemisphere of a 64-cell embryo of *Patella vulgata* showing the four groups of ciliated primary prototroch cells. The remaining animal micromeres form the molluscan cross (yellow). **5.32** shows a micrograph of an 82-cell embryo of *Lymnaea stagnalis* after micro-injection of one of the cells of the molluscan cross with Lucifer Yellow, a fluorescent low molecular-weight substance that can pass through gap junctions. Lucifer Yellow has been transferred to all other animal cells with the exception of the primary trochoblasts (×300. **5.31** adapted from Serras, Dictus and van den Biggelaar, 1990; **5.32** adapted from Serras and van den Biggelaar, 1987).

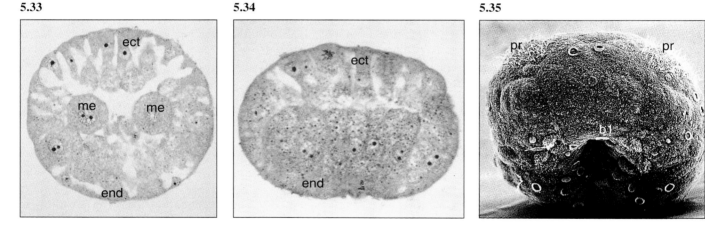

5.33–5.35 Gastrulae of *Lymnaea*. These figures are micrographs of histological sections of gastrulae of *Lymnaea stagnalis* (haematoxylin-eosin stain). **5.33** is a frontal-section showing the position of two centrally placed large derivatives of the mesodermal stem cells located between the ecto- and endodermal cells. **5.34** is a longitudinal section through the flattened gastrula. **5.35** is a ventral view of a late gastrula of *Lymnaea stagnalis* with the vegetally located blastopore. Note the rings at the surfaces of the cells, they are stages in the process of pinocytosis of capsular fluid. (×250. Key: **bl**, blastopore; **ect**, ectoderm; **end**, endoderm; **me**, mesoderm; **pr**, prototroch cells.)

The trochophore

During the onset of gastrulation, the blastopore lies diametrically opposite the animal pole. Mainly due to the extensive proliferation of the derivatives of the first somatoblast 2d, the stomodaeum (mouth region) is shifted anteriorly. Meanwhile, the ciliated trochoblasts form a transverse girdle, the prototroch (**5.36, 5.37** and **5.39–5.41**), which subdivides the ectoderm into pre-trochal and post-trochal domains. The former is derived from the first quartet of micromeres, while the latter is developed from the second and third quartet of micromeres. The pre-trochal region separates into the larval cells around the animal pole, which become involved in the formation of an apical tuft or apical plate, and the two cephalic plates lateral to it which are composed of small-celled ectoderm (**5.36**). From each of these plates a tentacle, an eye, a cerebral ganglion and half of the epidermal ectoderm develop. The post-trochal ectoderm predominantly stems from the dorsal region, mainly from 2d, and forms the somatic plate which, in turn, gives rise to the shell gland, the mantle epithelium and the foot (**5.37, 5.39**). The ectoderm around the anteriorly displaced blastopore develops into the stomodaeum from whose ventral region are derived the anlage of the radula and the radular sac.

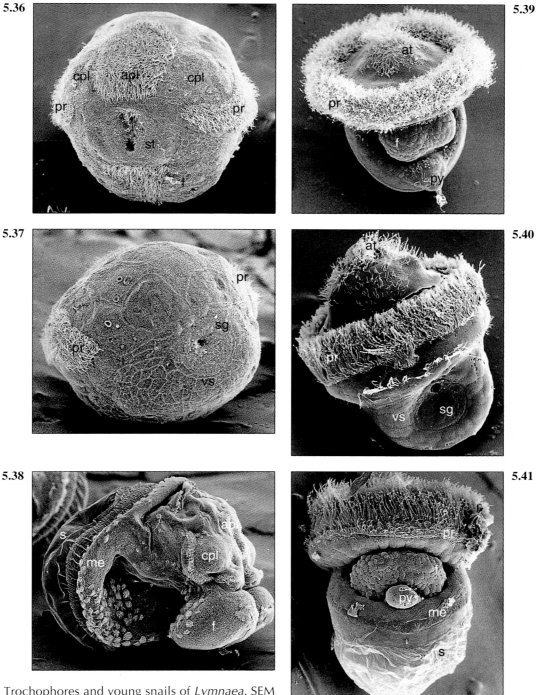

5.36–5.38 Trochophores and young snails of *Lymnaea*. SEM micrographs of embryos of *Lymnaea stagnalis*. **5.36** is a front view of a trochophore. The stomodaeum can be distinguished as can the prototroch and the apical plate with the two interspersed cephalic plates. The narrower, lower part of the embryo represents the foot anlage. **5.37** is a dorsal view of a trochophore showing the invaginated shell gland. The dorso-posterior part of the embryo is already enclosed by the embryonic shell. **5.38** is a lateral view of a so-called hippo stage embryo. The head is almost complete, having eyes and tentacle rudiments. The shell almost completely surrounds the visceral sac. The foot is well developed. (×300. Key: **apl**, apical plate; **cpl**, cephalic plate; **f**, foot; **me**, mantle edge; **pr**, prototroch; **py**, pygidium; **s**, shell; **sg**, shell gland; **st**, stomodaeum; **vs**, visceral sac.)

5.39–5.41 Trochophores of *Patella*. SEM micrographs of trochophores of *Patella vulgata*. **5.39** is a ventral view. Enclosed in the ring of ciliated prototroch cells is the cephalic region with the central, apical tuft. In the post-trochal part, the bilobed anlage of the foot and the pygidium with telotroch can be seen. The visceral sac behind the foot anlage is not yet completely surrounded by the shell. **5.40** is a dorsal view showing the invaginated shell gland. **5.41** is a late trochophore stage with further developed post-trochal part and a visceral sac almost completely covered by a shell. (×350. Key: **at**, apical tuft; **f**, foot; **me**, mantle edge; **pr**, prototroch; **py**, pygidium; **s**, shell; **sg**, shell gland; **st**, stomodaeum; **te**, telotroch; **vs**, visceral sac).

The veliger

Depending upon the species, the trochophore may metamorphose directly into a young adult (e.g., a snail), or develop into a second larval form, the veliger. In the latter case, the lateral parts of the prototroch may be transformed into large semicircular, velar folds, provided with strongly developed cilia. A veliger already has eyes, tentacles, a well-developed foot, shell gland, a shell mass (**5.38, 5.41**), and cerebral, pedal and buccal ganglia. The stomodaeum is differentiated into a mouth cavity, pharynx, radular sac and radula and oesophagus, with the latter opening into the endodermal midgut. The anus then forms where the posterior part of the hindgut makes contact with the ventro-posterior part of the ectoderm (molluscs are Protostomata).

Metamorphosis

In Aplacophora and Polyplacophora, the trochophore larva metamorphoses directly into the adult form. In species with a veliger, metamorphosis takes place by the loss of such larval structures as the velum, while the larval heart, muscles and kidneys are replaced by their adult counterparts.

Molluscs: model systems in developmental biology

The most challenging problem in developmental biology is the regulation of patterns of gene expression as the body plan is laid down. In several respects, molluscs have been and are still useful model systems in developmental biology and have the following advantages which, in turn, suggest a range of important problems to be solved.

- The organization of the cytoskeleton, the egg cortex and the ooplasm in the ovarian oocyte and zygote is directly related to the organization of the later embryo. The vegetal region of the zygote appears to have determinants (probably maternal RNAs) for the development of the dorso-ventral body axis. A molecular analysis of the informational molecules, especially in the polar lobe, thus appears to be a promising area of research.
- The consistency of the cleavage pattern as the blastomeres form, together with the cell cycle arrests as the larval trochoblasts differentiate, indicate that the molluscan embryo is a suitable model system for the study of the developmental regulation of the cell cycle.
- The molluscan embryo progressively separates into non-communicating compartments with specific developmental fates, and, presumably, with specific proliferation rates. This phenomenon lends itself to an analysis of the causal relation between cell proliferation, cell communication and determination of developmental fate and differentiation. In particular, a molecular analysis of the relation between the early mitotic arrest and the early differentiation of the larval cells is well worth undertaking.
- The great range of molluscan species shows many differences in the rate of cell proliferation in the various cell lineages and these lead to differences in the speed with which different parts of the embryo develop. These differences may be fundamental in the evolution of different species and a comparative analysis of the early development of various species may therefore shed new light on the relation between ontogeny and phylogeny (van den Biggelaar, 1993).
- Although molluscs have been a favourite system for cell lineage studies since the end of the last century, these studies have been limited by the absence of any reliable cell marker other than position, and this becomes less useful as cell numbers increase. A thorough cell lineage analysis of the commoner species using good markers is urgently needed.
- Molluscs are the best system for analysing the molecular mechanism of spiral cleavage which characterizes many of the animal phyla.
- The invariance of the molluscan cleavage pattern is a suitable problem for mathematical modelling.
- In contemporary developmental biology, considerable emphasis is put on elucidating the mechanisms of gene regulation involved in the development of the body pattern and in cell differentiation. The species mainly being studied are restricted to a few 'state horses' such as *Drosophila, Xenopus, Caenorhabditis* and the mouse. In our opinion a thorough understanding of morphogenesis in general and of the relation between ontogeny and phylogeny in particular requires a continued interest in the embryonic development of other major animal phyla. For these reasons a detailed analysis of the molecular basis of molluscan development needs to be undertaken.

References

Arnolds, W.J.A., van den Biggelaar, J.A.M. and Verdonk N.H. (1983). Spatial aspects of cell interactions involved in the determination of dorsoventral polarity in equally cleaving gastropods and regulative abilities of their embryos, as studied by micromere deletions in *Lymnaea* and *Patella*. *Roux's Arch. Dev. Biol.*, **192**, 75–85.

Boring, L. (1989). Cell-cell interactions determine the dorsoventral axis in embryos of an equally cleaving opisthobranch mollusc. *Dev. Biol.*, **136**, 239–253.

Boycott A.E. and Diver C. (1923). On the inheritance of sinistrality in *Limnaea peregra* (Mollusca, Pulmonata). *Phil. Trans. Roy. Soc.*, **95B**, 207–213.

Collier, J.R. (1983). The biochemistry of molluscan development. In *The Mollusca*, Vol 3 Development (Verdonk N.H., van den Biggelaar, J.A.M., and Tompa, A.S., eds) pp. 253–297, Academic Press, New York.

Collier, J.R. (1989). Cytoplasmic regulation of translation during *Ilynassa* embryogenesis. *Development*, **106**, 263–269.

Conklin, E.G. (1897). The embryology of *Crepidula*. *J. Morph.*, **13**, 1–226.

Dohmen, M.R. and van der Mey, J.C.A. (1977). Local surface differentiations at the vegetal pole of the eggs of Nassarius reticulatus, *Buccinum undatum*, and *Crepidula fornicata* (Gastropoda, Prosobranchia). *Dev. Biol.*, **61**, 104–113.

Dohmen, M.R. and Verdonk, N.H. (1974). The structure of a morphogenetic cytoplasm, present in the polar lobe of *Bithynia tentaculata* (Gastropoda, Prosobranchia). *J. Embryol. Exp. Morph.*, **31**, 423–433.

Draetta, G., Luca, F., Westendorf, J., Brizuella, L., Ruderman, J. and Beach, D. (1989). cdc2 Protein kinase is complexed with both cyclin A and B: evidence for proteolytic inactivation of MPF. *Cell*, **56**, 829–838.

Dubé, F. and Dufresne, L. (1990). Release of metaphase arrest by partial inhibition of protein synthesis in blue mussel oocytes. *J. Exp. Zool.*, **256**, 323–332.

Dubé, F., Dufresne, L., Coutu, L. and Clotteau, G. (1991). Protein phosphorylation during activation of surf clam oocytes. *Dev. Biol.*, **146**, 473–482.

Freeman, G. and Lundelius, J.W. (1982). The developmental genetics of dextrality and sinistrality in the gastropod *Lymnaea peregra*. *Roux's Arch. Dev. Biol.*, **191**, 69–83.

Guerrier, P. (1968). Origine et stabilite de la polarité animale vegetative chez quelques Spiralia. *Ann. Embryol. Morphol.*, **1**, 119–139.

Guerrier, P. (1970). Les caractères de la segmentation et la détermination de la polarité dorsoventrale dans le développement de quelques Spiralia. **III**. *Pholas dactylus* et *Spisula truncata* (Mollusques Lamellibranches). *J. Embryol. Exp. Morph.*, **23**, 667–692.

Guerrier, P., Colas, P. and Néant, I. (1990). Meiosis reinitiation as a model system for the study of cell division and cell differentiation. *Int. J. Dev. Biol.*, **34**, 93–109.

Haszprunar, G. (1988). On the origin and evolution of major gastropod groups, with special reference to the Streptoneura. *J. Moll Stud.*, **54**, 367–441.

Kühtreiber, W.M., van Til, E.H., and van Dongen, C.A.M. (1988). Monensin interferes with the determination of the mesodermal cell line in embryos of *Patella vulgata*. *Roux's Arch. Dev. Biol.*, **197**, 10–18.

Martindale, M.Q., Doe, C.Q. and Morrill, J.B. (1985). The role of animal-vegetal interaction with respect to the determination of dorsoventral polarity in the equal cleaving spiralian, *Lymnaea palustris*. *Roux's Arch. Dev. Biol.*, **194**, 281–295.

Meshcheryakov, V.N. (1990). The common pond snail *Lymnaea stagnalis*. In *Animal species for developmental studies*, Vol **1** Invertebrates, (TA Dettlaff and SG Vassetzky, eds) pp. 69–132, Consultants Bureau, New York.

Meshcheryakov, V.N. and Beloussov, L.V. (1975). Asymmetrical rotations of blastomeres in early cleavage of Gastropoda. *Roux's Arch. Dev. Biol.*, **177**, 193–203.

Raven, C.P. (1945). The development of the egg of *Limnaea stagnalis* L. from oviposition till first cleavage. *Arch. Néerl. Zool.*, **7**, 91–119.

Ries, E. and Gersch, M. (1936). Die Zelldifferenzierung und Zellspezialisierung wahrend der Embryonalentwicklung von Aplysia limacina L. Zugleich ein Beitrag zu Problemen der vitalen Fårbung.

Robert, A. (1902). Recherches sur le développement des troques. *Arch. Zool. Exp. Gén.*, **10**, 269–359.

Serras, F., Dictus, W.J.A.G. and van den Biggelaar, J.A.M. (1990). Changes in junctional communication associated with cell cycle arrest and differentiation of throchoblasts in embryos of *Patella vulgata*. *Dev. Biol.*, **137**, 207–216.

Serras, F. and Speksnijder, J.E. (1990). F-actin is a marker of dorsal induction in early *Patella* embryos. *Roux's Arch. Dev. Biol.*, **199**, 246–250.

Serras, F. and van den Biggelaar J.A.M. (1987). Is a mosaic embryo also a mosaic of communication compartments? *Dev. Biol.*, **120**, 132–138.

Swenson, K.I., Farrell, K.M. and Ruderman, J.V. (1986). The clam embryo protein cyclin A induces entry into M phase and the resumption of meiosis in *Xenopus* oocytes. *Cell*, **47**, 861–870.

Tompa, A.S., Verdonk, N.H. and van den Biggelaar, J.A.M. (1984). In *The Mollusca*, Vol 7 Reproduction, Academic Press New York.

van den Biggelaar, J.A.M. (1977). Development of dorsoventral polarity and mesentoblast determination in *Patella vulgata*. *J. Morph.*, **154**, 157–186.

van den Biggelaar, J.A.M. (1993). The cleavage pattern in embryos of *Haliotis tuberculata* (Archaeogastropoda) and gastropod phylogeny. *J.Morph.* in press.

van den Biggelaar, J.A.M. and Guerrier, P. (1979). Dorsoventral polarity and mesentoblast determination as concomitant results of cellular interactions in the mollusk *Patella vulgata*. *Dev. Biol.*, **68**, 462–471.

van den Biggelaar, J.A.M. and Guerrier, P. (1983). Origin of spatial organization. In *The Mollusca* Vol **3**, Development (N.H. Verdonk, J.A.M. van den Biggelaar, and A.S. Tompa eds) pp. 179–213, Academic Press, New York.

van Dongen, C.A.M., Wes, J.H., Goedemans, J.H. and Reijenga, J.C. (1985). Composition of the nucleotides pool in a morphogenetic compartment in eggs of *Nassarius reticulatus* (Mollusca) analysed by capillary isotachophoresis. *Exp. Cell. Res.*, **161**, 406–420.

van Loon, A.E., Colas, P., Goedemans, H.J., Néant, I., Dalbon, P. and Guerrier, P. (1991). The role of cyclins in the maturation of *Patella vulgata* oocytes. *EMBO J.*, **10**, 3343–3349.

Verdonk, N.H., Geilenkirchen, W.L.M. and Timmermans, L.P.M. (1971). The localization of morphogenetic factors in uncleaved eggs of *Dentalium*. *J. Embryol. Exp. Morph.*, **25**, 57–63.

Wilson, E.B. (1892). The cell lineage of *Nereis*. *J. Morph.*, **6**, 361–481.

6. The Leech

David A. Weisblat

Introduction

Leeches are segmented worms (**6.1**), and are generally treated as the class (Pearse and Buchsbaum, 1987) or subclass (Barnes *et al.*, 1988) Hirudinea in the phylum Annelida, along with oligochaetes (earthworms) and polychaetes. In leech development, individually identified, experimentally accessible cells can be examined from the embryo through adulthood.

The work to be summarized here is rooted jointly in studies carried out independently by Charles O. Whitman (1878) and Gustav M. Retzius (1891). Whitman used leech embryos for the first studies of developmental cell lineages to test Haeckel's Law ('ontogeny is a rapid and shortened recapitulation of phylogeny'; see Maienschein, 1978), while Retzius, a neuroanatomist, found that segmental ganglia in leech contain large cells that can be individually identified (**6.2–6.4**). Modern

6.1 Dorsal view of a small adult leech, of the glossiphoniid species *Helobdella triserialis*, on a US dime (anterior to right). This species, which reaches a maximum size of approximately 3cm, provided the embryos for many of the studies described in this chapter (×19).

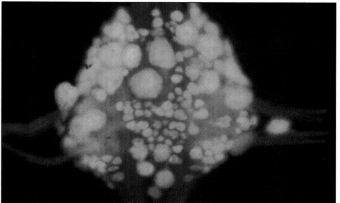

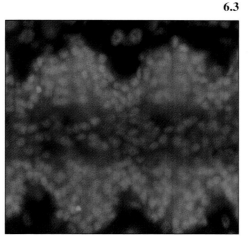

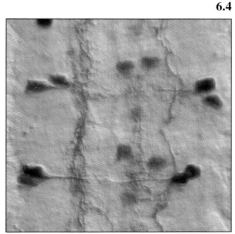

6.2–6.4 Segmental ganglia contain individually identifiable neurons. **6.2** (×100) shows a single ganglion from *Haemopis marmorata* (a hirudinid species) stained with a monoclonal antibody that labels all neuronal cell bodies (courtesy of B. Zipser and R. McKay). **6.3** (×550) shows two ganglia in which the cell nuclei are stained from a Stage 9 *Helobdella* embryo; the connective nerves have not yet formed, and the cells are all the same size (courtesy of S. Zackson). However, staining similar ganglia with an antibody to serotonin (**6.4**, ×550), reveals that at least some neurons have already begun to differentiate (courtesy of R. Ho).

electrophysiological explorations of the leech nervous system were initiated by Stephen Kuffler, David Potter and John Nicholls in the 1960s; they soon found that the identifiable cell bodies correspond to neurons and glia with distinct physiological and morphological characteristics (see Nicholls, 1987). Today, the nervous system of the European medicinal leech, *Hirudo medicinalis*, is among the best known in terms of explaining animal behaviour through the properties and interconnections of individual neurons (for reviews, see Muller *et al.*, 1981; Friesen, 1989; Stent *et al.*, 1992).

In the 1970s Gunther Stent recognized that the leech nervous system constitutes a detailed 'endpoint' upon which to focus embryological studies. But, because zygotes of *Hirudo* and other species in its family (Hirudinidae) are small and develop inside yolk-filled cocoons, their early development is hard to study. The family Glossiphoniidae, in contrast, features large, easily cultured embryos and it was these that Whitman chose for his cell lineage studies (**6.5**). He provided an outline of leech development, finding that the embryos, like the adult nervous system, contain individually identifiable cells. The patterns of early cell divisions that he described are extensively conserved throughout the annelid phylum (see Sandig and Dohle, 1988).

In the past 15 years, the description provided by Whitman has been amended slightly and extended significantly, often using microinjected 'cell lineage tracers' to mark the progeny of selected blastomeres (**6.6–6.8**); for a review of this technique, see Stuart *et al.*, 1990. Here, after a brief overview of leech development which concentrates on glossiphoniid species, some areas of leech development under active investigation are reviewed.

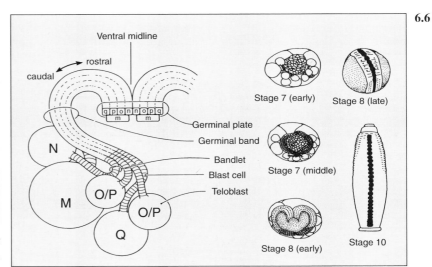

6.5 Size comparison of leech embryos. Embryos of hirudinid leeches such as *Hirudo* (upper left) are less than 100µm in diameter, while those of glossiphoniid species range from 400µm in *Helobdella* (upper right) to almost 2mm in *Haementeria* (bottom, ×10).

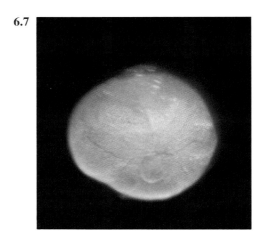

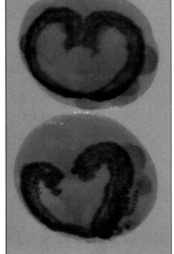

6.6–6.8 Lineage tracers confirm the 'central dogma' of leech development. During cleavage (**6.6**), five bilateral pairs of teloblasts arise from the D quadrant. During Stages 7–8, teloblasts make blast cells in coherent columns called bandlets. Ipsilateral bandlets form left and right germinal bands and these zipper together along the ventral midline to form the germinal plate, from which the definitive segments arise. Views of Stage 7 and early Stage 8 embryos are from the future dorsal side, while those of late Stage 8 and Stage 10 are from future ventral side. In a living embryo at early Stage 8 (**6.7**, ×100), the teloblasts, blast cells and germinal bands are difficult to discern. **6.8** (×100) shows similar embryos fixed and stained for horseradish peroxidase (HRP) two days after the ectodermal proteloblast (cell DNOPQ) was injected with HRP.

Normal development
Cleavage

The 0.5mm eggs of *Helobdella triserialis* (a common glossiphoniid leech) are fertilized internally, but arrest in meiosis until after they are laid (for obtaining leech embryos, see Fernandez and Olea, 1982; Jellies *et al.*, 1987; Wedeen *et al.*, 1990). Zygotes are deposited on the ventral aspect of the parent hermaphrodite in transparent cocoons, from which they can be isolated and cultured in simple solutions (**6.9–6.12**). Cleavages are stereotyped and many are unequal; early blastomeres can be individually identified by size and position in the embryo, by birth order, and by inheritance of yolk-deficient cytoplasm known as teloplasm (**6.13–6.16**) which arises at the poles of the embryo prior to first cleavage and is enriched in mitochondria and maternal mRNAs (Weisblat *et al.*, 1990).

Three classes of blastomeres arise during cleavage (Stages 1–6; Fernandez, 1980). Four macromeres and four micromeres are generated at the highly unequal third cleavage (Stage 4a). By the end of Stage 6, the three remaining macromeres

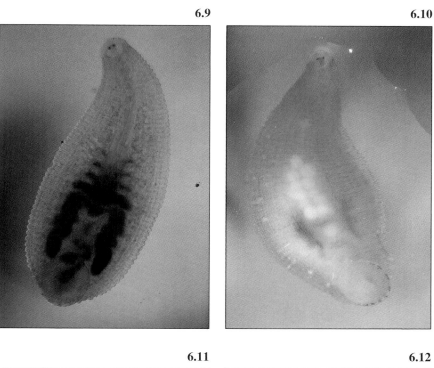

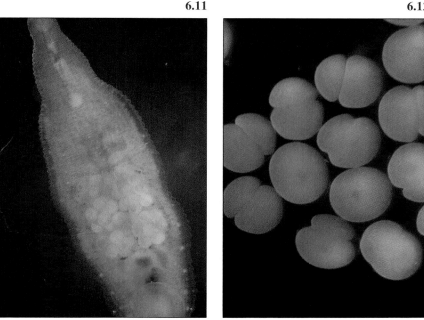

6.9–6.12 Egg production in *Helobdella*. These figures show ventral views of adult leeches (anterior is up). **6.9** shows freshly fed leeches (on physid snails), crop secae are dark; **6.10**, gravid leeches, whitish egg mass is visible through the body wall; **6.11**, brooding; embryos are laid in transparent cocoons attached to the parental venter; and **6.12** shows embryos can be removed from the cocoons and develop normally in simple salt solutions; first cleavage gives a smaller cell AB and a larger cell CD. (**6.9–6.11** ×10, **6.12** ×40; courtesy of *Scientific American*.)

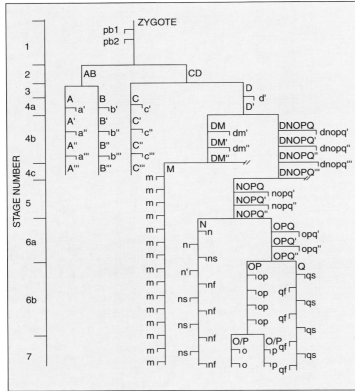
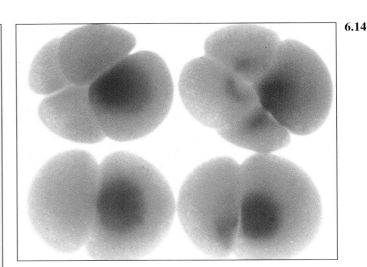
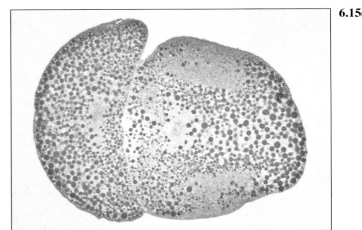
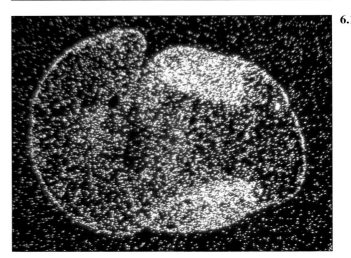

6.13–6.16 A cell lineage diagram of the early cleavage divisions is seen in **6.13**. Zygotes are in meiotic arrest until they are laid; polar bodies (pb) are produced and teloplasm forms prior to first cleavage. Micromeres are denoted by lower case, and primed letters and blast cells are denoted by lower case, unprimed letters. Only one side of the bilaterally symmetric divisions leading from cells DM″ and DNOPQ‴ is shown. **6.14** shows vegetal (left) and animal (right) views of 2-cell and 4-cell *Helobdella* embryos stained with avidin HRP to reveal the distribution of yolk-deficient cytoplasm; cells CD and D are to the right (courtesy of B. Holton). **6.15** is a brightfield view of a methylene blue-stained section through a 2-cell embryo, showing the predominant yolk (dark blue spots) and the yolk-deficient cytoplasm (telobplasm) at the animal and vegetal ends of cell CD (right). Prior to sectioning, the embryo had been hybridized with tritiated polyuridylic acid to localize maternally derived mRNAs. **6.16** is a darkfield view of the same section, showing silver grains concentrated over the cell cortex and teloplasm (**6.14** ×90, **6.15–6.16** ×180; courtesy of B. Holton and C. Wedeen).

(A‴–C‴; for lineage details, see **6.13**) are the largest cells in the embryos and provide a framework for the morphogenetic movements of embryogenesis. The third class of blastomeres comprises five bilateral pairs of teloblasts, stem cells derived (in Stages 4c–6) from macromere D′ of the 8–cell embryos and the progenitors of all segmentally iterated cells in the leech. Additional micromeres are also generated during Stages 4–6 and a total of 25 arise during cleavage, clustering in the micromere cap at the animal pole.

Genesis of the germinal plate

Over most of the length of the leech, mesoderm and ectoderm are organized into 32 segments which arise over an extended period (Stages 7–10) from a posterior growth zone composed of the ten teloblasts. One pair, the mesoteloblasts (M), generates segmental mesoderm and the other four pairs, ectoteloblasts (N, O/P, O/P and Q), generate segmental ectoderm. During Stages 7 and 8, each teloblast makes several dozen highly unequal divisions at the rate of about one per hour, generating a column (bandlet) of primary blast cells. Ipsilateral bandlets form parallel arrays called germinal bands within which they occupy stereotyped positions. On each side, the mesodermal bandlet (m) lies on the surface of the macromeres, and the ectodermal bandlets (n, o, p and q) lie atop and adjacent to the m bandlet. According to the dictates of the germ layer theory (e.g., Gilbert, 1988, p.6), the macromeres constitute endoderm.

The two germinal bands contact each other via their distal ends at the future head of the embryo and are separated from each other along the future dorsal side by a temporary epithelium derived from the cells of the micromere cap which also covers the germ bands. As more blast cells bud off the teloblasts, the germinal bands lengthen and move across the surface of the embryo, coalescing progressively from anterior to posterior along the future ventral midline into a structure called the germinal plate (**6.17**, **6.18**; Stage 8). The micromere-derived epithelium expands concomitantly, continuing to cover the germ bands and the area behind them with a squamous epithelium (**6.19–6.22**). Beneath the micromere-derived epithelium lie muscle fibres of mesodermal origin and these two cell layers together constitute the provisional integument which serves as a temporary body wall, pending the generation of the definitive body wall by the proliferation of cells in the germinal plate.

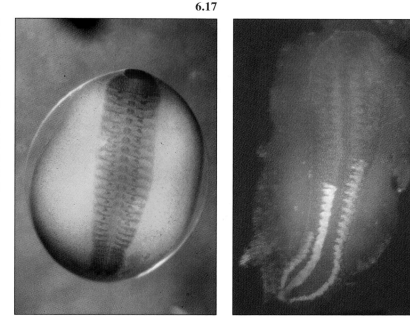

6.17, 6.18 A vegetal view of a Stage 8–9 embryo of the giant glossiphoniid leech *Haementeria ghilianii*, stained with haematoxylin to reveal the segmentally iterated structures in the germinal plate is seen in **6.17** (×30, photograph by R. Sawyer, courtesy of *Scientific American*). **6.18** shows a dissected germinal plate of a much smaller, *Helobdella triserialis* germinal plate, stained with Hoechst 33258 (blue) to reveal the segmentally iterated structures. Three teloblasts in this embryo had been injected with fluorescent cell lineage tracers during Stage 7 (×30; courtesy of D. Lans).

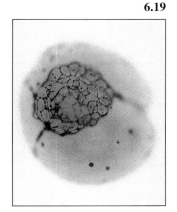 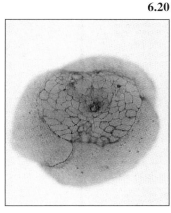 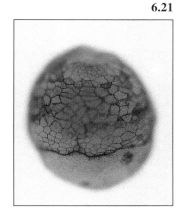 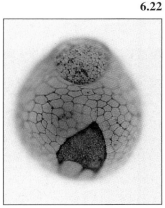

6.19–6.22 Epiboly of the micromere-derived epithelium. Views from the animal pole (prospective dorsal aspect) of silver stained *Helobdella* embryos, showing the spreading of the micromere-derived epithelium during Stages 7–8. (× 90; courtesy of R. Kostriken).

Morphogenesis of segments

During Stage 9, segments become apparent first through the cavitation of mesodermal hemisomites to create coelom and later through the appearance of segmental ganglia, both in a rostrocaudal progression (**6.17, 6.18** and **6.23–6.25**). The ventral nerve cord contains 32 segmental ganglia and connects anteriorly to a non-segmental dorsal ganglion via circumoesophageal connectives. During Stage 10, the edges of the germinal plate expand at the expense of the provisional integument and finally meet along the dorsal midline, closing the body tube. Segmental tissues approach their mature form during Stage 11 (**6.26**).

Each segmental ganglion forms in close contact with its anterior and/or posterior neighbour(s); later, as axons grow out, intersegmental connective nerves arise throughout the midbody and segmental nerves extend from the ganglia into the body wall. But ganglia in the rostral four (designated R1–R4) and the caudal seven (C1–C7) segments remain fused to form specialized ganglia that innervate suckers. Even among the 21 midbody segments (designated M1–M21), segment-specific properties can be identified in neural and non-neural tissues, especially in M5 and M6, which innervate the reproductive organs (e.g., see Stewart *et al.*, 1986).

Beginning in Stage 9, the yolk-filled macromeres and remnants of teloblasts become enclosed within the developing

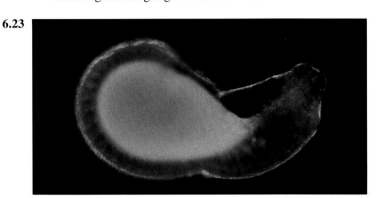

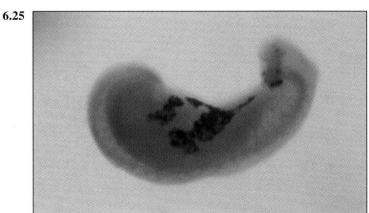

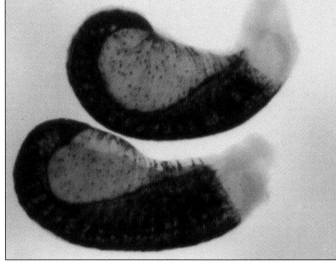

6.23–6.25 Lateral view of a living Stage 9 *Helobdella* embryo (anterior is right; dorsal up) is shown in **6.23**. The embryo is just emerging from the fertilization membrane, which is constricting the embryo just in front of the solid pink yolk mass. Segmentally iterated coelom is demarcated by the faint grey septa in the germinal plate, which wraps around the yolk at the posterior end of the embryo. **6.24** shows similar embryos stained for HRP after an M teloblast had been injected with HRP early in Stage 7. Note the sharp anterior boundaries between stained posterior segments and unstained anterior segments derived from blast cells produced prior to the injection. Cells dorsal to the lateral edge of the darkly stained germinal plate are M-derived muscle fibres of the provisional integument. **6.25** shows a similar embryo, in which a micromere had been injected with HRP in the 8-cell embryo. Progeny of the micromere are confined to nonsegmental tissues, including the supraoesophageal ganglion and epithelial cells in the provisional integument (all ×93).

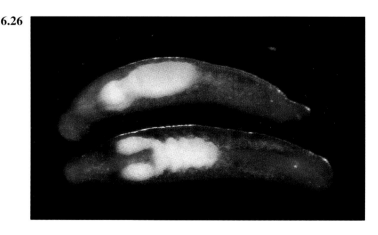

6.26 This figure shows lateral (top) and dorsal (bottom) views of Stage 11 *Helobdella* embryos. By this point, the anterior (right) and posterior suckers have formed and the gut enclosing the remaining yolk (pink) begins to differentiate into the crop secae seen in the adult (× 38).

gut and are digested. By the end of Stage 11 the yolk has been exhausted and the juvenile leech is ready for its first meal. Juvenile *Helobdella* are normally carried to their first meal still attached to the ventral body wall of the parent, but individuals reared from isolated eggs can be fed on small freshly killed snails and thus raised to adulthood.

Embryos of hirudinid leeches diverge significantly from the pattern described above, so that a different staging system has been devised for them (see Stent *et al.*, 1992). Since the hirudinid egg is so small, there is no surface across which the germinal bands might move and no need for such a movement; instead, the germinal plate forms directly along the ventral midline. The increasing length of the germinal plate is accommodated by a rapid growth of the embryo, made possible by the formation of larval structures which swallow and digest the yolk in the cocoon. By the time the germinal plate has formed in hirudinid embryos, they are large enough to be dissected routinely (by skilled individuals!), and neurons at the earliest stages of neurite outgrowth can be visualized under the compound microscope with DIC optics and impaled with microelectrodes.

Experimental case studies
Embryonic origins of definitive cell types

Whitman postulated that the stereotyped cleavages in the leech embryo segregated 'determinants' among the teloblasts, thereby specifying each pair of teloblasts as progenitors of a particular type of tissue. Since the n bandlets come into apposition at the ventral midline as the germinal bands coalesce, parsimony led him to propose that the N teloblasts were the progenitors of the leech nervous system (**6.27**). But this prediction was untestable because it was not possible to follow the fates of the many small cells in the germinal plate. Hence the first experiments undertaken once lineage tracers became available were to follow cell fates by micro-injecting a selected teloblast with tracer during Stages 5–7 and then observing the distribution of labelled progeny during Stages 7–10, either alone or in conjunction with other labelling techniques.

From such studies, we now know that Whitman's notion of tissue type determinants was too simple. Each of the five bandlets contributes a distinct subset of segmentally iterated progeny, including ganglionic neurons, to the mature leech (**6.23–6.25**, **6.28–6.34** and **6.53–6.55**; for review, see Stent *et al.*, 1992). Gangliogenesis thus entails medial migration of neuroblasts from the laterally situated p and q bandlets, a process that can be examined even in fixed preparations because of the rostrocaudal gradient of development (**6.30**, **6.31**; Weisblat *et al.*, 1984; Torrence and Stuart, 1986). In normal development, individually identified neurons derive from particular teloblasts and segmental tissues arise in a highly determinate manner by the stereotyped timing and orientation of mitosis in primary blast cells and their descendants (e.g., see Bissen and Weisblat, 1989); older blast cells in each bandlet contribute to more anterior segments, accounting for the pronounced rostrocaudal temporal gradient seen throughout development. Some definitive nonsegmental tissues, including the supraoesophageal ganglion, proboscis, and anterior epidermis, arise from micromeres, but the origins of the gut and germ line remain unknown.

By labelling teloblasts that had already begun making blast cells and then determining the number of possible positions for the boundary between unlabelled anterior cells and labelled posterior cells in the resultant embryos, it has been possible to deduce the number of blast cells required to produce one segment's worth of definitive progeny. In the n and q bandlets, two blast cells are required to make one segmental complement, and here two classes of blast cells, nf and ns (qf and qs), arise in exact alternation, each contributing a specific subset of N or Q progeny to a segment (**6.32–6.34** and **6.38**;

6.27 A Stage 8 embryo, fixed and counterstained with Hoechst 33258, in which an N teloblast had been injected with rhodamine lineage tracer at Stage 6a. The n bandlet can be seen emerging from the teloblast (centre left), entering the left-hand germinal band (bottom left), and then in apposition to its unlabelled right hand counterpart within the anterior portion of the germinal plate (top; ×155; courtesy of S. Zackson).

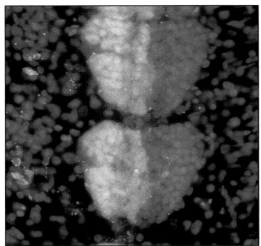 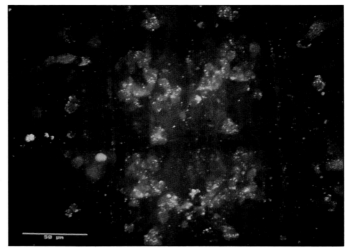

6.28, 6.29 Lineage tracers reveal distinct segmentally iterated progeny distribution patterns (anterior is up). **6.28** (×330) shows a medial portion of a *Theromyzon* rude germinal plate, counterstained with Hoechst 33258 to show segmental ganglia M2 and M3 (ventral midline is in the centre). Cells derived from the left-hand n bandlet are labelled with rhodamine lineage tracer (courtesy of D. Stuart). **6.29** (×350) is a similar view showing segments M3 and M4 in a different *Theromyzon* embryo, in which both m bandlets were labelled with rhodamine lineage tracer and both o bandlets were labelled with fluorescein lineage tracer. Note that in this embryo segment M4 lacks a left-hand nephridial tubule (irregular red columns extending medially from both edges of the photo in segment M3 and from the right side in segment M4), but does contain a male genital primordium (irregular red mass). As a result of an anteriad translocation of the left-hand m bandlet, the blast cell that normally contributes to segment M5 was shifted to segment M4 (courtesy of L. Gleizer).

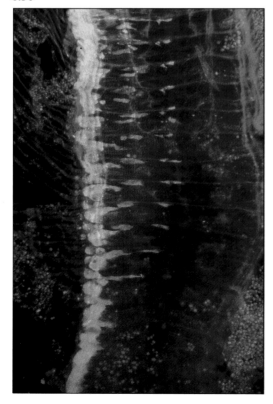 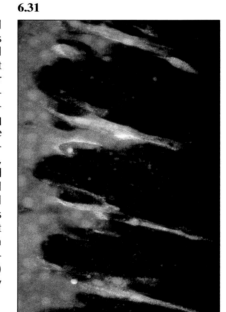

6.30, 6.31 Migration in the Q cell line (anterior is up). **6.30** (×70) shows a Stage 8–9 *Theromyzon* germinal plate in which left-hand q bandlet contains rhodamine lineage tracer and all the muscles are labelled immunohistochemically with fluorescein. In caudal (lower) segments, q cells are confined to the lateral edge of the germinal plate, but in developmentally advanced segments (top), small numbers of cells have migrated medially, some reaching the ventral midline. **6.31** (×310) is a confocal micrograph showing four segments in which q bandlet cells (red) are just initiating their medial migration (as in caudal segments in **6.30**), in apposition to circular muscle fibres (blue) derived from the m bandlet (courtesy of S. Torrence).

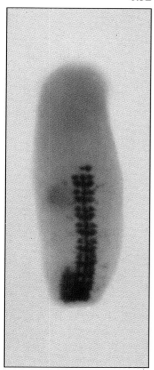 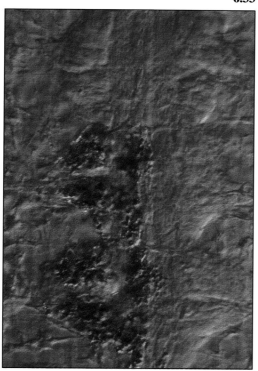 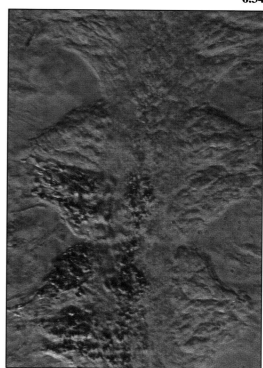

6.32–6.34 A ventral view of a Stage 10 *Helobdella* embryo in which both N teloblasts were injected with HRP in mid Stage 7, after blast cell production had already begun, is shown in **6.32** (×63); anterior segments (top) are thus unlabelled. Left-hand stain boundary lies at segment boundary; right-hand boundary falls midway through a segment. **6.33** and **6.34** show views of three segmental ganglia from similar embryos, in which just one N teloblast had been injected. The boundary between stained and unstained tissues falls between two ganglia in one specimen (**6.33**, ×750) side but through the middle of the ganglion in the other (**6.34**, ×750), reflecting the fact that two n blast cells are required to make one segment's worth of progeny.

6.35 The five kinship groups (M, N, O, P, and Q) of segmentally iterated progeny. For each pattern half of one segment is shown, with the ventral midline and ganglion at the left, and dorsal midline at the right (dashed lines). Hatching and the black diagonal line (in M) lineage represent muscles. Stippling and open contours represent epidermal cells; solid black contours represent individual neurons or clusters of neurons (and nephridia in M).

Weisblat and Shankland, 1985; Bissen and Weisblat, 1987). In the N cell line, the anterior and posterior portions of a segmental complement are almost entirely generated from ns-derived and nf-derived progeny, respectively (Bissen and Weisblat, 1987). The nf and ns clones within a segment intermingle slightly (Braun and Stent, 1989a).

In contrast to the situation described above, each m, o and p blast cell generates an entire segmental complement of M, O or P progeny, but individual m, o and p clones extend over more than one segment, interdigitating extensively with anterior and posterior clones of the same lineage. In consequence, we cannot regard the set of cells derived from any teloblast in a given segment as a clone and, for this reason, these segmentally iterated sets of cells are referred to as the M, N, O, P or Q kinship groups (**6.35**). It is important to note that the experiments underlying this data used minimally perturbed embryos and, although they help describe normal development, do not address the issue of the plasticity of cell fates, something that can only be assessed after substantial experimental perturbation (see also Weisblat, 1988).

Cell cycle composition and zygotic transcription

Using tritiated thymidine triphosphate (or its immunohistochemically detectable analogue, bromodeoxyuridine triphosphate) to label S phase nuclei in carefully staged embryos, it has been possible to determine the lengths and compositions of the cell cycles of identifiable cells throughout early development in *Helobdella* (**6.36, 6.37**; Bissen and Weisblat, 1989). This work showed that the stereotypey of cell lineages in the embryo extends even to the composition of embryonic cell cycles. For example, the dramatic lengthening of the cell cycles in going from the teloblasts (approximately 1hr cell cycle) to the primary blast cells (20–28hr cell cycle), is primarily accomplished by lengthening the G2 phase (**6.38**). In contrast to *Drosophila* and *Xenopus*, which seem to be specialized for rapid development, cell cycles in leech embryos include postreplicative gap (G2) phases from the first cell cycle, while the G1 phase is first observed in some of the secondary blast cells. Other cell-specific differences in cell cycle composition are also seen, so indicating that the control of embryonic cell cycles is not limited to a single transition point (e.g., G1/S transition) in the cell cycle.

The onset of zygotic transcription has also been examined in *Helobdella* embryos, using autoradiographic detection of tritiated uridine incorporated in the presence or absence of low concentrations of α-amanitin (Bissen and Weisblat, 1991). By this technique, RNA synthesis was first detected after the second cleavage, and α-amanitin-sensitive RNA synthesis was first detected during the formation of the teloblasts (Bissen and Weisblat, 1991). More recently, zygotic transcription has been detected in the 8-cell embryo using probes for a specific mRNA (Kostriken and Weisblat, 1992). During Stages 7–8, RNA synthesis increases significantly, with the bulk of the α-amanitin-sensitive synthesis occurring in the blast cells and in progeny of the micromeres. The dramatic increase in both cell-cycle duration and transcription in primary blast cells relative to teloblasts suggests that this period of leech development may correspond to the 'mid-blastula' transition in *Xenopus* and *Drosophila* (e.g., Yasuda *et al.*, 1991).

The significance of the early α-amanitin-sensitive RNA synthesis in *Helobdella* has been examined by observing the developmental consequences of α-amanitin injections. Zygotes injected with α-amanitin cleaved normally through Stage 4c, but then underwent aberrant cleavages while the control embryos were generating teloblasts in Stages 5–6 (**6.39–6.41**); these data thus suggest that early mRNA synthesis is required

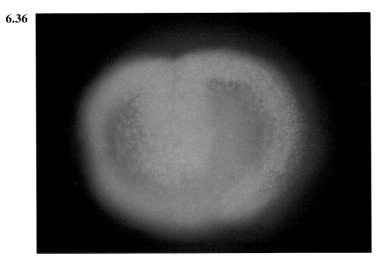
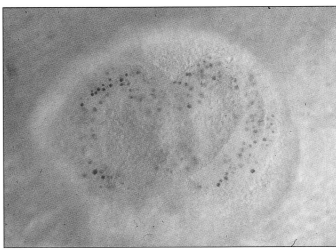

6.36, 6.37 Cell cycle analysis. **6.36** is a fluorescence micrograph of a Stage 8 *Helobdella* embryo, stained with Hoechst 33258 to reveal all the nuclei. **6.37** is a brightfield view of a similar embryo, showing immunocytochemically identified S phase nuclei (×170; courtesy of S. Bissen).

for the formation of normal teloblasts. But, when embryos were injected with α-amanitin after the teloblasts had formed, the production of primary blast cells was hardly affected (although the divisions of the blast cells and the micromere-cap cells were blocked as expected), so indicating that the rapid divisions of teloblast cells are not dependent on new mRNA synthesis.

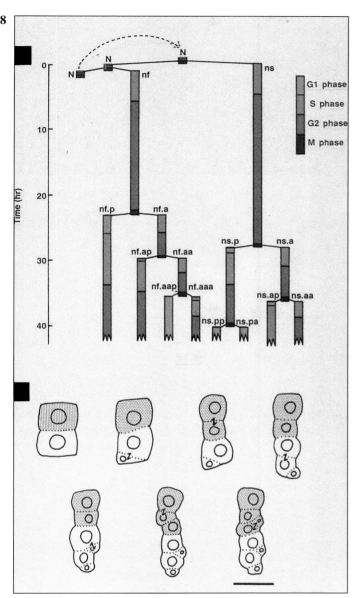

6.38 Reconstruction of cell cycle composition of the nf (white cells) and ns (stippled cells) blast cell lineages (courtesy of S. Bissen).

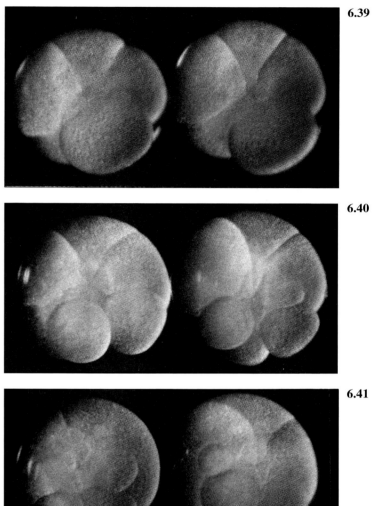

6.39–6.41 Teloblast formation requires zygotic transcription. Embryos injected with either salt solution (left) or α-amanitin (right) develop normally through Stage 4c (6.39). But subsequently (6.40, 6.41), the amanitin-injected embryos exhibit abnormal cleavages, failing to generate normal teloblasts (×93; photograph by S. Bissen, courtesy of *Development*).

Genesis of the teloblasts

By what process is the D' macromere designated as the teloblast precursor? The unequal distribution of teloplasm during cleavage was an obvious starting point for investigating this process. Normally the first two cell divisions are unequal, with the larger cell (cell CD at first cleavage and cell D at second cleavage) inheriting the bulk of the teloplasm. Experiments in which the normal distribution of teloplasm during cleavage was experimentally altered by centrifugation (Astrow et al., 1987), compression (Nelson and Weisblat, 1992) or blastomere isolation (Symes and Weisblat, 1992) have shown

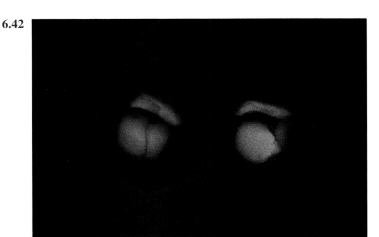
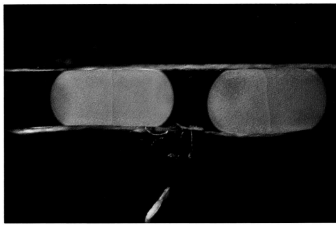
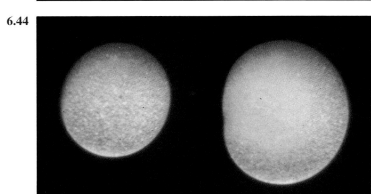

6.42–6.44 Experimental manipulations leading to altered teloplasm distribution. **6.42** (×52) shows centrifugation: two *Helobdella* embryos that had been centrifuged gently at the 2-cell stage and then stained with acridine orange at the 3-cell stage, to reveal the teloplasm (courtesy of S. Astrow). **6.43** (×80) shows compression: two *Helobdellla* zygotes compressed within a Sylgard 'ditch', to elongate them along their animal vegetal axis (courtesy of B. Nelson). **6.44** (×100) shows a blastomere isolation: a *Helobdella* zygote was removed from its fertilization membrane, and then, at first cleavage, the two daughter blastomeres were separated from one another by pressing gently on the cleavage furrow with a fine glass needle (courtesy of K. Symes).

that teloplasm content, not size, is critical in causing a blastomere to act as a proteloblast in early cleavage (**6.42–6.44**).

During cleavage, the ectoteloblasts and mesoteloblasts arise from blastomeres DNOPQ and DM, the animal and vegetal daughters of macromere D'. How are the distinct fates of these two sister cells determined? One possibility is that the animal and vegetal teloplasms contain distinct determinants for ectoderm and mesoderm, but this seems unlikely for several reasons. First, animal and vegetal teloplasms mix during normal development in the period between second and fourth cleavage (Holton *et al.*, 1989). Next, even when the animal and vegetal teloplasm are prematurely mixed by centrifugation, the DNOPQ and DM blastomeres that arise follow distinct fates (Astrow *et al.*, 1987). Finally, embryos from which either animal or vegetal teloplasm is removed (**6.45–6.47**) can generate both mesodermal and ectodermal proteloblasts (Nelson and Weisblat, 1991).

Further insight into this process was obtained by compressing zygotes to re-orient the first cleavage plane (Nelson and Weisblat, 1992). When the animal and vegetal hemispheres are separated by re-orientation of the first cleavage from meridional to equatorial, the ectodermal fate tends to co-segregate with the animal hemisphere and the mesodermal fate with the vegetal hemisphere. But complete segregation of ectodermal and mesodermal fates requires a cleavage orientation that separates the animal and vegetal cortical regions as well as the two teloplasms.

From these results, a model has been proposed according to which teloblast-producing lineages are determined in a multi-step process: teloplasm is generated by microtubule-

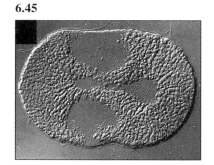
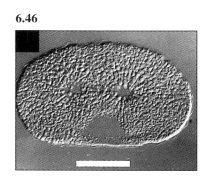
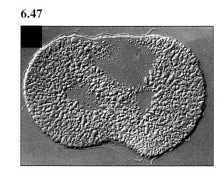

6.45–6.47 Teloplasm extrusion. Mid-saggital sections through three *Helobdella* zygotes. **6.45** is normal; **6.46** is after extrusion of animal teloplasm; and **6.47** is after extrusion of vegetal teloplasm (×115; photographs by B. Nelson, courtesy of *Science*; copyright 1991 by the AAAs).

dependent cytoplasmic re-arrangements prior to first cleavage and then segregated to the D quadrant during the initial, unequal cleavages (Astrow et al., 1987, 1989). The mechanisms by which these unequal cleavages are generated remains largely mysterious, but it appears that mechanical deformation by blastomere AB plays a role in determining the unequal division of blastomere CD at second cleavage (Symes and Weisblat, 1992). Inheritance of teloplasm above some critical threshold designates a blastomere as a proteloblast. But whether mesodermal or ectodermal teloblasts are generated depends on determinants localized to a cortical domain in the animal hemisphere of the zygote. It appears that these determinants normally segregate to the ectodermal precursor and interact over short distances with factors in teloplasm to transform the fate of this cell from a mesodermal 'ground state' to one in which ectodermal teloblasts are generated (Nelson and Weisblat, 1992).

Expression of an engrailed-class gene in leech

A comparative approach is required to identify general mechanisms of development, and to learn how animal diversity arises by developmental changes during evolution. The segmental body plans of arthropods (e.g., *Drosophila*) and annelids are, for example, presumed to have arisen from a segmented common ancestor, yet segmentation in these groups can differ extensively at the cellular level. Most obviously, *Helobdella* development proceeds via complete cleavages, whereas *Drosophila* undergoes 13 initial rounds of syncytial nuclear proliferation.

One way to compare *Drosophila* and *Helobdella* is by using 'reverse genetics' to identify potential *Helobdella* homologues to developmental regulatory genes in *Drosophila* and then compare the role of these genes in the two species. In doing so, we recognize three different kinds of 'homology' (Weisblat et al., 1992). *Sequence homology* refers to the degree of similarity between two or more nucleic acid coding sequences and reflects the evolutionary distance(s) between the compared sequences. *Biochemical homology* refers to the conservation of biophysical properties and as a result, the most proximal function of two or more sequence homologs (e.g., DNA-binding for the proteins encoded by homeobox genes). Neither type of homology, however, says anything about the embryological significance of the genes. A third, developmentally relevant level of homology is designated *syntagmal homology* (Weisblat et al., 1992) as Antonio Garcia-Bellido has coined the term *syntagma* to refer to an evolutionarily conserved set of interacting genes that subserve some developmental function.

In leech development, the reverse genetic approach is being used in various ways that include analysing presumptive gene complexes homologous to the antennapedia (ANT-C) and bithorax complexes (BX-C) in *Drosophila* (**6.48**; Wysocka-Diller et al., 1989; Nardelli-Haefliger and Shankland, 1992). It is also being used to characterize a leech homologue (*ht-en*) of the *Drosophila* gene *engrailed*, a homeobox gene expressed in the posterior compartment of each segment during segmentation and later in a subset of neurons or neuroblasts in the developing arthropod nervous system (see Patel et al., 1989).

Polyclonal antibodies raised against a fusion protein of β-galactosidase and a cloned fragment of *ht-en* have been used to study *ht-en* expression at the cellular level (**6.49, 6.50**). By this technique, expression is first apparent during late Stage 7, and occurs as a dynamic, rostrocaudally progressing series of segmentally iterated patterns (Wedeen and Weisblat, 1991).

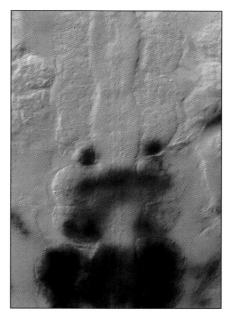

6.48 Segment-specific expression of *lox2*, a leech homologue of *abdominal A* (a *Drosophila* segmentation gene), revealed by non-radioactive *in situ* hybridization. Nomarski optics view of 4 ganglia in a Stage 10 embryo (anterior is up); note the stepwise transition in the number of stained cells between segments 5 (top) and 8 (bottom; ×460; courtesy of M. Shankland, D. Nardelli-Haefliger and M. Perkins).

Double-label experiments which combine immunohistochemical stain with micro-injected lineage tracers show that the segmentally iterated patterns comprise specific cells or small subsets of cells in each of the five bandlets at different times in development (Lans et al., 1993).

During Stages 8–9, before ganglia become apparent (Torrence and Stuart, 1986), *ht-en* immunoreactive nuclei appear as segmentally iterated transverse stripes extending across the central portion of the germinal plate (**6.51, 6.52**). Each stripe comprises a subset of the cells in the nf primary blast cell clones (**6.49, 6.50**). As development proceeds, *ht-en* disappears from all but two pairs of cells in each stripe, which eventually differentiate into identified peripheral neurons. Subsequently, new *ht-en* expression appears in a subset of

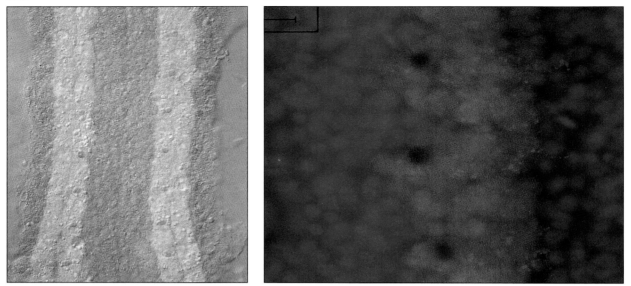

6.49, 6.50 Segmentally iterated expression of *ht-en*, revealed immunohistochemically. **6.49** (×700) shows combined brightfield and fluorescence view of the germinal plate in a Stage 8 embryo: o and p bandlets on both sides contain rhodamine lineage tracer; *ht-en* is revealed as dark nuclei in the o (medial, labelled) and q (lateral, unlabelled) bandlets. **6.50** (×1575) shows an early Stage 9 embryo: left-hand N lineage is rhodamine-labelled (ventral midline at right edge of lineage tracer); at this stage, each nf blast cell clone (more laterally extensive cell cluster) contains one cell expressing *ht-en* (courtesy of C. Wedeen and D. Lans).

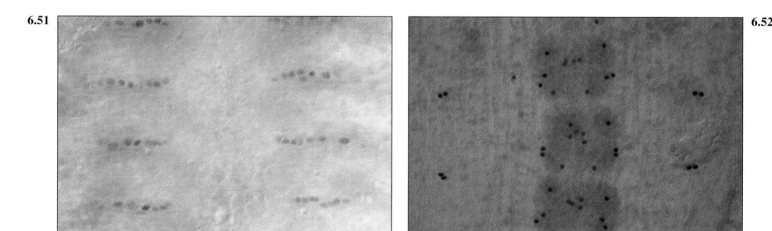

6.51, 6.52 Later expression of *ht-en*. **6.51** (×555) shows segmentally iterated stripes of nuclei in late Stage 9 (courtesy of D. Lans). Segmentally iterated subsets of ganglionic and extraganglionic neurons express *ht-en* in Stage 11 in **6.52** (×256; courtesy of C. Wedeen).

ganglionic and extraganglionic cells derived from various classes of blast cells. Most of these cells maintain their expression at least through Stage 11 (**6.51, 6.52**).

The segmentally iterated expression of *ht-en* in stripes prior to neurogenesis, and then later in neurons, is strikingly reminiscent of the expression patterns of *en* class proteins in arthropods and supports the notion that *ht-en* and *engrailed* are syntagmal homologues. Moreover, the fact that the stripes of expressing cells in *Helobdella* are descended from just the nf (posterior) class of n blast cells supports the notion that the anterior/posterior compartments of *Drosophila* and the nf/ns blast cells in *Helobdella* may also be evolutionary homologues. Finally, these similarities suggest that the 'dual expression' of *en* class protein was present in an ancestor of annelids and arthropods, and therefore may occur in other protostomes as well.

Terminal differentiation

Since cells become different from one another from the first cell division in leech, it is inappropriate to refer to any particular process as 'differentiation' and we therefore refer to the way in which cells acquire definitive morphological and biochemical phenotypes as 'terminal differentiation'. A fundamental issue here is distinguishing between lineage-dependent and interaction-dependent aspects of terminal differentiation. It is important to realize that observations on normal development alone are, *a priori*, insufficient to make this distinction. Instead, it is necessary to observe the effects (or lack thereof) on cell fate following experimental perturbations that alter normal interactions between cells.

O-P equivalence group

The O/P teloblasts are so designated because various experimental perturbations have shown that their primary blast cells are capable of generating either the O or the P pattern of definitive progeny, and thus constitute the *O-P equivalence group*. In *Helobdella*, for example, if the primary blast cells in the p bandlet are missing (due to ablating the teloblast or photolesioning blast cells in the bandlet, as described in the next section) the primary o blast cells 'transfate' and generate P instead of O kinship group cells (**6.53–6.55**). By ablating the p bandlet further anterior in the germinal band, after the o and p blast cells have undergone one or more divisions, it has been shown that the commitment of higher order o blast cells to their normal 'O' fate is a stepwise process (see also Shankland and Weisblat, 1984; Shankland, 1987).

Several sorts of interactions influence fates of o and p blast cells. For example, disrupting the provisional epithelium overlying the germinal bands results in partial transfating of o blast cell clones to the 'P' fate, even when the p bandlet is intact (Ho and Weisblat, 1987). Conversely, even when the nominal p blast cells are missing, in embryos of *Theromyzon rude*, the nominal o blast cells transfate only if they shift their position within the germinal band (**6.56**; Keleher and Stent, 1990). Since interactions with the n and q bandlets are known not to be important in determining the O and P fates (Zackson, 1984), this result suggests that interactions with the underlying mesodermal bandlet are also critical in determining the definitive fate of o and p blast cells.

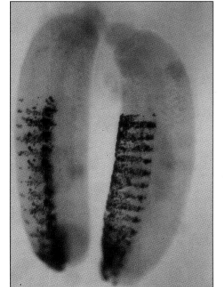

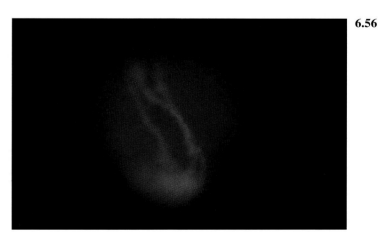

6.53–6.55 O-P equivalence group. **6.53** (×83) shows two Stage 7 embryos in which both the o and the p bandlets are labelled on one side. In the right-hand embryo, the bandlets cross between the teloblasts and the germinal band, making cell identifications ambiguous. **6.54** (×70) and **6.55** (×70) show three Stage 10 *Helobdella* embryos, showing the O pattern (left), the P pattern (centre) and a hybrid pattern (right) in which a single bandlet switched from making O pattern anteriorly (top) to P pattern posteriorly, following the ablation of an O/P teloblast.

6.56 Wandering o bandlet in Stage 7–8 *Theromyzon* embryo. Fluorescence micrograph of a germinal band in which the n and q bandlets are labelled with blue. The p bandlet is missing, and the o bandlet (red) lies next to the n bandlet anteriorly (top) and next to the q bandlet posteriorly (bottom; ×60; courtesy of G. Keleher).

Segment-specific differences

In contrast to the O-P equivalence group, m, n and q blast cells do not transfate, even after baroque combinations of cell ablations. But distinctions between lineage and positional effects in terminal differentiation cell lines have also been examined in the context of generating of *segment-specific* differences within the cell lines. Lacking techniques for transplanting individual blast cells, a key technique here is translocating segments of one bandlet relative to all the others, so that cells end up out of their normal segmental register (Shankland, 1984). For this purpose, the bandlet of interest is rendered photosensitive by micro-injecting the parent teloblast with fluoresceinated dextran lineage tracer. One or more primary blast cells in the labelled bandlet at the posterior end of the germinal band are then photolesioned using a microbeam of light (**6.57–6.59**).

This treatment has no direct effect on other cells in the labelled bandlet, but the section of bandlet posterior to the lesion delays entering the germinal band, so that its cells become ectopically situated. This translocation technique was originally useful for ectodermal bandlets alone, but a modification has now been worked out for the m bandlet (Gleizer and Stent, 1990) in which individual blast cells much closer to the teloblast are killed directly by DNAse injection. Curiously, the posterior section of the m bandlet may shift either forward or back after such treatment.

Blast cell translocations have been used to show that some segment-specific differences, such as death of supernumerary blast cells at the caudal end of the germinal plate (Shankland, 1984) and survival of the cell from the O cell line that lies at the distal tip of the nephridial duct (Martindale and Shankland, 1988), are determined by the environment of the blast cell or its proliferating descendants. But other differences, including aspects of neuropeptide expression in neurons of the N cell line (**6.60, 6.61**; Shankland and Martindale, 1989; Martindale and Shankland, 1990a) and the formation of nephridia by the M cell line (Gleizer and Stent, 1990; **6.28, 6.29**), are inherent properties of the blast cell.

It thus appears that blast cells are born with unique segmental identities or acquire them well before entering the germinal band, but that the position- or lineage-dependent basis of any given segment-specific difference must be determined case by case (see Shankland, 1991). Ablation and transplantation of gonadal tissues have, for example, also been used to show that interactions between neurons and target tissues are important in determining segment-specific properties of a pair of large serotonergic neurons called the Retzius cells (**6.60, 6.61**; Jellies *et al.*, 1987; Loer and Kristan, 1989a,b,c). At the finest level of resolution, individual neurons in adjacent segments have been shown to interact in determining the axonal morphology of a number of other individually identified neurons (see Gao and Macagno, 1988). Most of these experiments dealing with the terminal differentiation of identified neurons have been carried out in *Hirudo* embryos, which are particularly appropriate because results can be compared directly with the many physiological and morphological studies carried out on the same neurons in adults.

The photosensitizing technique described above for inducing blast cell translocation has also been used later in development to identify cells that guide migration of neurons and neuroblasts (e.g., see Torrence, 1991) and the growth of axons in the body wall (Braun and Stent, 1989b), and to investigate cell interactions regulating the bilaterally asymmetric differentiation of peptidergic neurons (Martindale and Shankland, 1990b; Blair *et al.*, 1990). To kill cells in later development, when the lineage tracer has been diluted, eosin, a more potent photosensitizing agent, can be used instead of fluorescein as the dextran-linked chromophore.

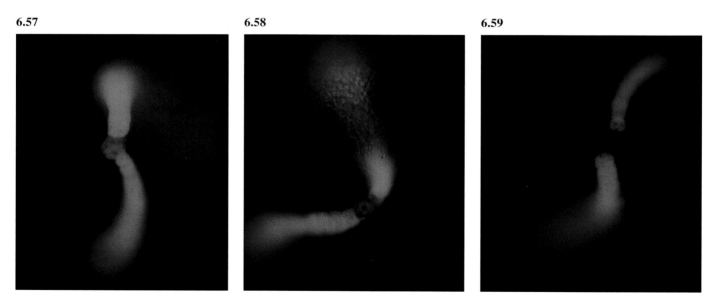

6.57–6.59 Photolesion-induced blast cell translocation. Fluorescence micrographs of germinal bandlets labelled with fluoresceinated lineage tracer in which 1–2 blast cells had been irradiated with intense light at about the fluorescein absorption maximum. 0.5hr (**6.57**) and 2hr (**6.58**) afterwards, the irradiated blast cells show bleaching, and vacuole formation. 5hr (**6.59**) afterwards, the bandlet has broken and a gap is forming between the anterior (top) and posterior (lower) portions of the bandlet (×625; courtesy of M. Shankland).

6.60, 6.61 Segment-specific neuronal differences (anterior is up). **6.60** (×20) shows lucifer yellow-filled Retzius neurons in a 12-day-old *Hirudo* embryo. Retzius neurons in segmental ganglia M5 (top) and M6 (centre) exhibit distinct physiological and morphological properties, including unique branches that innervate the reproductive tissues. Throughout the rest of the midbody, Retzius neurons extend long processes along the connectives and into the body wall, as in ganglion M7 (bottom; courtesy of J. Jellies). **6.61** (×160) shows a fluorescence micrograph of segmental ganglia M18 (top) and M19 (bottom) in *Theromyzon*, showing neurons that exhibit Small Cardioactive Peptide-like immunoreactivity. Note the large, unpaired neurons (CAS neurons) on the right of M18 and the left of M19 (photographs by M. Martindale and M. Shankland, courtesy of Academic Press).

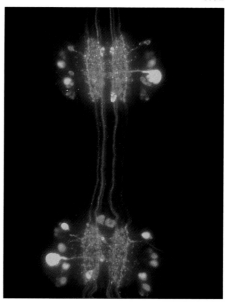

Scaffold cells

A technique of importance in studying terminal differentiation is that of micro-injecting individual cells with small, rapidly diffusing fluorescent dyes, so that the morphology of the injected cell can be examined by fluorescence microscopy. For this technique, which requires relatively large postmitotic cells, embryos of a giant glossiphoniid leech, *Haementeria ghilianii*, have been used, as well as those of *Hirudo*. Investigations using this technique have revealed a class of embryonic cells that seems to serve as scaffolding to guide the outgrowth of definitive neurons or muscles. Such cells were first implicated in the formation of the dorsal posterior segmental nerve in *Haementeria* (Kuwada, 1985) and each segmental hemiganglion contains the cell bodies for two pressure-sensitive, mechanosensory neurons (Nicholls and Baylor, 1968; Kramer and Goldman, 1981), which seem to be the first neurons to extend processes out from the ganglion during development (Kuwada and Kramer, 1983; Kuwada, 1985). One of these (neuron PV) primarily innervates ventral ipsilateral epidermis (Kramer and Kuwada, 1983) and, during axonogenesis, processes of this neuron establish two branches of the anterior segmental nerve by entering the ventral body wall and branching as soon as they exit the ganglion. The other (neuron PD) primarily innervates dorsal epidermis (Nicholls and Baylor, 1968; Kramer and Goldman, 1981) and, during axonogenesis, it establishes the dorsal branch of the posterior segmental nerve by growing without branching across the prospective ventral territory, in intimate contact with the non-neuronal DV cell to form the dorsal posterior nerve (Kuwada, 1985). This DV cell, presumably of mesodermal origin, features flat thin processes that extend from the posterior edge of the ganglion to the prospective dorsal territory of the germinal plate.

Similar cells, called ARC cells, have been implicated by ablation experiments in the formation of the 'sex nerves' that extend rostrally from the anterior segmental nerves of ganglion M6 in *Hirudo* to innervate the male reproductive tissue (**6.62** Jellies and Kristan, 1988a). The ARC and DV cells resemble one another in morphology, function, and in the fact that they stain with a monoclonal antibody (Lan 3–14; Zipser and McKay, 1981) that recognizes leech muscle cells. These cells also resemble the 'muscle pioneer' cells that seem to organize myogenesis and some aspects of neuron outgrowth in arthropods (Ho *et al.*, 1983). In leech, it seems likely that the DV cell also prefigures the dorsoventral muscle through which the posterior dorsal nerve runs.

More is known about the formation of the body wall muscles, particularly the oblique muscles, which form around the processes of a truly remarkable scaffold cell, called the comb cell (C–cell; **6.63**, **6.64**; Jellies and Kristan, 1988b, 1991). A pair of these initially spindle-shaped cells arises bilaterally in each segment, roughly midway between the ventral midline and the prospective nephridiopore. Each C-cell normally elongates along the anterior/posterior axis until it contacts the growing

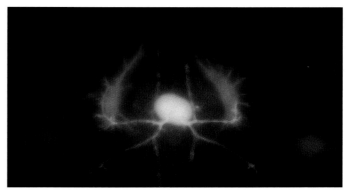

6.62 The ARC cells. Fluorescence micrograph showing a Lucifer yellow-filled Retzius neuron in segmental ganglion M6 of a *Hirudo* embryo on embryonic day 10. On each side, the neurites that will innervate the male reproductive tissues are growing out across the (sulphorhodamine-filled) ARC cells (×220; courtesy of J. Jellies).

ends of its adjacent homologues (**6.63, 6.64**), to which it forms transient gap junctions. These contacts may be important in signalling the cessation of elongation because, if one C-cell is ablated early in development, its homologues will extend into its territory. As the C-cell extends longitudinally, it elaborates dozens of laterally directed growth cones that arrange themselves into parallel arrays, half directed posteromedially and half anterolaterally, at roughly 45° to the long axis of the embryo. These growth cones extend slowly at first and the older ones then pause until all processes have reached about the same medial and lateral extent; they then enter a period of rapid growth, crossing the ventral midline and the processes of the contralateral C-cell to form a multi-toothed helix that reaches from dorsal midline to dorsal midline. Myoblasts aggregate along this scaffold and differentiate into the definitive oblique muscles (**6.65**). The C-cell is not contractile, never expresses the Lan 3–14 antigen and degenerates late in embryogenesis. The remarkable morphology and fascinating growth of this cell should provide an excellent window into the cell biology of many aspects of terminal differentiation.

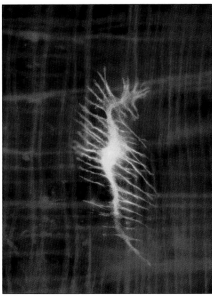

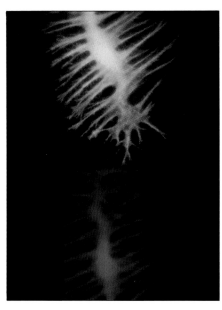

6.63, 6.64 The comb cell (anterior is up). **6.63** (×700) shows a single, Lucifer-yellow-filled comb cell in a 9–10-day-old *Hirudo* embryo, viewed against a background of immunohistochemically stained longitudinal and circular muscles (courtesy of J. Jellies). **6.64** (×800) shows two comb cells in adjacent segments of a 9 day embryo; longitudinal growth of the cells ceases when their axially oriented growth cones meet (courtesy of J. Jellies).

6.65 Beginning on embryonic day 12, the comb cells organize myoblasts to form the oblique muscles: this micrograph shows three oblique muscle fibres injected with Lucifer yellow, two circular and one longitudinal muscle fibre injected with sulphorhodamine 101 (red), and the comb cell injected with 8-hydroxy-1,3,6-pyrenetrisulphonic acid, trisodium salt (blue; ×600; photograph by J. Jellies and W. Kristan, Jr., courtesy of *J. Neuroscience*).

The future

From a technical point of view, developmental biology is highly derivative, relying on concepts and techniques developed in cell and molecular biology, biochemistry and biomechanics. Intellectually, however, it occupies centre stage, in the sense that the central fact of biology is the evolution of species and that the morphological changes associated with evolutionary processes arise by modification of developmental processes. Our goal in studying development is thus twofold: first, to obtain as satisfying an understanding as possible of the development of whatever species we choose to study; and second, to see how modifications of developmental processes might generate species diversity.

For both purposes, leeches have much to offer. First, it should be evident from the case studies presented here that leeches, developmentally and neurobiologically well-characterized annelids, provide a useful point of comparison with representatives of other phyla. Second, as subjects for the analysis of development in its own right, leeches offer the advantages of a simple body plan, combined with the experimental accessibility of the cells throughout development. One area in which the study of leech development is particularly useful is in understanding the connections between early developmental events and later processes during neurogenesis; another is in devising biochemical and molecular genetic explanations for macroscopic developmental phenomena such as cell lineage-based restrictions on developmental potential, morphogenetic cell movements and cell-division patterns. Good progress is being made in generating probes with which to describe normal development at the molecular level and to assess molecular consequences of such classic embryological perturbations as cell ablation and translocation. An important next step is to generate methods of manipulating gene expression in the embryo directly, so as to observe embryological consequences of molecular perturbations with the same cell level resolution that has characterized previous work in the leech.

Many people regard leeches with a mixture of horror and disgust so intense as to prevent their serious consideration as objects for biological investigation. But with contempt as the starting point, it might be hoped that the familiarity provided by this brief treatment will lead to some appreciation of leech embryos as a source of beauty and truth in understanding developmental processes.

Acknowledgements

Research in the author's laboratory has been supported by grants from the National Science Foundation, the National Institutes of Health and the March of Dimes Birth Defects Foundation.

References

Astrow, S.H., Holton, B. and Weisblat, D.A. (1987). Centrifugation redistributes factors determining cleavage patterns in leech embryos. *Dev. Biol.*, **120**, 270–283.

Astrow, S.H., Holton, B. and Weisblat, D.A. (1989). Teloplasm formation in a leech, *Helobdella triserialis*, is a microtubule-dependent process. *Dev. Biol.*, **135**, 306–319.

Barnes, R.S.K., Calow, P. and Olive, P.J.W. (1988). *The Invertebrates: a New Synthesis*. Blackwell Scientific, Oxford.

Bissen, S.T. and Weisblat D.A. (1987). Early differences between alternate n blast cells in leech embryos. *J. Neurobiol.*, **18**, 251–269.

Bissen, S.T. and Weisblat, D.A. (1989). The durations and compositions of cell cycles in embryos of the leech, *Helobdella triserialis*. *Development*, **106**, 105–118.

Bissen, S.T. and Weisblat, D.A. (1991). Transcription in leech: mRNA synthesis is required for early cleavages in *Helobdella* embryos. *Dev. Biol.*, **146**, 12–23.

Blair, S.S., Martindale, M.Q. and Shankland, M. (1990). Interactions between adjacent ganglia bring about the bilaterally alternating differentiation of RAS and CAS neurons in the leech nerve cord. *J. Neurosci.*, **10**, 3183–3193.

Braun, J. and Stent, G.S. (1989a). Axon outgrowth along segmental nerves in the leech: I. Identification of candidate guidance cells. *Dev. Biol.*, **132**, 471–485.

Braun, J. and Stent, G.S. (1989b). Axon outgrowth along segmental nerves in the leech: II. Identification of actual guidance cells. *Dev. Biol.*, **132**, 486–501.

Fernandez, J. (1980). Embryonic development of the glossiphoniid leech *Theromyzon rude*: Characterization of developmental stages. *Dev. Biol.*, **76**, 245–262.

Fernandez, J. and Olea, N. (1982). Embryonic development of glossiphoniid leeches. In *Development Biology of Freshwater Invertebrates* (F.W. Harrison and R.R. Cowden, eds) pp. 317–361, A.R. Liss, Inc., NY.

Friesen, W.O. (1989). Neuronal control of leech swimming movements. In *Neuronal and Cellular Oscillators* (J.W. Jacklet, ed.) pp. 269–316, Marcel Dekker, NY.

Gao, W.Q. and Macagno, E.R. (1988). Axon extension and retraction by leech neurons: severing early projections to peripheral targets prevents normal retraction of other projections. *Neuron*, **1**, 269–277.

Gilbert, S.F. (1988). *Development Biology*. Sinauer Assoc. Sunderland, MA.

Gleizer, L. and Stent, G.S. (1990). Control of segment identity in the leech embryo. *Soc. Neurosci. Abstr.*, **16**, 650.

Ho, R.K., Ball, E.E. and Goodman, C.S. (1983). Muscle pioneers: large mesodermal cells that erect a scaffolding for developing muscles and motorneurons in grasshopper embryos. *Nature*, **301**, 66–69.

Ho, R.K. and Weisblat, D.A. (1987). A provisional epithelium in leech embryo: Cellular origins and influence on a developmental equivalence group. *Dev. Biol.*, **120**, 520–534.

Holton, B., Astrow, S.H. and Weisblat, D.A. (1989). Animal and vegetal teloplasms mix in the early embryo of the leech, *Helobdella triserialis*. *Dev. Biol.*, **131**, 182–189.

Jellies, J. and Kristan, W.B. Jr. (1988a). An identified cell is required for the formation of a major nerve during embryogenesis in the leech. *J. Neurobiol.*, **19**, 153–165.

Jellies, J. and Kristan, W.B. Jr. (1988b). Embryonic assembly of a complex muscle is directed by a single identified cell in the medicinal leech. *J. Neurosci.*, **8**, 3317–3326.

Jellies, J. and Kristan, W.B. Jr. (1991). The oblique muscle organizer in *Hirudo medicinalis*, an identified embryonic cell projecting multiple parallel growth cones in an orderly array. *Dev. Biol.*, **148**, 334–354.

Jellies, J., Loer, C.M. and Kristan, W.B. Jr. (1987). Morphological changes in leech *Retzius* neurons after target contact during embryogenesis. *J. Neurosci.*, **7**, 2618–2629.

Keleher, G.P. and Stent, G.S. (1990). Cell position and developmental fate in leech embryogenesis. *Proc. Nat. Acad. Sci.*, **87**, 8457–8461.

Kostriken, R.G. and Weisblat, D.A. (1992). Expression of a *Wnt* gene in embryonic epithelium of the leech. *Dev. Biol.*, **151**, in press.

Kramer, A.P. and Goldman, J.R. (1981). The nervous system of the glossiphoniid leech *Haementeria ghilianii*. I. Identification of neurons. *J. Comp. Physiol.*, **144**, 435–448.

Kramer, A.P. and Kuwada, J.Y. (1983).

Formation of the receptive fields of leech mechanosensory neurons during embryonic development. *J. Neurosci.*, **3**, 2474–2486.

Kuwada, J.Y. (1985). Pioneering and pathfinding by an identified neuron in the embryonic leech. *J. Embryol. Exp. Morph.*, **86**, 155–167.

Kuwada, J.Y. and Kramer, A.P. (1983). Embryonic development of the leech nervous system: Primary axon outgrowth of identified neurons. *J. Neurosci.*, **3**, 2098–2111.

Lans, D., Wedeen, C.J., and Weisblat, D.A. (1993). Cell lineage analysis of the expression of an *engrailed* homolog in leech embryos. *Development,* **117**, in press.

Loer, C.M. and Kristan, W.B. Jr. (1989a). Peripheral target choice by homologous neurons during embryogenesis of the medicinal leech. I. Segment-specific preferences of *Retzius* cells. *J. Neurosci.*, **9**, 513–527.

Loer, C.M. and Kristan, W.B. Jr. (1989b). Peripheral target choice by homologous neurons during embryogenesis of the medicinal leech. II. Innervation of ectopic reproductive tissue by nonreproductive *Retzius* cells. *J. Neurosci.*, **9**, 528–538.

Loer, C.M. and Kristan, W.B. Jr. (1989c). Central synaptic inputs to identified leech neurons determined by peripheral targets. *Science*, **244**, 64–70.

Maienschein, J. (1978). Cell lineage, ancestral reminiscence, and the biogenetic law. *J. Hist. Biol.*, **11**, 129–158.

Martindale, M.Q. and Shankland, M. (1988). Developmental origin of segmental differences in leech ectoderm: survival of the distal tubule cell is determined by the host segment. *Dev. Biol.*, **125**, 290–300.

Martindale, M.Q. and Shankland, M. (1990a). Intrinsic segmental identity of segmental founder cells of the leech embryo. *Nature*, **347**, 672–674.

Martindale, M.Q. and Shankland, M. (1990b). Neuron competition determines the spatial pattern of neuropeptide expression by identified neurons of the leech. *Dev. Biol.*, **139**, 210–226.

Muller, K.J., Nicholls, J.G. and Stent, G.S. eds (1981). *Neurobiology of the Leech.* New York, Cold Spring Harbor Lab.

Nardelli-Haefliger, D. and Shankland, M. (1992). *Lox-Z* a putative leech segment identity gene, is expressed in the same segmental domain in different stem lineages. *Development,* **116**, 697–710.

Nelson, B.H. and Weisblat, D.A. (1991). Conversion of ectoderm to mesoderm by cytoplasmic extrusion in leech embryos. *Science*, **253**, 435–438.

Nelson, B.H. and Weisblat, D.A. (1992). Cytoplasmic and cortical determinants interact to specify ectoderm and mesoderm in the leech embryo. *Development*, **115**, (in press).

Nicholls, J.G. (1987). *The Search for Connections: Study of Regeneration in the Nervous System of the Leech.* Sinauer Associates, Inc. Sunderland, MA.

Nicholls, J.G. and Baylor, D.A. (1968). Specific modalities and receptive fields of sensory neurons in the CNS of the leech. *J. Neurophysiol.*, **31**, 740–756.

Patel, N.H., Kornberg, T.B. and Goodman, C.S. (1989). Expression of *engrailed* during segmentation in grasshopper and crayfish. *Development*, **107**, 201–212.

Pearse, V., Pearse, J., Buchsbaum, M. and Buchsbaum, R. (1987). *Living Invertebrates.* Blackwell Scientific, Palo Alto, CA.

Retzius, G. (1891). Zur Kenntnis des centralen Nervensystems der Wurmer. *Biol. Unters.* (NF). **2**, 1–28.

Sandig, M. and Dohle W. (1988). The cleavage pattern in the leech *Theromyzon tessulatum* (Hirudinea, Glossiphoniidae). *J. Morph.*, **196**, 217–252.

Shankland, M. (1984). Positional control of supernumerary blast cell death in the leech embryo. *Nature*, **307**, 541–543.

Shankland, M. (1987). Differentiation of the O and P cell lines in the embryo of the leech: II. Genealogical relationship of descendant pattern elements in alternative developmental pathways. *Dev. Biol.*, **123**, 97–107.

Shankland, M. (1991). Leech segmentation: Cell lineage and the formation of complex body patterns. *Dev. Biol.*, **144**, 221–231.

Shankland, M. and Martindale, M.Q. (1989). Segmental specificity and lateral asymmetry in the differentiation of developmentally homologous neurons during leech embryogenesis. *Dev. Biol.*, **135**, 431–448.

Shankland, M. and Weisblat, D.A. (1984). Stepwise commitment of blast cell fates during the positional specification of the O and P cell fates during serial blast cell divisions in the leech embryo. *Dev. Biol.*, **106**, 326–342.

Stent, G.S., Kristan, W.B. Jr., Torrence, S.A., French, K.A. and Weisblat, D.A. (1992). Development of the leech nervous system. *Int. Rev. Neurobiol.*, **33**, 109–193.

Stewart, R.R., Spergel, D. and Zipser, B. (1986). Segmental differentiation in the leech nervous system: The genesis of cell number in the segmental ganglia of *Haemopis marmorata.* *J. Comp. Neurol.*, **253**, 253–259.

Stuart, D.K., Torrence, S.A. and Stent, G.S. (1990). Microinjectable probes for tracing cell lineage in development. *Methods in Neuroscience*, **2**, 375–392.

Symes, K. and Weisblat, D.A. (1992). An investigation of the specification of unequal cleavages in leech embryos. *Dev. Biol.*, **150**, 203–218.

Torrence, S.A. (1991). Positional cues governing cell migration in leech neurogenesis. *Development*, **111**, 993–1005.

Torrence, S.A. and Stuart, D.K. (1986). Gangliogenesis in leech embryos: migration of neural precursor cells. *J. Neurosci.*, **6**, 2736–2746.

Wedeen, C.J., Price, D.J. and Weisblat, D.A. (1990). Analysis of the life cycle, genome, and homeo box genes of the leech, *Helobdella triserialis.* In *The Cellular and Molecular Biology of Pattern Formation.* (D.L. Stocum and T.L. Karr, eds) pp. 145–167, Oxford U. Press, NY.

Wedeen, C.J. and Weisblat, D.A. (1991). Segmental expression of an *engrailed*-class gene during early development and neurogenesis in an annelid. *Development*, **113**, 805–814.

Weisblat, D.A. (1988). Equivalence groups and regulative development. In *From Message to Mind: Directions in Development Neurobiology.* (S.S. Easter, K.F. Barald and B.M. Carlson, eds) pp. 209–223, Sinauer Press, Sunderland, MA.

Weisblat, D.A., Astrow, S.H., Bissen, S.T., Ho, R.K. and Liu, K. (1990). Role of cytoplasmic factors that affect cleavage in leech embryos. In *Cytoplasmic Organization Systems* (G.M. Malacinski, ed.) pp.243–261, McGraw-Hill, NY.

Weisblat, D.A., Kim, S.Y. and Stent, G.S. (1984). Embryonic origins of cells in the leech *Helobdella triserialis.* *Dev. Biol.*, **104**, 65–85.

Weisblat, D.A. and Shankland, M. (1985). Cell lineage and segmentation in the leech. *Phil. Trans. R. Soc. Lond.*, **313**, 39–56.

Weisblat, D.A., Wedeen, C.J. and Kostriken, R.G. (1992). Evolutionary conservation of developmental mechanisms: comparisons of annelids and arthropods. In *Soc. for Dev. Biol. Symp.,* **50** (A. Spradling, ed.) pp 125–140. Wiley-Liss, New York,.

Whitman, C.O. (1878). The embryology of Clepsine. *Quart. J. Microscop. Sci.*, **18**, 215–315.

Wysocka-Diller, J.W., Aisemberg, G.O., Baumgarten, M., Levine, M. and Macagno, E.R. (1989). Characterization of a homologue of bithorax-complex genes in the leech *Hirudo medicinalis.* *Nature*, **341**, 760–763.

Yasuda, G.K., Baker, J. and Schubiger, G. (1991). Temporal regulation of gene expression in the blastoderm *Drosophila* embryo. *Genes Dev.*, **5**, 1800–1812.

Zackson, S.L. (1984). Cell lineage, cell-cell interaction, and segment formation in the ectoderm of a glossiphoniid leech embryo. *Dev. Biol.*, **104**, 143–160.

Zipser, B. and McKay, R. (1981). Monoclonal antibodies distinguish identifiable neurons in the leech. *Nature*, **289**, 549–554.

7. *Drosophila*

Maria Leptin

Introduction

In the first part of this century, while the embryology of vertebrates, especially amphibia, was intensely investigated, *Drosophila* was mainly used as an organism for studying the laws of genetics. Experimental work on the embryonic development of insects was carried out on other species whose eggs are larger, or easier to manipulate. Important insights into such problems as anterior-posterior polarity of the developing embryo came from these experiments well before similar mechanisms were studied in *Drosophila*. It was, however, the genetic approach to *Drosophila* embryology, and the molecular analysis of genes identified by classical genetics that led to the extensive understanding of developmental mechanisms that we have today.

The realization that mutations in certain genes (some of them originally identified because of their effects on adult structures) resulted in specific defects in the developing embryo – such as the deletion of defined parts of the body, or their transformation into other body parts – led to the systematic application of the genetic method to study *Drosophila* embryonic development. This was greatly helped by the genetic tools that had been developed over decades of *Drosophila* genetic research, and of course by the advantages that *Drosophila* had been chosen for in the first place (short generation time, ease of keeping in the lab, etc.).

In order to find the genes required for embryonic development, a number of laboratories conducted saturation screens for zygotic and maternal mutations causing embryonic lethality. A subset of the mutations found led to visible defects in the cuticle of the fully developed embryo (the cuticle of the mature embryo or freshly hatched larva with its numerous specializations and landmarks is used as a read-out for regional specification and differentiation; see **7.1–7.3**). For example, about 30 maternal effect genes seem to be required for the establishment of the embryonic axes, while some 120 zygotically active genes are needed for the embryo to develop with all pattern elements of the cuticle present and in the correct place and orientation. These genes probably represent the majority of those *specifically* required for pattern formation in the embryo. Genes for components like general cytoskeletal elements, ribosomes etc., are of course also needed for embryonic development.

The techniques of molecular biology and biochemistry, as well as those of classical genetic experiments, have made it possible to work out the function of many of these genes during embryogenesis and a large number of them have now been cloned and sequenced. In many cases, homologies to other known genes – such as transcriptional regulators or cell surface receptors – gave clues to their mechanism of action. The development of the technique of *in situ* hybridization allowed the expression patterns of cloned genes in the embryo to be revealed, a further important step towards understanding their function, and in establishing the hierarchy of gene activities during development. As a result, we now have the answers to some fundamental questions of embryology. We understand how anterior-posterior and dorso-ventral polarity are set up in the embryo and how the initially broad patterns of gene expression along the axes are interpreted and refined to create the patterns that are reflected in the structures of the living organism.

Apart from describing *Drosophila* embryogenesis, this chapter will also attempt to give an account of what we know about the genes directing embryonic development and how they work. The principle of the screens that identified these genes will be explained, and, as an example of the developmental mechanisms revealed by the analysis of genes discovered in these screens, one system of pattern formation, polarity and segmentation along the anterior-posterior axis, will be discussed. The molecular basis of many more aspects of embryonic development is understood or becoming clear (see Lawrence, 1992, or the *Drosophila* chapter in Slack, 1991, for excellent accounts), but their detailed description is not the purpose of this chapter. Instead, the functions of the main genes that determine cell fates and direct differentiation and organogenesis will be pointed out in the paragraphs describing these processes.

Normal development

The *Drosophila* egg is surrounded by two protective shells, the inner vitelline membrane and the outer chorion (**7.1–7.3**). The chorion with its two long appendages at the anterior end is filled with air, while the space between the vitelline membrane and the egg plasma membrane is filled with liquid which contains molecules involved in setting up the body axes of the embryo, probably including ligands for receptors in the plasma membrane.

The egg has a clearly recognizable antero-posterior (a-p)

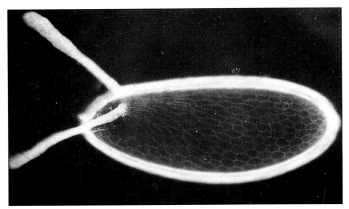

7.1 Egg in its chorion. The structured surface of the chorion shows the imprints of the somatic follicle cells which build the egg membranes during oogenesis and then die (×130). (Photograph by Trudi Schüpbach.)

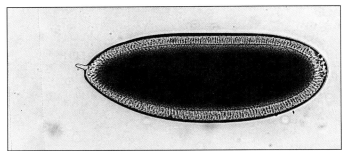

7.2 Live embryo at the blastoderm stage after removal of the chorion. The embryo is now only surrounded by the vitelline membrane. The small protrusion at the anterior end is the micropyle through which the sperm enters the egg. Cell membranes are invaginating from the surface of the egg to divide the layer of nuclei into individual cells. The advancing front of the membranes is visible as a dark line close to the yolk. The vitelline membrane has to be removed before embryos can be stained and is therefore not present in any of the following photographs. (×130; photograph by Kristina Straub.)

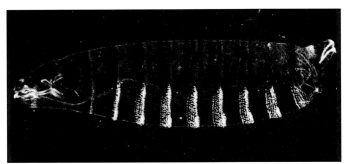

7.3 At the end of embryogenesis the first instar larva hatches from the egg membranes with its cuticle covered with denticles and sensory structures. Since these differentiations vary between segments and along the dorso-ventral axis, they are useful landmarks for the identification of specific body regions and are very helpful for the interpretation of transformations or deletions in the epidermis of mutant embryos (for example, homeotic or segmentation mutants) (×130).

and dorso-ventral (d-v) asymmetry. To refer to particular regions, the distance from the posterior pole along the a-p axis, and from the ventral midline along the d-v perimeter is used, in each case measured as a percentage. Thus, 100% egglength is the anterior tip of the embryo, and 100% perimeter, the dorsal midline.

The sperm enters the egg through the micropyle, a specialized structure in the egg shell at the anterior end (visible in **7.2**). Since the axes of the egg are determined before fertilization, sperm entry plays no role in axis determination. The development of the fertilized egg to an adult fly takes 10 days (at 25°C), of which embryogenesis occupies the first 24 hr, ending with the first instar larva (**7.3**) hatching from the egg shell.

The larva grows during three larval phases (instars) separated by moults. The first and second instar each last one day, the third instar two days. The larva then encloses itself in a puparium in which it undergoes metamorphosis. The pupal stage lasts five days, after which the adult fly emerges and is able to reproduce within a day or two. Although larval and pupal development are not part of embryogenesis, they are of course part of the development from fertilized egg to mature organism, and a number of interesting processes occur during this period, that include the growth, differentiation and morphogenesis of the imaginal discs, the precursors of most of the adult body structures.

Embryogenesis can be divided roughly into three phases: the formation of the blastoderm; gastrulation and organogenesis; and differentiation. Various systems of finer subdivisions have been used. Zalokar and Erk (1976) and Foe and Alberts (1983) have defined stages of development up to the blastoderm, while Bownes (1975) and Campos-Ortega and Hartenstein (1985, see also Wieschaus and Nüsslein-Volhard, 1986) describe staging systems from fertilization to the end of embryogenesis. The latter, based on a comprehensive histological analysis as well as observations on living embryos, is now the most commonly used and is the one that will be followed here. These phases will be described below. The numbering system and the main visible events occurring at each stage are given in the caption to **7.4**.

Stages 1–5: cleavage divisions and blastoderm formation

Fertilization occurs in the body of the mother. Depending on the environmental conditions, the egg can be laid immediately after fertilization, or begin to develop in the body of the mother. The nucleus of the zygote undergoes 13 divisions without cytokinesis, each division cycle lasting about 10min, with the first seven divisions taking place in the centre of the egg (Foe and Alberts, 1983). Up to this point, the egg has a homogeneously opaque appearance due to the homogeneous distribution of yolk droplets. The nuclei then begin to move towards the periphery of the egg while they continue to divide and, at the same time, the egg becomes less opaque at the periphery. This is the beginning of a process that separates the yolk components (opaque) from cytoplasmic components (clear) and continues until the formation of the cellular blastoderm. The first nuclei to reach the egg cortex (after cycle 8; end of Stage 2) are those at the posterior pole; these form the pole cells, the precursors of the germ-line. When these nuclei reach the posterior pole, the plasma membrane above them forms visible protuberances, the pole buds (Stage 3). By cycle 10, the other nuclei (apart from about 26 nuclei which remain at the centre of the egg to form vitellophages) have reached the egg cortex where they divide three more times (Stage 4; syncytial blastoderm). When the nuclei reach the cortex, the actin cytoskeleton underlying the plasma membrane becomes re-organized (reviewed in Karr and Alberts, 1986). During interphase, actin filaments form a cap above each nucleus, which rearranges into a hexagonal network of filaments between the nuclei at metaphase (**7.5**, **7.6**). The plasma membrane is pulled down between the nuclei along this network, but is released again at each interphase. At interphase 14, the synchronous nuclear cleavages cease.

The membrane invaginations now continue to deepen past the nuclei (Stage 5; **7.4**). At a depth of 30μm, they reach the yolk at the interior of the egg and close off the compartments of cytoplasm that they have created around each nucleus, thus forming cells (although these cells remain connected with the interior of the egg through a bridge of cytoplasm). At this stage (cellular blastoderm), the embryo consists of an epithelium of approximately 5000-identical looking columnar cells surrounding the yolk, and about 35 round pole cells attached to the surface of the epithelium at the posterior pole. Practically no random cell mixing occurs after the cellular blastoderm stage, and the fates of blastoderm cells can be mapped quite precisely, as shown in the fate map in **7.4**.

Stages 6–9: gastrulation

Gastrulation includes several independent, but simultaneously occurring, processes. The most important of these are: the invagination of the mesoderm, the invagination of the anterior and posterior midgut (endoderm formation), germ-band extension and neuroblast delamination. In the first phase of gastrulation, several transient folds are formed, the most prominent being the head fold or cephalic furrow, an invagination behind the head region. The colours mentioned in the following paragraphs refer to the fates shown in **7.4**.

The first morphogenetic movement is the invagination of the mesoderm which derives from a wide band of cells along the ventral side of the egg, reaching from approximately 10–70% egglength (yellow). At the end of Stage 5, these cells begin to change shape to form an indentation in the epithelium, creating the ventral furrow which deepens until it invaginates, making a tube of prospective mesoderm (**7.7–7.14**). This tube then disperses into single cells which divide twice and spread out on the underlying ectoderm as a single cell layer (Stages 8–10). Mesoderm development is directed by two genes, *twist* and *snail*, that are expressed in the ventral region of the embryo. Without the twist and snail proteins (both transcription factors), none of the later mesodermal genes, like muscle *actin* and *myosin*, *Drosophila MyoD* etc., is expressed and no mesoderm develops (reviewed in Leptin *et al.*, and Costa *et al.*, 1992).

All blastoderm cells that lie dorsal of the prospective mesoderm will become ectoderm (blue and white). The region adjacent to the prospective mesoderm on each side is the ventral neurogenic region (dark blue and white). A subset of cells from this region enlarges and delaminates just after the prospective mesoderm has invaginated (Stage 9) and come to lie between the ectoderm and the mesodermal cell layer. These cells are neuroblasts and will proliferate and their progeny differentiate and later condense to form the central nervous system. The cells of the neurogenic region that do not delaminate form the ventral epidermis. The next region dorsal to the neurogenic region will form the dorsal epidermis and most of the peripheral nervous system (light blue and white). The most dorsal band of blastoderm cells makes the amnioserosa (green), a thin extra-embryonic membrane which covers the dorsal side of the egg until the embryo proper closes dorsally.

The endoderm is made by the anterior and posterior midgut invaginations (dark red). The posterior midgut arises from the epithelium underlying the pole cells at the posterior pole of the embryo. The cells in this region undergo shape changes very similar to those in the ventral furrow (Stage 6), and an indentation is formed (the amnioproctodeum) which gradually deepens (Stage 7) and carries the pole cells with it into the interior of the embryo (Stages 8–10). The anterior midgut invaginates as a transverse fold at the anterior end of the ventral furrow (Stage 7). It later fuses with the stomodeum, an invagination in front of the anterior midgut invagination (bright red).

Very soon after the ventral furrow has begun to invaginate, and while the anterior and posterior midgut are forming, the

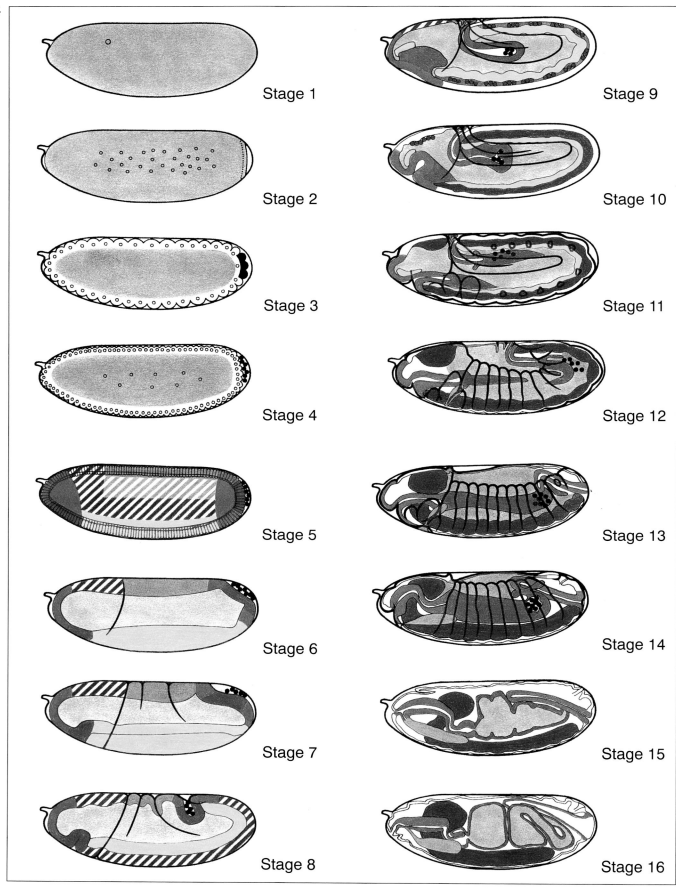

7.4 This figure shows 16 stages (15hr) of embryonic development. In all drawings and photographs, anterior is to the left and dorsal is up, unless stated otherwise. The diagrams show midsagittal sections. The broad lines in Stages 6–14 are the outlines of shapes and folds on the surface of the embryo to show the main morphogenetic movements. Stage 5 has the blastoderm fate map superimposed on it, with the different colours representing the following fates: dark red, midgut; bright red, hindgut and foregut; orange, salivary gland; green, amnioserosa; yellow, mesoderm; blue, central nervous system; light blue, peripheral nervous system; white, epidermis; grey, yolk; black dots, germ line cells. Several different methods have been used to determine which parts of the larva are derived from which parts of the blastoderm. These include observations of cell movements, or inference of their movements from comparisons of very narrowly staged fixed embryos; ablation of cells and mapping of the resulting deletions in the fully developed embryo; and dye and enzyme marking of cells.

Stage 1 (25min): From fertilization to the second nuclear cleavage. The egg looks homogeneously opaque.

Stage 2 (40min): Nuclear cleavages 3–8. During Stage 2 the egg cytoplasm transiently retracts at the anterior and posterior poles, leaving visible clear spaces.

Stage 3 (15min): Nuclear cleavage 9. The clear cytoplasmic layer at the periphery of the egg becomes wider, and begins to look less homogeneous as the somatic nuclei begin to arrive at the egg surface. The most easily visible event during this stage is the formation of the pole buds above the nuclei at the posterior pole.

Stage 4 (50min): Syncytial blastoderm stage. Nuclear cleavages 9–13 of the somatic nuclei. The cytoplasmic rim at the periphery is still becoming wider, and nuclei are now clearly visible in it. The pole buds, which have divided three times, pinch off to form the pole cells.

Stage 5 (40min): Cellularization. The nuclei begin to elongate and the cytoplasmic region at the periphery of the egg becomes subdivided into cells by the invagination of membranes between the nuclei. The advancing front of the membranes is visible. Cellularization is slower on the dorsal side than on the ventral side, and at the end of this stage, even before cellularization is complete dorsally, gastrulation begins in the ventral region of the embryo.

Stage 6 (10min): The ventral furrow begins to invaginate. At the same time, a fold behind the head region (the cephalic fold) becomes visible at 65% egglength. The posterior midgut primordium with the pole cells begin to shift dorsally.

Stage 7 (10min): The anterior and posterior midgut begin to invaginate. The cephalic fold deepens and extends more dorsally and ventrally. Germ-band extension begins and transient deep dorsal folds appear.

Stage 8 (30min): The posterior midgut invagination deepens into the interior of the embryo, curving to point backwards and carrying the pole cells with it. The surrounding epithelium is drawn into the invagination, forming the proctodeum (which will give rise to the hindgut). Germ-band elongation continues rapidly, bringing the posterior midgut invagination to about 60% egglength.

Stage 9 (40min): The germ-band slowly continues to elongate (gut invagination at 70% EL at the end of this stage). Neuroblasts in the germ-band and in the head delaminate. At the end of this stage, the stomodeal cell plate appears on the ventral side of the anterior tip of the egg.

Stage 10 (60min): The germ-band extends until the proctodeum reaches 70%EL. The stomodeum invaginates. The AMG grows posteriorwards and flattens out. The pole cells leave the lumen of the posterior midgut. First signs of visible segmentation of the germ-band, invagination of tracheal pits.

Stage 11 (120min): The germ-band remains extended throughout this stage. A clear space appears at the posterior end of the egg, where the embryo retracts from the vitelline membrane. Segmentation becomes more obvious with clear furrows marking the boundaries between parasegments (the ectoderm later becomes rearranged with respect to these furrows, such that the furrows mark the segmental, rather than the parasegmental, boundaries). Buds for the head appendages appear in segments 1–3 (mandibular, maxillary and labial buds). Nervous system: the primordium of the optic lobe becomes morphologically distinguishable as a placode in the dorsal epithelium of the head region and invaginates. Neuroblasts in the head proliferate and displace the yolk.

Stage 12 (120 min): The germ-band begins to retract and the amnioserosa unfolds to cover the yolk. The clear space at the posterior end is still visible. The posterior midgut has grown to the posterior end of the yolk sac and now bends anteriorly. Later in Stage 12, after the germ-band has retracted further, anterior and posterior midgut fuse and form two bands along the sides of the yolk. The tracheal pits fuse to make a continuous tracheal tree.

Stage 13 (60min): The germ-band completes its retraction. The posterior end of the ventral nerve cord lies at the posterior tip of the egg. The edges of the germ-band begin to grow over the yolk towards the dorsal midline (beginning of dorsal closure). The two bands of midgut epithelium extend dorsally and ventrally until their edges meet at the ventral (in Stage 13) and dorsal (in Stage 14) midlines to enclose the yolk completely. Salivary gland ducts meet at the midline and fuse to create a single opening. An epithelial ridge appears dorsally behind the head (dorsal ridge).

Stage 14 (60min): Head involution begins as the anterior edge of the germ-band begins to move anteriorly over the head region. Dorsal closure of the gut and the germ-band continues. The hindgut grows and assumes its typical shape, ascending dorsally from the anus to about 50% EL, where it bends backwards to join the midgut.

Stage 15 (30min): Dorsal closure of gut and epidermis are completed. The three midgut constrictions appear. Head involution continues. The ventral nerve cord begins to contract.

Stage 16 (180min): Head involution continues. The ventral nerve cord continues to contract, its posterior end moving to 40% EL. Secretion of cuticle begins. Proventriculus and gastric caeca appear.

A little later, in Stage 17, the ventral cord completes contraction and the gut continues to constrict until it is a long thin tube. Towards the end of embryogenesis, the gut begins to move and the embryo twitches, signs that the musculature is beginning to work. The tracheae fill with air and are easily visible.

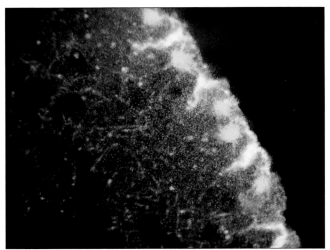

7.5, 7.6 Early events: distribution of actin and tubulin in the syncytial blastoderm (metaphase of nuclear division cycle 13). **7.5** shows optical section through the blastoderm. **7.6** shows surface view. Tubulin (orange) is mainly found in mitotic spindles, while actin (green) forms a network associated with plasma membranes that invaginate from the surface of the blastoderm. The invagination is transient during nuclear cleavages 8–13, disappearing during interphase, but deepens during interphase 14 to separate the syncytial blastoderm into individual cells. (×200; photographs by Bill Theurkauf.)

process of germ-band extension (or elongation) begins. The germ-band is the part of the embryo that will make the segmented part of the body (i.e. everything except the head region, the posterior region of the egg and the amnioserosa). Cells in the germ-band intercalate along the dorso-ventral axis, so that the germ-band elongates. This results in the posterior midgut invagination being pushed onto the dorsal side, and then anteriorwards. Simultaneously, the epithelium on the dorsal side begins to fold, and later, to become very thin as it makes the amnioserosa (Stage 8). Germ-band elongation continues until the posterior midgut invagination lies directly behind the head, and the germ-band is completely folded back upon itself (Stage 10). The amnioserosa is squeezed into thin folds between the sides of the germ-band and between the head and the posterior midgut invagination.

During gastrulation, most cells go through their 14th division cycle. However, the synchrony of the first 13 divisions now ceases. This does not mean that cells now divide at random or in an unorganized fashion. Rather, defined groups of cells ('mitotic domains') divide simultaneously at specified times (**7.15–7.18**) (Foe, 1989).

Stages 10–17: organogenesis and differentiation

General morphogenetic movements and segmentation

The main general morphogenetic movements after gastrulation are germ-band retraction, dorsal closure and head involution. The final phase of germ-band elongation brings the posterior midgut invagination to lie directly behind the head fold (Stage 10). The germ-band remains in its extended state for about 2hr and then begins to retract, bringing the hindgut opening back to the posterior end of the embryo. The amnioserosa, previously folded up between the head and the sides of the germ-band, now spreads over the dorsal side of the yolk sac. The function of germ-band extension and retraction are not understood.

While the germ-band is extended, it becomes divided into a series of bulges whose appearance is the first visible sign of segmentation, but the indentations between the bulges do not correspond to the final segment boundaries. They lie at the boundaries between the anterior and posterior compartments within each segment. The regions defined by the bulges have therefore been called parasegments (Martinez-Arias and Lawrence, 1985). Parasegments are not merely a transient morphological feature, but are the genetic building blocks of the embryo, since they are the realms of action of the primary segmentation genes and the homeotic genes. The final subdivision of the embryo into segments becomes visible during germ-band retraction.

The retracted germ-band lies on the ventral side of the embryo, with the amnioserosa covering the dorsal side. The edges of the retracted germ-band grow dorsally (Stage 13) until they meet and fuse at the dorsal midline (dorsal closure), enclosing the amnioserosa (Stage 15).

The region anterior to and within the cephalic furrow mainly gives rise to the nervous system of the head and to the most anterior parts of the alimentary duct, i.e. the pharynx, oesophagus, the mouth parts and the salivary glands. The primordia of these structures are brought into the interior of the embryo by a process called head involution, during which the epidermis of the germ-band moves anteriorly over the head region (Stages 14–17).

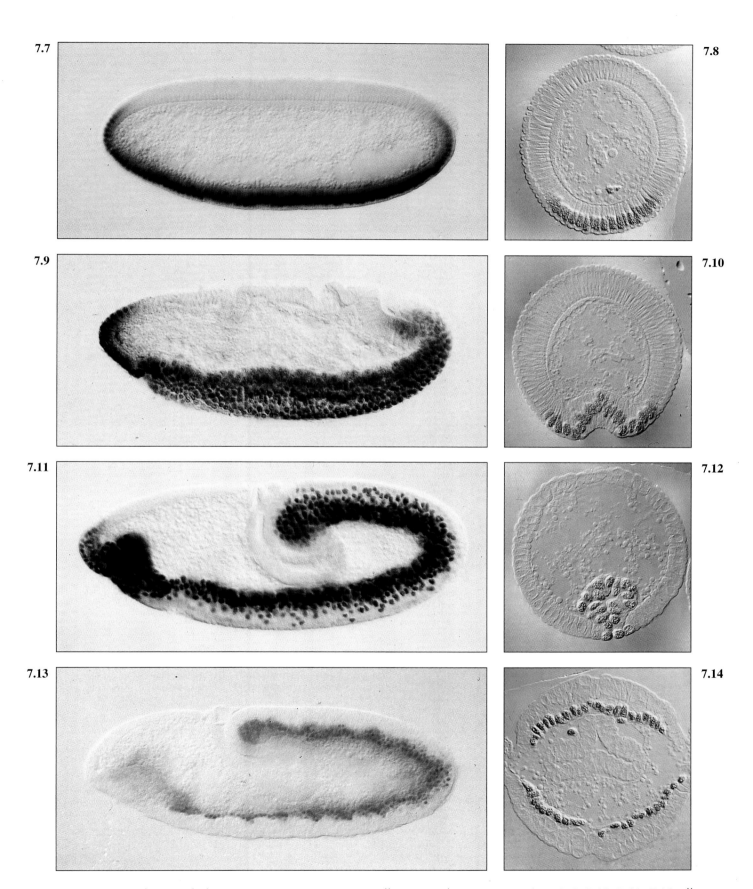

7.7–7.14 Gastrulation: whole mounts (**7.7, 7.9, 7.11, 7.13**, all ×180) and transverse sections (**7.8, 7.10, 7.12, 7.14**, all ×230) of gastrulating embryos stained with antibodies against twist, a nuclear protein expressed in prospective mesodermal and some endo- and ectodermal cells (compare to fate map in **7.4**). See text for details. (Photograph **7.11** by Rolf Reuter.)

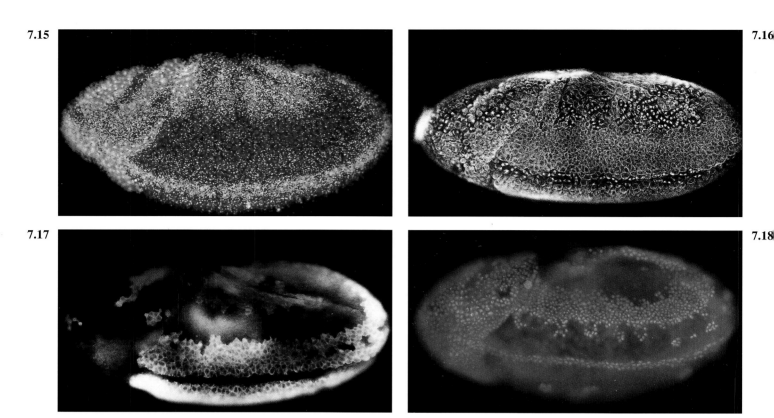

7.15–7.18 Mitotic domains: embryos during successive stages of cell cycle 14. Groups of cells that enter mitosis 14 synchronously (mitotic domains) start expressing the gene *string* synchronously (**7.15**, *in situ* hybridization showing the distribution of *string* RNA). A few minutes later, mitotic figures are observed in these cells (**7.16**, embryo stained with anti-tubulin antibody) and cyclin A is degraded (**7.17**, embryo stained with anti-cyclin A antibody; degradation of cyclin A seen as black patches of unstained cells). Immediately after mitosis, cells enter the next S phase and synthesize DNA (**7.18**, BrdU incorporation visualized with an anti-BrdU antibody). (×150; photographs by Bruce Edgar.)

Organogenesis

Gut

The gut consists of foregut, midgut and hindgut and has a number of appendages (salivary glands, gastric caeca and malpighian tubules). The foregut is further subdivided into pharynx, oesophagus and proventriculus.

Gut development begins with the invagination of the anterior and posterior midgut primordia during gastrulation. These separately developing anlagen grow towards each other and eventually fuse and form the midgut (**7.19–7.22**). In the late stages of embryogenesis, the midgut becomes constricted into a long convoluted tube with the site of each constriction being determined by the activity of different homeotic genes (*Antp*, *Ubx* and *Abd-A*; **7.23**) in the overlying visceral mesoderm. In the most anterior region of the midgut, the gastric caeca, four tube-like appendages develop. Their morphogenesis depends on the activity of another homeotic gene (*Scr*) expressed in the visceral mesoderm (Immergluck *et al.*, 1990; Reuter and Scott, 1990).

The development of the foregut begins with the invagination of the stomodeal plate, a depression in front of the anterior midgut invagination (Stage 9), which grows, extends backwards and fuses with the anterior midgut invagination. During head involution, much of the surrounding ectoderm is also internalized and contributes to the foregut. The salivary glands, which will eventually open into the pharynx, originally arise as two invaginations of the ectoderm in the labial region during Stage 11. The outer parts of the salivary tubes later fuse to form a single duct. This is pushed into the oral opening during head involution such that it comes to lie at the bottom of the pharynx.

The hindgut is derived from the blastoderm epithelium which follows the posterior midgut into the interior of the embryo during late germ-band extension. During the extended germ-band stage, four bulges appear on the hindgut, which will elongate (by proliferation of a stem cell at the tip of each; Skaer, 1989) and develop into the malpighian tubules.

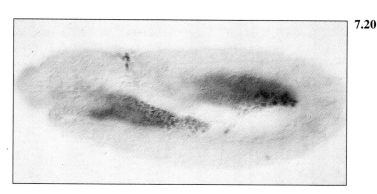
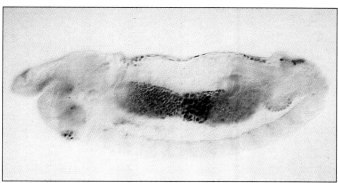
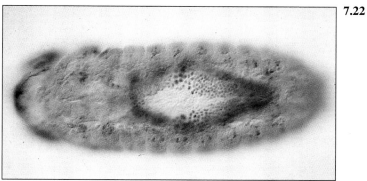

7.19–7.22 Midgut development: embryos from an enhancer trap line that expresses ß-galactosidase in the anterior and posterior midgut. After having invaginated during gastrulation (**7.19**), anterior and posterior midgut grow towards each other (**7.20**) by stretching out two bands of cells on the right and left of the yolk mass in the space between yolk and mesoderm. During germ-band retraction, the tips of the two arms of the AMG extending from the front meet those of the PMG extending from the back and they fuse (**7.21**). The midgut epithelium grows around the yolk until it is completely enclosed (**7.22**, dorsal view). (×150; photographs by Rolf Reuter.)

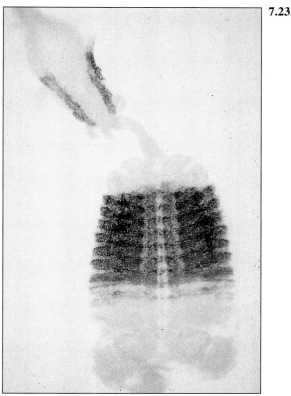

7.23 Homeotic genes from the bithorax complex. Expression of two homeotic genes, *Ultrabithorax* (*Ubx*; brown) and *abdominal-A* (*abdA*; blue) in the epidermis and gut of a 12 hour-old dissected embryo (Karch *et al.*, 1990). The embryo has been opened along the dorsal midline, the epidermis flattened out, and the internal organs have been removed, except for the midgut and hindgut, which have been folded out of the embryo. *Ubx* is expressed from parasegment 5 to the posterior end, *abdA* from parasegment 7 back. The function of these genes is to give parasegments 5–14 and 7–14, respectively, identities different from the more anterior parasegments which do not express *Ubx* or *abdA*. A third bithorax complex gene, *AbdB*, is responsible for making parasegments 10–14 different from the more anterior parasegments. The expression of *Ubx* and *abdA* in the mesoderm surrounding the gut is important for the formation of the gut constrictions. These arise at the anterior boundary of *abdA* and in the middle of the *abdA* domain, and are absent in *Ubx* (second constrictions) *abdA* mutants (second and third constriction). (×130; photograph by Welcome Bender and François Karch.)

Muscles

The musculature is derived from the mesoderm. During germ-band retraction, the mesoderm separates into visceral mesoderm, which will produce the muscles that attach to the gut, and the somatic mesoderm which produces the musculature that attaches to the body wall. The visceral musculature grows as a sheath over the gut epithelium, and plays an important role in gut morphogenesis (as described above, see **7.23**). Some mesoderm cells are used for the gonads, heart and fat body.

The somatic mesoderm is divided into segmental groups of

7.24

7.25

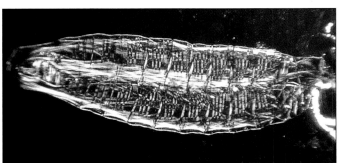

7.26

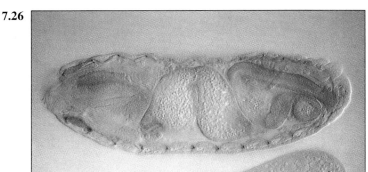

7.27

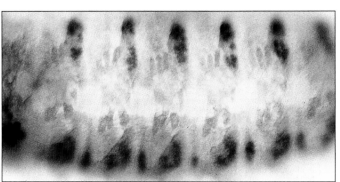

7.28

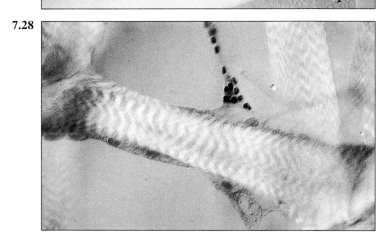

7.29

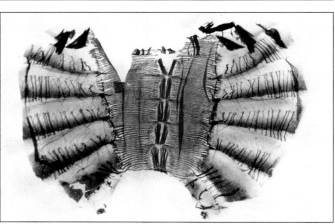

7.24–7.29 Muscles. **7.24** (×150) is an embryo stained with antibodies against muscle myosin to mark the segmentally repeated muscle pattern. **7.25** (×120) shows muscles in the mature first instar larva, visualized with polarized light. The longitudinal muscles are seen particularly well in this preparation. **7.26** (×150) is an embryo stained with antibodies against the α-subunit of the PS2 integrin, which is needed for the attachment of muscles to the body wall and the gut. The accumulation of this protein at the points where the longitudinal muscles attach to sites at the segmental boundaries is visible in this focal plane. The protein is also seen at high levels in the pharyngeal muscle and in the visceral mesoderm attached to the hindgut epithelium. **7.27** (×350) shows an embryo stained with antibodies against muscle myosin (brown) and the homeodomain protein S59, which is only expressed in a subset of the muscles in each segment, and probably specifies their fates. **7.28** (×350) shows *twist* – expressing cells lying near a muscle in a third instar larva. At this stage, embryonic (now larval) muscles and other embryonic mesodermal derivatives have ceased to express *twist*, except for a few cells like those illustrated here, which are the precursors of adult muscles. When these cells are ablated, gaps appear in the adult muscle pattern. **7.29** (×40) displays muscles in the adult abdomen. The abdomen has been dissected and spread out to illustrate the muscles (stained blue). (Photographs by Rolf Reuter (**7.24**), Michael Bate (**7.25, 7.28** and **7.29**), Thierry Bogaert and Michael Wilcox (**7.26**), and Manfred Frasch (**7.27**).)

cells, subsets of which fuse to form the syncytial muscle fibres whose pattern is precisely defined for each segment. Most muscles do not extend beyond one segment (**7.24–7.29**). The fates of different sets of muscles probably depend on transcription factors expressed only in specific muscle founder cells (**7.28**; Dohrmann *et al.*, 1990). The developing muscles attach to specific attachment sites within segments and at the segmental boundaries (the apodemes, Stage 14), a process mediated by two integrins, one expressed in the gut epithelium and the epidermis, the other in the mesoderm (**7.24–7.29**). (Bogaert *et al.*, 1987; Leptin *et al.*, 1989).

Some mesodermal cells do not contribute to the musculature of the larva, but are set aside to provide the muscles of the adult fly. When such precursors are eliminated, subsets of muscles are absent in the adult (**7.24–7.29**, Bate *et al.*, 1991; Broadie and Bate, 1991).

Nervous system

The nervous system, which consists of peripheral (PNS) and central nervous system (CNS), arises from two distinct sets of precursors which arise from the ectoderm at different times (dark and light blue striped regions in **7.4**). The dorso-ventral limits of these regions are determined by the maternal morphogen *dorsal* together with the zygotically expressed genes *twist* and *snail* on the ventral side and *dpp* on the dorsal side (**7.13** and **7.46–7.48**).

Central nervous system: The CNS consists of the brain and the ventral nerve cord and is mainly derived from the neuroblasts that delaminate in three longitudinal rows on each side of the ventral midline from the neuroectoderm during gastrulation (**7.30**). While all cells in the neuroectoderm have the potential to develop into neuroblasts, a process of lateral inhibition, mediated by the transmembrane receptors *Notch* and *Delta*, selects only a few to do so (reviewed by Campos-Ortega and Jan, 1991; Cabrera, 1992; and Doe, 1992). When the *Notch* and *Delta* genes are mutated, all cells from the neurogenic region develop as neuroblasts. The opposite effect, loss of neuroblasts, is caused by mutations in the *achaete-scute*-complex, a family of genes coding for bHLH type transcription factors that are essential for neuroblast development.

The neuroblasts divide as stem cells to produce a series of ganglion mother cells which proliferate further and produce neurons. The location and behaviour of these neurons are very reproducible, and some have been mapped and named as 'identified neurons', (originally in the grasshopper; Bate, 1976, but the arrangement is the same in the fly; Thomas *et al.*, 1984; **7.34–7.39**). A number of transcription factors are known to be responsible for differences in the fates of these neurons. They are expressed in subsets of neuronal precursors (**7.34–7.39**) and their absence or misexpression can lead to transformations in cell fates.

The differentiating ganglion cells condense to form the ventral nerve cord. During late embryogenesis, the ventral nerve cord contracts in length to approximately half the length of the embryo. Axon outgrowth begins during germ-band retraction (Stage 12). Axons in the central nervous system are bundled together in fascicles, forming two bundles along the length of the ventral nerve cord, connected by two commissures in each segment. A set of cell-surface proteins, the fasciclins (first found in grasshoppers, later also isolated in *Drosophila*; Bastiani *et al.*, 1987; Patel *et al.*, 1987) are expressed only on certain fascicles, such that the part of an axon that is in the fascicle expresses the protein, but as soon as the axon leaves the fascicle, it ceases to carry the protein on its surface.

Mesectoderm: A further population of cells that contributes to the CNS, the mesectoderm cells, arises in two lines

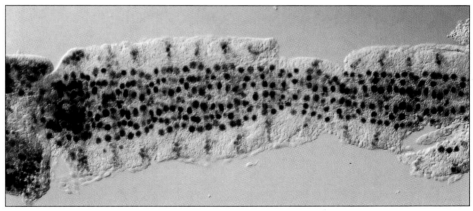

7.30 Development of the central nervous system. Neuroblast pattern in the neurogenic region of the ectoderm at the extended germ-band stage. A Stage 10 extended germ-band embryo (from an enhancer trap line expressing β-galactosidase in neuroblasts) was stained with anti-engrailed antibodies (brown) to mark segments, and antibodies against β-galactosidase (black) to mark neuroblasts. The germ-band was then unfolded and spread out for photography (see **7.53–7.56** for an intact embryo at this stage, stained only with anti-engrailed antibody). (×180; photograph by Marc Haenlin and Jose Campos Ortega.)

bordering the mesoderm. When the mesoderm invaginates, the mesectodermal cells meet at the ventral midline. They then delaminate from the surface to form the midline precursors, and differentiate into neurons and glia.

Brain: The brain develops from neuroblasts in the procephalic neurogenic region (in **7.4**, blue and white region in the head; also visible in **7.30**) which delaminate around the same time as the neuroblasts in the germ-band. The optic lobes develop from placodes in the dorsal ectoderm of the head region.

Peripheral nervous system: The PNS consists of sensory organs, the peripheral nerves, and glia. There are three main types of sense organs; the external sensory organs (es-organs), with visible differentiations in the cuticle on the surface of the embryo, the chordotonal organs, which are internal stretch receptors (**7.31–7.33**) and the less well characterized multipolar sensory neurons (Bodmer and Jan, 1987). These sense organs arise from groups of cells in the dorso-lateral epidermis of each segment which delaminate and proliferate from the epidermis from Stage 10 onwards. The cells then differentiate into sensory organs and extend axons towards the CNS, which become joined into two fascicles per segment on each side. The difference between these two types of sense organs depends on the homeodomain-protein *cut*, which is normally expressed in es-organ precursors. When it is deleted, the es-organ precursors develop as chordotonal organs. When it is misexpressed in chordotonal organs, they develop as es-organs (Blochlinger *et al.*, 1991).

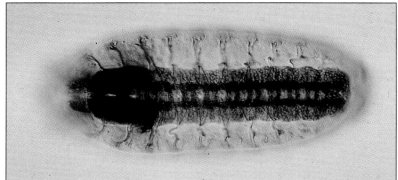

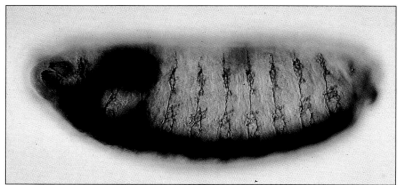

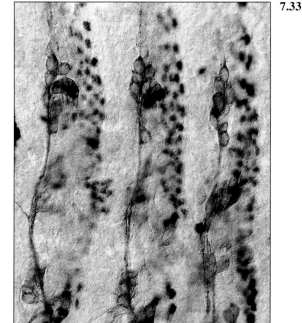

7.31–7.33 Nervous system. **7.31** (×180) and **7.32** (×180) shows ventral and lateral views of embryos stained with antibodies against horseradish peroxidase which stain all peripheral and central neurons (Jan and Jan, 1982). In the ventral nerve cord, the longitudinal axonal bundles and the commissures connecting them in each segment are visible (see also **7.30**). **7.33** (×500) shows part of the peripheral nervous system. Three segments (2nd and 3rd thoracic and first abdominal) of an embryo stained with antibodies against the product of the segmentation gene *engrailed* (blue; this marks the posterior compartment of each segment) and an antibody that stains all peripheral neurons (brown). (Photographs by Yuh Nung Jan (**7.31** and **7.32**) and Nipam Patel (**7.33**).)

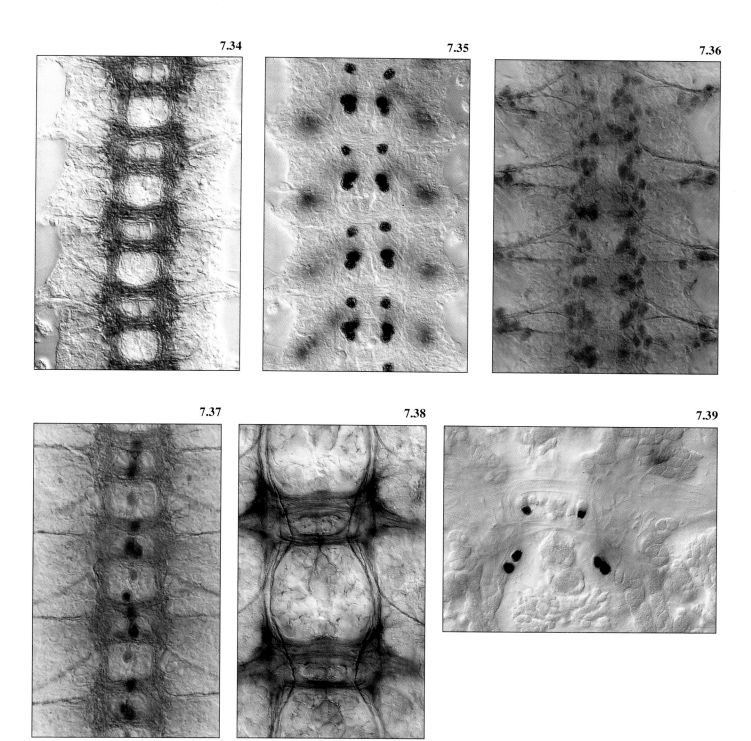

7.34–7.39 Pattern in the central nervous system. Part of the ventral nerve cord (compare to **7.30**) is stained with antibodies to visualize different subsets of neurons and axons. **7.34** (×700) is an antibody that stains all CNS axons in the ventral nerve cord, marking the longitudinal and commissural axon bundles. **7.35** (×700) shows the segmentation gene *even-skipped* (expressed in stripes in every other segment at the blastoderm stage, **7.53–7.56**) and is expressed in the nuclei of certain identified neurons (RP2, aCC and pCC) in each segment. The same neurons express *even-skipped* in the grasshopper (**7.39**). **7.36** and **7.37** (both ×700) are glial cells (blue staining) in the CNS: midline (**7.36**) and longitudinal and exit glial cells (**7.37**) in the CNS, counterstained with the antibody against CNS axons. **7.38** (×1250) are subsets of CNS axons express fasciclin II (brown) or fasciclin IV (black). **7.39** (×500) is a segment of the grasshopper embryo stained with antibodies against the grasshopper *even-skipped* protein. (Photographs by Nipam Patel, A. Kolodkin, Christian Klämbt, and Corey Goodmann.)

Gonads

The gonads are composed of somatic cells (derived from the mesoderm) and germ-line cells (**7.40–7.43**), the progeny of the pole cells (black dots in **7.4**) which were carried into the interior of the embryo in the posterior midgut invagination. They leave the midgut by migrating through its epithelium and come to lie between the midgut epithelium and the mesoderm in abdominal segments a5–a7. During late germ-band retraction, they collect in segment a5, where they are joined by the mesodermal cells that form the somatic gonadal sheath (Stage 13).

The development of the pole cells depends on determinants present in the cytoplasm at the posterior pole of the egg (Illmensee and Mahowald, 1974), which are made and localized under the control of a group of maternal effect genes (reviewed in St. Johnston and Nüsslein-Volhard, 1992. **7.49–7.52**).

7.40
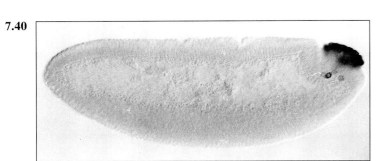

7.41

7.42
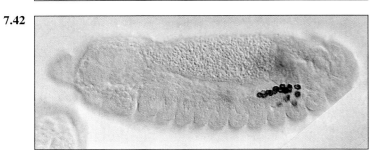

7.43
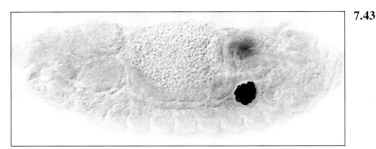

7.40–7.43 Germ cells and gonad. Embryos stained with antibodies against the vasa protein. The pole cells develop at the posterior end of the egg under the control of a set of maternal effect genes, among them *vasa*. The pole cells are carried dorsally (**7.40**) and then into the interior of the embryo with the posterior midgut invagination. They leave the midgut by migrating through its epithelium (**7.41**) and come to lie between the midgut epithelium and the mesoderm in abdominal segments a5–a7 (**7.42**). During late germ-band retraction they collect in segment a5, where they are joined by the mesodermal cells that form the somatic gonadal sheath (**7.43**). The embryo in **7.43** is seen from a slightly dorsal angle and the right gonad is therefore also faintly visible. (×150; photographs by Bruce Hay.)

Tracheae

The tracheae develop from invaginations (known as tracheal pits) which become visible at Stage 11 in the lateral ectoderm of every segment from t2 to a9. The invaginations branch out and fuse during germ-band retraction to make the continuous trunk of the tracheal tree (**7.44, 7.45**). The openings where tracheae take up air are the anterior and posterior spiracles.

7.44
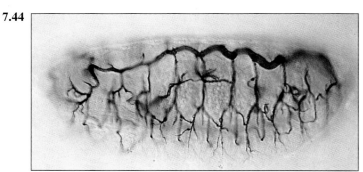

7.45
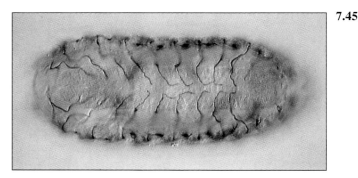

7.44, 7.45 Tracheae. **7.44** shows the tracheal branches throughout the embryo that provide the body with oxygen. In a lateral view, the tracheal trunks running along the dorsal side can be seen. These are connected to the outside at the anterior and posterior ends of the embryo in structures called spiracles, visible in the cuticle (**7.1–7.3**). **7.45** is a ventral view showing branches penetrating the ventral nerve cord. (×150; photographs by Nipam Patel.)

Larval cuticle

The cuticle begins to be secreted at Stage 16 by epidermal cells and the epithelia of the hindgut, foregut and tracheae. It has numerable visible specializations including dorsal hairs, ventral denticles and sense organs (**7.3**).

Imaginal discs

Imaginal discs are epithelial invaginations of the epidermis that evaginate during metamorphosis and develop into most of the external structures of the adult head and thorax during metamorphosis. There is a disc for each appendage (wings, halteres and legs), eye and antenna and one for the genitalia. The discs are derived from small groups of cells that are set aside during embryogenesis and proliferate extensively during the larval stages. They are hardly visible in the embryo, but are large conspicuous organs in the third larval instar. Their complicated differentiation programmes are reflected in complex patterns of gene activity.

Experimental manipulations

The *Drosophila* embryo is small (0.5mm long) and, at least until the end of gastrulation, so fragile that it cannot be removed from the vitelline membrane without destroying it. It does not therefore lend itself to much experimental manipulation and the main method of interfering with embryogenesis has been by using mutations. Nevertheless, a few techniques, both genetic and mechanical, have been developed to manipulate the development of the embryo.

Mechanical manipulations

Microinjection

One of the main manipulations used to study *Drosophila* embryogenesis is microinjection. Cells, nuclei and cytoplasm can be transplanted between embryos, and RNA, DNA, proteins, drugs and other molecules can be injected into the embryo, as well as into the perivitelline space surrounding the embryo.

- *Transplantation of cytoplasm*
  Cytoplasm can be drawn out of an embryo into a microcapillary and then injected into another embryo. Transplantation experiments have demonstrated two very important principles of early embryogenesis: the existence of germ-cell determinants in the posterior cytoplasm (Illmensee and Mahowald, 1974) and the presence of localized cytoplasmic components that direct axis formation in the egg (Nüsslein-Volhard et al., 1987; Anderson, 1987).
- *Injection of DNA, RNA, drugs and labelled molecules*
  The most important application of this method is the injection of DNA into the pole plasm in order to transform germ-line cells (see below), but injection of DNA or RNA can be used for a variety of other purposes (e.g. assaying genetically engineered variatiants of genes). When fluorescently labelled actin, tubulin or histones are injected into the embryo, they become incorporated into the cytoskeleton and nuclei, and the behaviour of these structures can then be followed in living embryos by fluorescence microscopy (Minden et al., 1989).
- *Transplantation of cells and nuclei*
  Transplantation of cells or nuclei is used to create mosaic embryos, for example in order to test whether a gene acts in a cell autonomous way (i.e. whether a mutation affects the cell that normally expresses the gene, or neighbouring cells) or to observe how the progeny of individual cells populate the developing organism (lineage restrictions, compartment boundaries). Mosaics can, however, also be created genetically (see below).

Tissue culture

Most cell lines that exist (see Ashburner, 1989) are not very well characterized with respect to their embryonic origin. They are used mainly for biochemical purposes and for transfections to assay properties of molecules like cell-surface receptors or transcription factors, much in the same way as vertebrate cell lines are used (for example, one of the lines made by Imogen Schneider has been adapted to grow in mammalian tissue culture media; these are usually referred to as 'Schneider cells').

Older embryos can be dissected and continue to develop in culture. It is possible to fill neurons with dye and observe axon outgrowth in cultured embryos (Sink and Whitington, 1991).

Other manipulations

Cells in the embryo can be killed by laser ablation, an important method for constructing fate maps. The vitelline membrane can be made permeable to small molecules, including radioactive or otherwise-labelled amino acids and nucleotides, which allows easy *in vitro* labelling with such compounds for biochemical and histological studies.

Genetic manipulations

Mitotic recombination

In a fly or an embryo heterozygous for a mutation, clones of homozygous mutant cells can be created by using X-rays to induce somatic recombination in individual cells. Another way of making mosaics is through the creation of mutant clones in the developing female germ-line ('germ-line clones'). If a gene that is active during the development of the oocyte in the gonad is also required for the survival of the organism after fertilization, then the effect of recessive mutations in the germ-line cannot be analysed (because a homozygous mutant mother cannot develop). In such a case, clones of *homozygous* mutant germ-line cells have to be generated in the gonad of a *heterozygous* mother. Since as much as a third of all genes may be required both in the germ-line and during development of the zygote, this is a very important technique for analysing the role of these genes in early embryogenesis (Wieschaus and Noell, 1986).

Germ-line transformation

Recombinant DNA can be integrated into the germ-line of flies very easily by injecting appropriate constructs into the pole plasm of the embryo (Rubin and Spradling, 1982; Spradling and Rubin, 1982). The technique makes use of two components of a transposable element (the P element), namely a transposase, and the DNA segments recognized by the transposase for integration into the host DNA. DNA flanked by these segments is integrated into DNA efficiently if transposase is provided.

Random P element insertion can also be used for mutagenesis. The genes disrupted by the P element can then be cloned very easily by using the inserted P element as a tag for the adjacent genomic DNA.

Other molecular genetic techniques

When genes flanked by target sites for the yeast recombinase FLP are integrated into the *Drosophila* genome by germ-line transformation, they can be induced to be excised at high frequency at desired times of development, provided that the gene for FLP recombinase has also been integrated into the genome (usually under the control of a heat-shock promoter; Golic and Lindquist, 1989). This is another way of creating genetic mosaics.

A sophisticated method of cell ablation uses the directed expression of a temperature sensitive toxin (diphtheria or Ricin A chain) under the control of a cell or tissue-specific promoter to kill those cells or tissues by temperature shifts (Moffat *et al.*, 1992; Bellen *et al.*, 1992).

Developmental mutants: saturation screens

The description, or even listing, of all mutations affecting *Drosophila* development is beyond the scope of this chapter as systematic screens have identified several hundred genes specifically required for embryogenesis. Instead of discussing individual mutants, this chapter will describe the principles of the screens for mutants, and the general conclusions that can be drawn from them. The mutations isolated in these screens have provided the material upon which our present understanding of *Drosophila* embryogenesis is based, and a few examples will be discussed below (see also Chapter 6).

Background

A few mutations that, when homozygous, disrupt embryogenesis have been known for some time. In *Krüppel* mutant embryos, for example, the thorax and first five abdominal segments are missing (see Nüsslein-Volhard *et al.*, 1984). Such embryos die before they can hatch. The *Krüppel* gene is thus required for the proper development of these segments, and mutations in it cause embryonic lethality. Nüsslein-Volhard and her colleagues reasoned that other genes were probably required for the development of the other segments, and mutations in those genes should lead to the deletion of other segments. Similarly, there might be genes required for the establishment of regions along the d-v axis (one maternal effect gene, *dorsal*, that affected d-v fate was already known (Nüsslein-Volhard *et al.*, 1980)) or other parts of the embryo. with mutations in these genes being expected to be homozygous lethal. In order to find the genes specifically required to construct the embryo, Nüsslein-Volhard, Wieschaus and their collaborators therefore set out to search systematically for all mutations that caused embryonic lethality with visible defects in the pattern of the embryo (Jürgens *et al.*, 1984; Nüsslein-Volhard *et al.*, 1984; Wieschaus *et al.*, 1984).

Genetics and tools

Searches for mutations that do *not* cause lethality are relatively easy, because individuals carrying a mutation (e.g., one that causes eyes to be white rather than red) can be identified, and then be mated in order to propagate the chromosome carrying the mutation. The situation is more complicated in screens for mutations that cause lethality and a number of problems have to be considered. An organism carrying a lethal mutation is, dead (if it is homozygous for the mutation and therefore expresses the phenotype) and cannot be used for mating. Instead, its siblings that are heterozygous for the mutation (and therefore survive) have to be used for the propagation of the mutant chromosome. Furthermore, to generate homozygous embryos in which the lethal phenotype can be scored, one needs as parents two flies each carrying the mutated gene, which can be mated to produce homozygous offspring. Thus, the original mutagenized fly containing a mutated chromosome first has to be mated to multiply this chromosome and generate the parents of the embryo to be scored. Finally, heterozygotes carrying the mutated chromosome must be distinguishable from their sibs. The screen therefore has to be conducted in such a way that both the presence of a lethal mutation can be scored, and the chromosome carrying that mutation can be distinguished and propagated in heterozygotes.

The most important genetic tool used to achieve this are 'Balancer' chromosomes. These are chromosomes with three properties necessary for maintaining stocks carrying lethal mutations without losing or recombining the mutated chromosome (balanced stocks). First, balancers carry mutations that lead to the expression of visible markers in the fly (differences in eye colour and shape, body colour, wing shape, length of bristles etc.), which means that flies carrying these chromosomes can be distinguished unambiguously. Second, they carry recessive lethal mutations (unrelated to the mutations to be generated in the screen), which means that they can only be kept in heterozygosity, and therefore cannot out-compete chromosomes carrying the lethal mutations to be generated in the screen. Third, they contain many inversions, which means that they do not undergo meiotic recombination. An important additional tool for screens are dominant temperature-sensitive mutations which can be used to kill specifically those flies among the progeny that are not required for further breeding. This makes it unnecessary to sort out the desired flies manually in the back-crosses.

Mutagenesis and screen

Mutations are induced (usually in the germ cells of male flies) by chemical mutagens, by irradiation or by the insertion of transposable elements. Because of the use of balancers, one sets up separate screens for the different chromosomes. *Drosophila* has four chromosomes, with the X chromosome covering approximately 20% of the genome, the second and third chromosomes covering approximately 40% each, and the fourth chromosome less than 1%. Since each mutated chromosome has to be inbred to be able to score the mutations, one has to set up as many individual crosses as potential mutations are to be screened. The dose of mutagen is chosen to give on average at least one lethal mutation per chromosome. Each of the *Drosophila* autosomes is thought to contain about 2000 genes that can mutate to lethality. To have a good chance of obtaining at least one mutation in each of these genes (assuming they are about equally mutable) 5,000–10,000 individual crosses are usually screened.

When the whole *Drosophila* genome was screened in this way, mutations in 1200 different embryonic lethal genes were identified. Of these, more than 1000 caused lethality with no visible pattern defects in the embryo, while about 120 caused pattern deletions of the type described above, or pattern transformations, or other visible abnormalities. With a few exceptions, several independent mutations were recovered for each of these genes. The average number of hits per gene was five. This indicates that these screens missed few, if any, genes and that those identified represent all the genes specifically required for embryonic pattern formation.

Results and conclusions

Some of the phenotypes produced fell into obvious groups. For example, mutations in *hunchback, giant* and *knirps* produced deletions of segments in a similar fashion as *Krüppel* mutations, except that different segments were affected; these genes were called 'gap genes' (Nüsslein-Volhard et al., 1984). A large class of genes affected repeated elements of the pattern along the a-p axis, deleting either every other segment (pair-rule genes), or parts of each segment (segment-polarity genes). By their phenotypes, these 'segmentation genes' can be arranged in a hierarchy. Most of them have now been cloned and their expression patterns and molecular functions analysed, and these studies have shown that both the grouping of these genes and the postulated hierarchy were justified. The genes affect the same process, the subdivision of the embryo into repeat units, and they act in a sequential manner where the products of the early acting gap genes are transcriptional regulators of the later acting pair-rule genes which in turn regulate the expression patterns of the later acting segment polarity genes.

Other groups of genes are those that affect fates along the d-v axis, homeotic genes, genes affecting the development of the nervous system etc. It also appears that in many of these cases genes that have been grouped by their phenotypes turn out to code for molecules that interact directly, or act in a cascade to direct the development of the part of the embryo affected by the mutations. Whether such grouping will make biological sense or define developmental pathways for all possible phenotypes (e.g. germband retraction) remains to be seen. Nevertheless, the collection of groups of mutants affecting similar processes can be an important step towards establishing the molecular pathway of that process.

Molecular mechanisms

This section will describe the molecular mechanisms of two aspects of *Drosophila* embryogenesis that are now understood in principle, axis determination and pattern formation along the anterior-posterior axis.

Axis determination and early pattern formation

The primary axes of the embryo – anterior-posterior and dorsal-ventral, and thereby, as a necessary consequence left-right – are already determined when the egg is fertilized, and they are clearly recognizable in the egg shape. This is in marked contrast to vertebrate eggs where either none of the axes (e.g. in mammals) or only one axis (e.g. in amphibia) is determined in the unfertilized egg, and the others are determined by fertilization or by unknown mechanisms during early embryogenesis.

Since the axes of the *Drosophila* egg are determined before fertilization and the information for the determination of the axis is genetic, this information must come from the maternal genome. Indeed, mutations interfering with the establishment of polarity are mutations in maternal genes that are active during oogenesis (Nüsslein-Volhard *et al.*, 1987). Four groups of maternal-effect genes are required for setting up the axes of the egg and are defined by the phenotypes of the members in each group. Genes in the 'posterior group' are required for the development of the abdomen of the larva; the 'anterior group' contains only one member, *bicoid*, which directs the development of the head and thorax. Mutations in the 'terminal group' affect the terminal regions of the larva, acron and telson, while the genes in the 'dorsal group' are responsible for determining all cell fates along the dorso-ventral axis.

The activity of these genes causes the asymmetric distribution of transcription factors along the a-p and d-v axes during early embryogenesis. The anterior and posterior groups act through two tightly localized RNAs in the anterior and posterior tips of the egg (*bicoid* and *nanos*, respectively; **7.49–7.52**). In contrast, terminal structures and dorso-ventral cell fates develop under the control of transmembrane receptors (*torso* and *Toll*) that are distributed evenly in the plasma membrane surrounding the egg, but are activated only in the terminal or ventral regions of the egg by localized extracellular ligands (Stein and Stevens, 1991). The asymmetric information generated by these systems is then 'interpreted' by the zygotic genome, which in the first instance means that genes in the developing embryo are activated or repressed in particular places at particular times, according to the distributions and activities of maternal gene products (Driever and Nüsslein-Volhard, 1988a, b) (**7.44–7.52**).

The first of the axis-determining systems to be understood in molecular terms was the anterior system, and this will be described in more detail here.

The gene responsible for the development of the head is *bicoid* (Frohnhöfer *et al.*, 1986) which codes for a transcription factor containing a homeodomain (Frigerio *et al.*, 1986). Embryos lacking bicoid protein develop into larvae in which the head and thorax are missing. *bicoid* RNA is synthesized in the nurse cells of the ovary during oogenesis and transported into

7.46

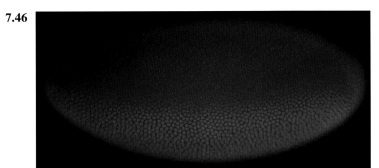

7.47

7.48

7.46–7.48 Pattern formation: dorso-ventral axis. **7.46** (×150) shows the distribution of the transcription factor *dorsal* along the dorso-ventral axis. The protein product of the maternal gene *dorsal* is initially distributed evenly throughout the egg cytoplasm, but, as nuclei populate the periphery, the protein enters ventral nuclei (red dots on ventral side), and is excluded from dorsal nuclei (red network of cytoplasmic staining on the dorsal side). In a narrow lateral zone, the protein is distributed evenly between nuclei and cytoplasm (homogeneous red staining in the middle region). **7.47** (×150) shows the RNA distribution of a zygotic gene expressed on the dorsal side of the embryo: *dpp* is transcribed in nuclei that do not contain *dorsal* protein (*dpp* RNA detected by *in situ* hybridization) **7.48** (×150) shows the expression of a zygotic gene on the ventral side of the embryo: *twist* is expressed in nuclei that contain high levels of *dorsal* protein (*twist* protein detected by antibodies). (Photograph **7.46** by Chris Rushlow.)

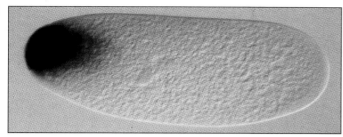

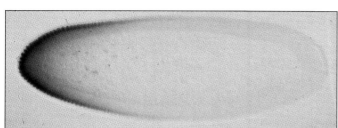

7.49–7.52 Pattern formation: anterior-posterior axis; maternal genes. **7.49** and **7.50** show *bicoid* RNA and protein; **7.51** and **7.52** show *nanos* RNA and protein. *bicoid* and *nanos* RNA (**7.49, 7.51**) are synthesized during oogenesis and localized to the anterior and posterior pole of the egg, respectively. When the egg is laid, the RNAs are translated, and the proteins diffuse away from the poles, forming gradients. The role of the proteins is to determine the expression patterns of the gap genes, including *hunchback* and *Krüppel* (**7.40–7.43**; for further details see text). (×150; photographs by Wolfgang Driever, Dominique Ferrandon and Christiane Nüsslein-Volhard (**7.49, 7.50**), Ruth Lehmann and Laura Dickinson (**7.51, 7.52**).)

the egg, where it becomes localized at the anterior tip (**7.49–7.52**). The RNA is translated after the egg is laid, and the protein begins to diffuse away from the anterior tip, forming a concentration gradient along the egg (Driever and Nüsslein-Volhard, 1988; Driever and Nüsslein-Volhard, 1988) and enters the nuclei which are proliferating in the egg. Nuclei close to the anterior tip thus receive higher concentrations of *bicoid* protein than those at more posterior positions. *Bicoid* now binds via its homeodomain to specific sites in the DNA of these nuclei, where it acts as a transcriptional regulator of zygotic genes. The trick is that it binds to different promoter sites with different affinities, and the way that it acts depends on its concentration (Driever *et al.*, 1989). Promoters containing low-affinity sites activate bicoid-dependent zygotic gene expression in a small anterior domain where high levels of *bicoid* protein are found, while high-affinity binding sites direct gene expression in a larger domain extending further towards the posterior. In this way, the expression of different zygotic genes is confined to different regions of the embryo (**7.53–7.54**). The product of one single maternal gene is thus used to divide the egg into several distinct domains in which different zygotic genes are activated and, in this way, complexity is generated and the initial maternal polarity of the egg is fixed.

Segmentation

The same principle of using a broad gradient of a transcription factor to set up expression patterns with sharper boundaries is used again at the next step of subdividing the a-p axis of the embryo, the division into double-segmental and segmental repeat units (reviewed in Hülskamp and Tautz, 1991, and Small and Levine, 1991). The genes responsible for the segmental subdivision of the embryo are the pair-rules genes which fall into two major classes. Pair-rules genes are expressed in patterns of seven stripes in alternating segments, each stripe being about one segment wide. The primary pair-rules genes *even-skipped*, *hairy* and *runt* depend only on the gap genes to be expressed in this stripe pattern, while the secondary pair-rules genes also require the function of the primary pair-rules genes and of each other to be expressed appropriately. The boundaries of the primary pair-rules stripes are set up by the transcription factors encoded by the gap genes *giant*, *hunchback*, *Krüppel* and *knirps*, which show graded distributions at the edges of their expression domains. One of the most surprising findings was that each of the stripes of the primary pair-rules genes is regulated independently, by different gap-gene products binding to specific elements in the upstream regulatory region of the gene (Riddihough and Ish-Horowicz, 1991,; Howard and Struhl, 1991). The promoters of the genes *even-skipped* and *hairy* can be divided into short stretches of DNA which separately control individual stripes. For example, a 480bp region in the promoter of *even-skipped* is necessary and sufficient to turn on *even-skipped* transcription in the area of the second stripe. This fragment contains sites to which the *giant*, *hunchback*, *Krüppel* and *bicoid* proteins bind. *Hunchback* and *bicoid* activate, while *Krüppel* and *giant* repress transcription from this region. This fragment of DNA thus allows expression of *even-skipped* in regions of the embryo where sufficiently high levels of *hunchback* and *bicoid* are present, but from which *giant and Krüppel* are absent. This is the case in precisely the region where the second *even-skipped* stripe is expressed. Similar rules operate for the other pair-rules genes (**7.53–7.56**).

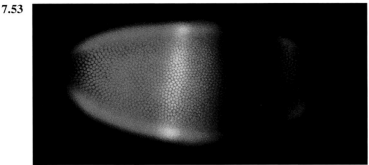
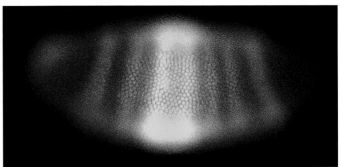
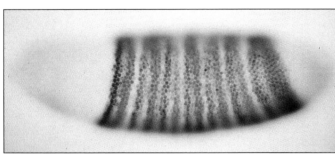
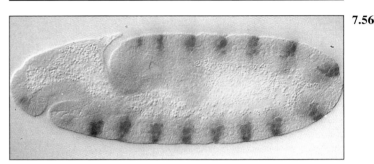

7.53–7.56 Pattern formation: anterior-posterior axis; zygotic genes. **7.53** shows an embryo stained with antibodies against the products of the gap genes *hunchback* (red) and *Krüppel* (green), two zinc-finger-containing transcription factors. Under the control of *bicoid* and *nanos*, these proteins are expressed in broad domains whose edges are not sharp, but show graded protein distributions. The posterior part of the hunchback domain overlaps the anterior part of the Krüppel domain (in the region where the green and red staining add up to yellow). These protein gradients and their overlaps determine the striped patterns of the primary pair-rule genes, including *hairy* and *even-skipped*. **7.54** is an embryo stained with antibodies against *Krüppel* (red) and the pair-rule gene product *hairy* (green) which is expressed in every other segment. Again, the overlap between the red and the green staining appears yellow. **7.55** shows expression of the pair-rule genes *even-skipped* and *fushi-tarazu* in alternating parasegments. The expression patterns of the *even-skipped* and *fushi-tarazu* proteins (transcription factors containing homeodomains) form a sharp line along the anterior border of each parasegment, the region where *engrailed* expression will be activated (**7.56**). **7.56** shows expression of *engrailed* at the extended germ-band stage. The engrailed protein (another homeodomain containing transcription factor) is expressed in the posterior compartment of each segment. (×150; photographs by Ken Howard (**7.53, 7.54**) and Peter Lawrence (**7.55, 7.56**).)

The pair-rule genes ultimately regulate the expression of the segment-polarity genes, which also interact among each other to determine and maintain their domains of expression. Most of the segment-polarity genes do not code for transcription factors, and the cell biology of their interactions and functions is not yet understood. Nor do we yet know how the their activities direct morphogenesis of each segment, e.g. how a furrow is formed at the segment boundary, how cuticle elements are arranged in the correct places and orientations etc.

Conclusion

Some basic principles of development and pattern formation have now been worked out in *Drosophila*. We understand how the primary axes of the egg are set up by maternal genes during oocyte development and how the activity of these genes is interpreted by the region-specific transcription of zygotic genes; we know how fields of similar cells in the embryos can then be divided into groups or regions with different fates by differences in the concentrations of transcription factors; and we are beginning to understand how cascades of gene activities regulate the morphogenesis and differentiation of organs.

Many of the genes that control these processes have homologues that play similar roles in other species. Homologues of homeotic genes determine regional differences along the anterior-posterior body axis in such diverse organisms as nematodes and vertebrates; *Drosophila* mesoderm-determining genes have homologues in frogs, fish and mice, which are also expressed in the early mesoderm; the segmentation genes *wingless* and *engrailed* regulate cell fate via cell-cell interactions between neighbouring groups of cells in the *Drosophila* blastoderm, and between neighbouring regions of the brain in vertebrates. These examples show that the analysis of *Drosophila* pattern formation can help to illuminate general biological principles used in building multicellular organisms.

Some general cell biological mechanisms – rearrangements of the cytoskeleton, RNA localization, or signal transduction – have been co-opted for specific developmental functions in the early egg. For example, a signal transduction cascade through the receptor tyrosine kinase *torso* and other elements also used in signal reception and transduction in vertebrate cells determines the terminal regions of the egg. Genetic studies of such processes in *Drosophila* can be supplemented by biochemical approaches, since large numbers of eggs can be obtained cheaply. In the future, therefore, *Drosophila* may well be increasingly often used to study cell biological phenomena of higher vertebrate cells, rather than just specific developmental questions.

If the general rules of pattern formation are now nearly understood, which are the developmental processes that remain to be studied? Apart from the development of the nervous system, one of the major unsolved problems is how morphogenesis is controlled. When cells have been allocated fates and specific positions in the body, how do they make three-dimensional structures and organs? Cell migration, cell shape changes and directed cell division are some of the mechanisms of morphogenesis. Some of them have been studied in vertebrate cells *in vitro*, and some of the proteins involved are known but we still do not know how they are integrated with the development of the whole organism, and which genes regulate them. Furthermore, some processes, such as the folding of epithelia, or cell intercalation, cannot easily be studied in isolation in cells separated from their environment. For such studies, the genetic approach in *Drosophila* will probably continue to provide answers that might be more difficult or impossible to obtain by other methods.

Acknowledgements

I thank Ken Howard, Peter Lawrence, Nipam Patel, François Karch, Bill Theurkauf, Bruce Edgar, Rolf Reuter, Michael Bate, Thierry Bogaert and Michael Wilcox, Manfred Frasch, Yuh Nung Jan, Marc Haenlin and José Campos Ortega, Bruce Hay, Helen Skaer, Christian Klämbt, Dominique Ferrandon, Ruth Lehmann, Chris Rushlow, Kristina Straub, and Trudi Schüpbach for contributing pictures; and Daniel St.Johnston, José Campos-Ortega and Michael Brand for comments on the manuscript.

References

Anderson, K. (1987). Dorsal-ventral embryonic pattern genes of *Drosophila. Trends Gen. ,3*, 91–97.

Ashburner, M. (1989). *Drosophila. A Laboratory Handbook.* Cold Spring Harbor Laboratory Press, New York.

Bastiani, M.J., Harrelson, A.L., Snow, P.M. and Goodman, C.S. (1987). Expression of fasciclin I and II glycoproteins on subsets of axon pathways during neuronal development in the grasshopper. *Cell ,48*, 745–755.

Bate, C.M. (1976). Embryogenesis of an insect nervous system. I. A map of the thoracic and abdominal neuroblasts in *Locusta migratoria. J. Embryol. Exp. Morphol.*, **35**, 107–123.

Bate, M., Rushton, E., and Currie, D.A. (1991). Cells with persistent *twist* expression are the embryonic precursors of adult muscles in *Drosophila. Development,* **113**, 79–90.

Bellen, H.J., D'Evelyn, D., Harvey, M. and Elledge, S.J. (1992). Isolation of temperature-sensitive diphtheria toxins in yeast and their effects on *Drosophila* cells. *Development,* **114**, 787.

Blochlinger, K., Jan, L.Y., and Jan, Y.N. (1991). Transformation of sensory organ identity by ectopic expression of *cut* in *Drosophila. Genes Dev.,* **5**, 1124–1135.

Bodmer, R. and Jan, Y.N. (1987). Morphological differentiation of the embryonic peripheral neurons in *Drosophila. Roux's Arch. Dev. Biol.*, **196**, 69–77.

Bogaert, T., Brown, N., and Wilcox, M. (1987). The Drosophila PS2 Antigen is and invertebrate integrin that, like the fibronectin receptor becomes localized to muscle attachments. *Cell*, **51**, 929–940.

Bownes, M. (1975). A photographic study of development in the living embryo of *Drosophila melanogaster. J. Embryol. Exp. Morphol.*, **33**, 789–801.

Broadie, K. and Bate, M. (1991). The development of adult muscles in *Drosophila*: ablation of identified muscle precursor cells. *Development,* **113**, 103–118.

Cabrera, C. (1992). The generation of cell diversity during early neurogenesis in *Drosophila. Development ,***115**, 893–901.

Campos-Ortega, J.A. and Haenlin, M. (1992). Regulatory signals and signal molecules in early neurogenesis of *Drosophila melanogaster. Roux's Arch. Dev. Biol.,* **201**, 1–11.

Campos-Ortega, J.A. and Hartenstein, V. (1985). *The Embryonic Development of* Drosophila melanogaster. Springer-Verlag, Berlin Heidelberg.

Campos-Ortega, J.A. and Jan, Y.N. (1991). Genetic and molecular basis of neurogenesis in *Drosophila melanogaster. Ann. Dev. Neurosci.,* **14**, 339–420.

Costa, M., Sweeton, D., and Wieschaus, E. (1992). Gastrulation in *Drosophila*: cellular mechanisms of morphogenetic movements. In *The Development of* Drosophila (Bate, M. and Martinez-Arias, A., eds) CSH Laboratory Press, New York.

Doe, C.Q. (1992). The generation of neuronal diversity in the *Drosophila* embryonic central nervous system. In *Determination of Neuronal Identity* (Shankland, M. and Macagno, E.) Academic Press, New York.

Dohrmann, C., Azpiazu, N., and Frasch, M. (1990). A new *Drosophila* homeobox gene is expressed in mesodermal precursor cells of distinct muscles during embryogenesis. *Genes Dev.,* **4**, 2098–2111.

Driever, W. and Nüsslein-Volhard, C. (1988a). The bicoid protein determines position in the *Drosophila* embryo in a concentration-dependent manner. *Cell,* **54**, 95–104.

Driever, W. and Nüsslein-Volhard, C. (1988b). A gradient of bicoid protein in *Drosophila* embryos. *Cell,* **54**, 83–93.

Driever, W., Thoma, G. and Nüsslein-Volhard, C. (1989). Determination of spatial domains of zygotic gene expression in the *Drosophila* embryo by the affinity of binding sites for the bicoid morphogen. *Nature,* **19**, 363–367.

Ephrussi, A., Dickinson, L.K., and Lehmann, R. (1991). *oskar* localizes the germ plasm and directs posterior localization of the posterior determinant *nanos. Cell,* **66**, 37–50.

Foe, V. and Alberts, B. (1983). Studies of nuclear and cytoplasmic behavior during the five mitotic cycles that precede gastrulation in *Drosophila*. *J. Cell Sci.*, **61**, 31–70.

Foe, V.E. (1989). Mitotic domains reveal early commitment of cells in *Drosophila* embryos. *Development*, **107**, 1–22.

Frigerio, G., Burri, M., Bopp, D., Baumgartner, S., and Noll, M. (1986). Structure of the segmentation gene *paired* and the *Drosophila* PRD gene set as part of a gene network. *Cell*, **47**, 735–746.

Frohnhöfer, H.G., Lehmann, R., and Nüsslein-Volhard, C. (1986). Organisation of anterior pattern in the *Drosophila* embryo by the maternal gene *bicoid*. *Nature*, **324**, 169–179.

Golic, K.G. and Lindquist, S. (1989). The FLP recombinase of yeast catalyzes site-specific recombination in the *Drosophila* genome. *Cell*, **59**, 499–509.

Hay, B., Ackermann, L., Barbel, S., Jan, L.Y., and Jan, Y.N. (1988). Identification of a component of *Drosophila* polar granules. *Development*, **103**, 625–640.

Howard, K.R. and Struhl, G. (1991). Decoding positional information – regulation of the pair-rule gene hairy. *Development*, **110**, 1223–1231.

Hülskamp, M. and Tautz, D. (1991). Gap genes and gradients – the logic behind the gaps. *Bioessays* **13**, 261–268.

Illmensee, K. and Mahowald, A.P. (1974). Transplantation of posterior pole plasm in *Drosophila*: induction of germ cells at the anterior pole of the egg. *Proc. Natl. Acad. Sci.* **71**, 1016–1020.

Immerglück, K., Lawrence, P., and Bienz, M. (1990). Induction across germ layers in *Drosophila* mediated by a genetic cascade. *Cell*, **62**, 261–268.

Jan, L.Y. and Jan, Y.N. (1982). Antibodies to horseradish peroxidase as specific neuronal markers in *Drosophila* and grasshopper embryos. *Proc. Natl. Acad. Sci.*, **70**, 2700–2704.

Jürgens, G., Wieschaus, E., Nüsslein-Volhard, C., and Kluding, H. (1984). Mutations affecting the pattern of the larval cuticle in *Drosophila melanogaster*. II. Zygotic loci on the third chromosome. *Roux Arch. Dev. Biol.*, **196**, 141–157.

Karch, F., Bender, W., and Weiffenbach, B. (1990). abdA expression in *Drosophila* embryos. *Genes and Development*, **4**, 1573–1587.

Karr, T.L. and Alberts, B.M. (1986). Organization of the Cytoskeleton in early *Drosophila* embryos. *J.Cell Biol.*, **102**, 1494–1509.

Klämbt, C. and Jacobs, J.R. (1991). The midline of the *Drosophila* central nervous system: A model for the genetic analysis of cell fate, cell migration and growth cone guidance. *Cell*, **64**, 801–815.

Lawrence, P. (1992). *The Making of a Fly*. Blackwell Scientific Publications, Oxford.

Lawrence, P.A. and Johnston, P. (1989). Pattern formation in the *Drosophila* embryo: allocation of cells to parasegments by *even-skipped* and *fushi-tarazu*. *Development*, **105**, 761–768.

Leptin, M., Bogaert, T., Lehmann, R., and Wilcox, M. (1989). The function of PS integrins during *Drosophila* embryogenesis. *Cell*, **56**, 401–408.

Leptin, M., Casal, J., Grunewald, B., and Reuter, R. (1992) Mechanisms of early *Drosophila* mesoderm formation. *Development*, Supplement, 23–31.

Martinez-Arias, A. and Lawrence, P. (1985). Parasegments and compartments in the *Drosophila* embryo. *Nature*, **313**, 639–642.

Minden, J.S., Agard, D.A., Sedat, J.W., and Alberts, B. (1989). Direct cell lineage analysis in *Drosophila melanogaster* by time-lapse, three-dimensional optical microscopy of living embryos. *J. Cell Biol.* **109**, 505–516.

Moffat, K.G., Gould, J.H., Smith, H.K., and O'Kane, C.J. (1992). Inducible cell ablation in *Drosophila* by cold-sensitive ricin A chain. *Development*, **114**, 681–687.

Nüsslein-Volhard, C., Frohnhöfer, H.G., and Lehmann, R. (1987). Determination of Anteroposterior Polarity in Drosophila. *Science*, **238**, 1675–1681.

Nüsslein-Volhard, C., Lohs-Schardin, M., Sander, K., and Cremer, C. (1980). A dorso-ventral shift of embryonic primordia in a new maternal-effect mutant of *Drosophila*. *Nature*, **283**, 474–476.

Nüsslein-Volhard, C., Wieschaus, E., and Kluding, H. (1984). Mutations affecting the pattern of the larval cuticle in *Drosophila melanogaster*. I. Zygotic loci on the second chromosome. *Roux Arch. Dev. Biol.*, **193**, 267–282.

Patel, N., Ball, E.E., and Goodmann, C.S. (1992). Changing role of even-skipped during the evolution of insect pattern formation. *Nature*, **357**, 339–342.

Patel, N.H., Snow, P.M., and Goodman, C.S. (1987). Characterisation and cloning of fasciclin III: a glycoprotein expressed on a subset of neurons and axon pathways in *Drosophila*. *Cell*, **48**, 975–988.

Reuter, R. and Scott, M.P. (1990). Expression and function of the homeotic genes *Antennapedia* and *Sex combs reduced* in the embryonic midgut of *Drosophila*. *Development*, **109**, 289–303.

Riddihough, G. and Ish-Horowicz, D. (1991). Individual stripe regulatory elements in the *Drosophila* hairy promoter respond to maternal, gap, and pair-rule genes. *Genes Dev.*, **5**, 840–854.

Rubin, G.M. and Spradling, A.C. (1982). Genetic transformation of *Drosophila* with transposable element vectors. *Science*, **218**, 348–353.

Rushlow, C.A., Han, K., Manley, J.L., and Levine, M. (1989). The graded distribution of the dorsal morphogen is initiated by selective nuclear transport in *Drosophila*. *Cell*, **59**, 1165–1177.

Sink, H. and Whitington, P. (1991). Pathfinding in the central nervous system and periphery by identified embryonic *Drosophila* motor axons. *Development*, **112**, 307–316.

Skaer, H. (1989). Cell division in Malpighian tubule development in *D. melanogaster* is regulated by a single tip cell. *Nature*, **342**, 566–569.

Slack, J.M.W. (1991). *From Egg to Embryo*. Cambridge University Press, Cambridge.

Small, S. and Levine, M. (1991). The initiation of pair-rule stripes in the Drosophila blastoderm. *Current Opinion in Genetics and Development*, **1**, 255–260.

Spradling, A.C. and Rubin, G.M. (1982). *Science*, **218**, 341–348.

St Johnston, D. and Nüsslein-Volhard, C. (1992). The origin of pattern and polarity in the *Drosophila* embryo. *Cell*, **68**, 201–219.

Stein, D.S. and Stevens, L.M. (1991). Establishment of dorsal-ventral and terminal pattern in the *Drosophila* embryo. *Current Opinion in Genetics and Development* **1**, 247–254.

Thomas, J.B., Bastiani, M.J., Bate, M., and Goodman, C.S. (1984). From grasshopper to *Drosophila*: a common plan for neuronal development. *Nature*, **310**, 203–207.

Wang, C. and Lehmann, R. (1991). *nanos* is the localized posterior determinant in *Drosophila*. *Cell*, **66**, 637–647.

Wieschaus, E. and Noell, E. (1986). Specificity of embryonic lethal mutations in *Drosophila* analyzed in germ line clones. *Roux's Arch. Dev. Biol.*, **195**, 63–73.

Wieschaus, E. and Nüsslein-Volhard, C. (1986). Looking at embryos. In Drosophila. *A practical approach*, (D. B. Roberts), IRL Press, Oxford

Wieschaus, E., Nüsslein-Volhard, C., and Jürgens, G. (1984). Mutations affecting the pattern of the larval cuticle in *Drosophila melanogaster*. I. Zygotic loci on the X chromosome and the fourth chromosome. *Roux Arch. Dev. Biol.*, **193**, 267–282.

Zalokar, M. and Erk, I. (1976). Division and migration of nuclei during early embryogenesis of *Drosophila melanogaster*. *J. microsc. Biol. Cell*, **25**, 97–106.

8. The Zebrafish

Walter K. Metcalfe

Introduction

The zebrafish (*Brachydanio rerio*) is a tropical freshwater cyprinid, closely related to goldfish, that is particularly well-suited to developmental and genetic analyses. Zebrafish are hardy and easy to care for in the laboratory, they readily produce many optically clear embryos which develop rapidly, and methods have been developed that facilitate genetic analyses, particularly screening for mutations among F1 offspring. There is thus considerable interest in the zebrafish as a model system for studying cellular, molecular, and genetic mechanisms of vertebrate development.

The first description of the zebrafish comes from Francis Hamilton's studies of the fishes of the Ganges river in 1822 when he wrote of "this beautiful fish I found in the Kosi river, where it grows to about two inches in length". These small schooling fish, which are available at pet stores throughout the world, are a favourite of many fish hobbyists who know them as the zebra danio. They first became common animals in embryological laboratories in the 1930s, when Creaser (1934) described simple methods for raising zebrafish to obtain eggs all year.

More recently, George Streisinger sought a model system in which to dissect neuronal development by the use of genetic strains and chose the zebrafish as a good model system for several reasons; the generation time is only about three months, large numbers of embryos are easily obtained which develop synchronously outside the mother, and the fish are small, hardy and easy to care for. Further, because of their rapid development and small size, large-scale screening for important developmental mutants, including many lethal mutants, is possible. Streisinger did, however, recognise the difficulty that heterozygosity posed to vertebrate developmental studies and therefore gave simple methods for producing homozygotes (Streisinger *et al.*, 1981).

At about the same time, Kimmel and co-workers began their studies of the development of identified neurons (Kimmel, 1982) and cell lineage (Kimmel and Law, 1985) in zebrafish embryos. Such studies are possible because zebrafish embryos are optically clear and develop very rapidly so that, within 24hr of fertilization, they look like fish, and their nervous and muscular systems are beginning to function, as evidenced by their response to touch. This rapid development offers an important advantage for experimental analysis of such features as pattern formation, cell movements and cell interactions, since, because of the exceptional clarity of the embryos, these events can be observed in real time throughout their development, even deep within the embryo. This combination of simple genetic methods and a detailed understanding of many features of early embryonic development in zebrafish have opened new possibilities for understanding how vertebrate genes function in early development.

In this chapter, the normal development of the embryo will be reviewed, some of the important genetic and molecular methods that have been developed will be described, and examples of experimental studies will be presented to illustrate how some of these methods have been used to help understand fundamental features of vertebrate development.

Obtaining embryos

Zebrafish are hardy animals, resistant to most diseases, and easy to care for and breed in the laboratory. A 45-litre aquarium is sufficient to hold about twenty fish which may produce 100 or more eggs daily. The fish breed shortly after the onset of daylight and this may be manipulated by the experimenter to provide eggs at convenient times. Since fertilization is external, the newly spawned eggs are available to the experimenter (**8.1**). Alternatively, large numbers (more than 1000) of synchronously developing eggs may be produced by *in vitro* fertilization and, as eggs are small, these numbers may be needed for isolating or analysing molecules that are developmentally regulated or are present in small quantities.

Methods for the care and use of zebrafish in the laboratory, along with procedures and recipes for experimental studies, have been compiled by Westerfield (1989). This manual, called *The Zebrafish Book*, is updated frequently by contributions from investigators around the world who use zebrafish in their laboratories, and is a valuable resource for researchers.

8.1 Zebrafish spawning in a tank. A female releases eggs (lower right) after being chased by several males. Marbles, visible in the bottom of the tank, are used to protect the eggs from being eaten by the fish.

Normal development

Staging series

Among staging series that have previously been published are those of Hisoaka and Battle (1958) and Kimmel and Westerfield (in Westerfield, 1989). Generally, the earliest stages are named by the number of cells in the blastula, gastrula stages by the advancement of the blastoderm over the yolk cell, stages through the formation of the embryonic axis by the number of developed somites, and some later stages by the advancement of the migrating primordium of the posterior lateral line (explained in more detail below). Developmental time is generally expressed as 'hr', referring to the number of hours after fertilization at 28.5°C.

The small eggs, approximately 1mm in diameter, are covered by a transparent and relatively soft chorion (**8.2**). For detailed observations of the beautifully transparent embryo, it is necessary to remove the chorion with fine forceps, although some observations can be made directly through this membrane. For accurate staging, especially of later stages, it is best to examine dechorionated embryos using a compound microscope with differential interference optics.

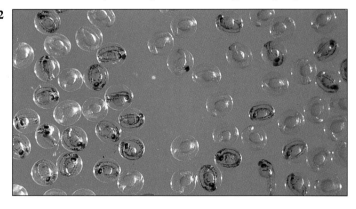

8.2 Wild-type and *golden* (*gol*) zebrafish embryos (×6). (Photograph by H. Howard, from Streisinger *et al.*, 1981.)

The egg and early cleavage stages

The cytoplasm of the unfertilized egg is initially distributed throughout the bulk of the yolk, but, following fertilization, it streams towards one pole of the egg, raising the blastodisc of the one-cell egg to form a cap on the surface of the yolk (**8.3**) and the egg is thus typically telolecithal. The first cleavage occurs approximately 40min after fertilization, and subsequent cleavages occur synchronously at 15min intervals (**8.4**) up to the 1024 cell stage, about 190min after fertilization. The 1024 cell stage marks the midblastula transition and is characterized by a lengthening of the cell cycle, nonsynchronous cell divisions, and new RNA synthesis (Kane and Kimmel, 1992).

The first five cleavages are vertical and incomplete, for, these first blastomeres are connected by a common basal cytoplasm overlying the uncleaved yolk. This cytoplasmic continuity means that one may micro-inject substances into the cytoplasm at the base of the blastoderm, and the injected material will then diffuse into each cell of the developing embryo (**8.5**). Such 'yolk injections' result in the complete labelling of the entire embryo, and have proven to be particularly useful for injecting rhodamine-dextran or fluorescein-dextran to prelabel donor cells for use in cell transplantation studies (Ho and Kane, 1990). Similarly, foreign DNA or RNA may be introduced intracellularly in this manner in order, for example, to study control of gene expression during development (Stuart *et al.*, 1988).

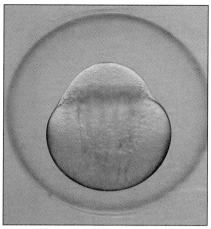
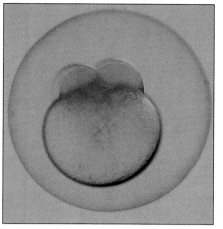

8.3 The one-celled egg, 30min after fertilization (×45).

8.4 The four-celled egg, 1hr after fertilization (1hr, ×45).

8.5 A micro-electrode is used to inject substances into early cleavage stage eggs. Here, a one-celled egg is being injected with an electrode containing dye (×45). (Photograph courtesy of Monte Westerfield.)

The blastula

Once the blasodisk contains about 2000 cells, the blastula undergoes a characteristic series of shape changes. At about 3.5hr, the blastoderm is perched high upon the yolk cell (**8.6**), but at 3.7hr the blastoderm has flattened out, so that the egg is spherical again (**8.7**). At these times, the interface between the yolk cell and blastoderm is rather flat, but by 4.3hr the yolk bulges into the blastoderm which begins to spread over the yolk cell (epiboly, **8.8**).

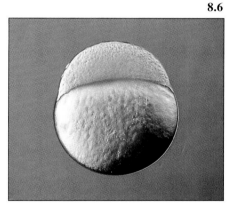
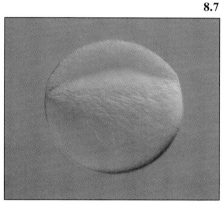
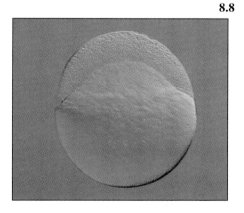

8.6 The high blastula (3.5hr, ×45).

8.7 Blastula at the sphere stage (4hr, ×45).

8.8 Blastula at a late dome stage, 40% epiboly (5hr, ×45).

Gastrulation

Once the blastoderm has spread over half of the uncleaved yolk cell (approximately 5.25hr), gastrulation begins. At its onset, the blastoderm is composed of an outer epithelial layer of cells called the enveloping layer and several layers of deep cells which undergo characteristic gastrulatory movements as epiboly continues to bring the blastoderm down over the surface of the yolk cell. In particular, deep cells near the blastoderm margin involute; that is, they approach the margin, turn under, and then migrate anteriorwards to form a deeper layer. Involution thus forms two layers of deep cells: the non-involuted epiblast cells and the deeper involuted cells of the hypoblast which together form a thickening, called the germ ring, around the circumference of the blastoderm. The hypoblast will later form mesoderm and endoderm, whereas the epiblast will give ectodermal derivatives (Warga and Kimmel, 1990). As involution and epiboly continue, deep cells converge towards the dorsal side of the gastrula and intercalate to form the embryonic axis and these movements continue until the blastoderm envelopes the entire yolk cell (about 9hr).

Somitogenesis

Shortly after closure of the yolk plug (10hr), the notochord becomes visible as a solid rod (**8.9**) while the nervous system is formed above it, developing as a solid neuro-epithelium, unlike many other vertebrates in which the neural tube forms from the closure of neural folds. The first somitic furrow also appears at about 10hr (**8.9**) and new somites will subsequently form caudally at 30min intervals until the final compliment of about 30 somites ispresent at 24hr (**8.10**).

During this period of somitogenesis, from 10–24hr, many of the major body parts undergo rapid development. At the 5-somite stage, the optic vesicles become visible as lateral expansions of the brain, while by 20 somites the optic cup has formed and the lens is clearly visible. At the 5-somite stage, the notochord extends caudally from the hindbrain into the caudal undifferentiated mass, and by 20 somites it consists of large vacuolated cells that extend to the tip of the tail. At the 15-somite stage, the otic placode has begun to invaginate, and by 25 somites otoliths are clearly visible within the otic vesicles. By about 17hr (19 somites), the nervous and muscular systems have begun to function as evidenced by spontaneous side-to-side body contractions that subside by 24hr.

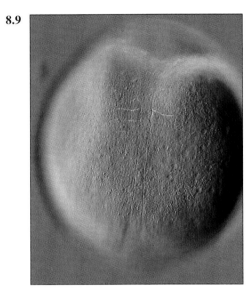

8.9 The first three somitic furrows are visible on either side of the notochord; dorsal view with the animal pole at the top (11hr, ×70).

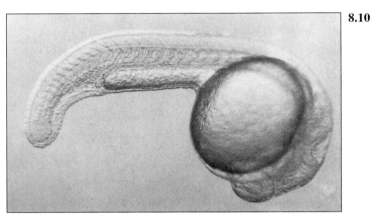

8.10 The zebrafish embryo one day after fertilization (24hr, ×40).

Later stages

Defining developmental stages according to the number of formed somites is only useful until 24hr when somitogenesis ends; other criteria are then needed for the later stages of development. The system that we have chosen for this is the location of the migrating primordium which later differentiates to form the posterior lateral line, a sensory system found in fish and aquatic amphibians that consists of groups of sensory hair cells within the skin, termed *neuromasts*, and that functions to detect water movements at the surface of the animal. The primordium originates as a postotic placode that begins a caudal migration within the epidermis at about the 18-somite stage and reaches the end of the tail at about 40hr (Metcalfe, 1985). The primordium migrates at a linear rate of about 2 somites/hr directly over the horizontal myoseptum of the muscles at the apex of the myotomal chevrons (**8.11**). In order to visualize the primordium in live embryos, careful

8.11 Scanning electron micrograph of the migrating primordium of the lateral line. The primordium (arrow) is a compact mass of cells that migrates within the skin, directly over the apex of the underlying myotomes which have been exposed by peeling back the skin (×400).

observation using a compound microscope with differential interference contrast optics is necessary.

The embryos hatch in about three days at 28.5°C, and reach maturity in approximately three months, averaging about 4–5cm in length (**8.12**).

8.12 Adult zebrafish. (Photograph from Carloina Biological Supply.)

Genetics

Analysis of mutants is a key method for learning about the roles that particular genes and gene products play in development. Although zebrafish have 50 diploid chromosomes, a variety of methods have been developed that make the zebrafish an attractive system in which to study the roles of genes in development.

Making mutant fish

Many interesting mutant strains have been recovered from wild-type stock (**8.13**) and these include pigmentation mutants such as *brass* (**8.14**), *albino* (**8.15**), *sparse* (**8.16**) and *golden* (**8.17**) as well as nonmotile mutants, such as *spadetail* (**8.18**). In addition, gamma irradiation and chemical and insertional mutagenesis have been used to make mutants.

Gamma-rays are thought to produce both chromosomal rearrangements and genetic deletions which can be quite large and it is of particular interest that gamma-rays applied during cleavage stages can produce mutations in pregonial cells (the precursors of germ cells in the cleaving egg). Calculations suggest that about five pregonial cells are present in mid-cleavage stages and, in one study, 5–50% of the progeny of a single mutagenized female exhibited the same mutant

8.13–8.17 Adult female pigment mutants: wild-type, **8.13**; *brass* (*brs*), **8.14**; *albino* (*alb*), **8.15**; *sparse* (*spa*), **8.16**; and *golden* (*gol*), **8.17**. (Photographs by Harrison Howard for George Streisinger.)

phenotype (Walker and Streisinger, 1983). In contrast, chemical mutagens such as ethyl-nitroso-urea usually cause point mutations and these have the advantage that, unlike gamma rays, the resulting phenotype is likely to have risen from a specific locus.

Insertional mutagenesis derives from injecting foreign DNA into the egg (Stuart *et al.*, 1988). When such DNA is incorporated into the host DNA, a transgenic fish is created and the injected DNA can later be detected if it includes a reporter gene, such as β-galactosidase (β-gal. **8.19**). If the DNA insert disrupts a functional gene in the host, then so-called insertional mutants result. The advantage of this system is that it is possible to recover the injected DNA together with some of its surrounding genetic material and, in such cases, the mutated host gene can be identified. In this way it is possible to correlate developmental mutant phenotypes with particular genes.

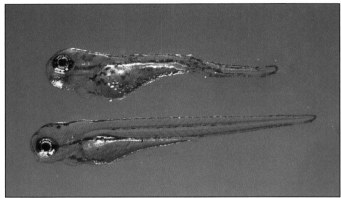

8.18 *Spadetail* (*spt*) and wild-type embryos two days after fertilization. Embryos containing lethal mutations often live long enough to study their early development (×20).

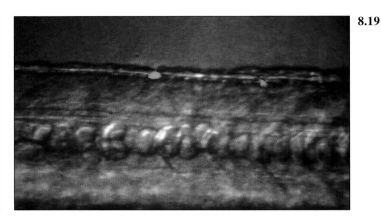

8.19 Two Rohon-Beard neurons (red spots) are visualized in the dorsal spinal cord of a live transgenic zebrafish embryo using a fluorescent substrate to β-gal. The human HOX-3.3 promoter was on the same injected DNA fragment and directed expression of the β-gal reporter gene (×360). (From Westerfield *et al.*, 1992.)

Mutant screens

Many of the interesting zebrafish mutations that have been studied so far have turned out to be recessive lethal mutations. Heterozygosity obviously complicates the analysis of such mutants, but simple methods have been developed that permit production of homozygous offspring directly from females that carry mutations (Streisinger *et al.*, 1981) and the simplest of these involves the production of haploid embryos. Such embryos are made by activating eggs of mutagenized females with sperm rendered genetically impotent by ultraviolet irradiation and thus directly reflect the genotype of the mother. As haploid embryos develop very much like diploid ones until hatching stages, the detection and characterization of developmentally important mutations, including recessive lethal mutations, are possible if the mutant phenotype is expressed in the embryo. The female fish that provided the eggs may be maintained and crossed to wild-type stock to perpetuate the mutation, while further characterization of the mutation in diploid embryos that are homozygous for the mutation is possible by intercrossing and screening the heterozygous stock.

As there can be some variablility in the development of haploid embryos, it is often preferable to screen homozygous diploid embryos for subtle mutant phenotypes. Such embryos may be produced by activating eggs with ultraviolet-treated sperm as described above, and preventing the first mitotic division by applying heat shock (at 41.4°C for 2min, beginning 13min after fertilization). Alternatively, hydrostatic pressure can be used to inhibit the second meiotic division (5min at 8000psi, beginning 1.4min after fertilization). The embryos resulting from pressure treatment will be diploid, with all of their genes derived from the mother, but they will not be homozygous because of recombination during the first meiotic division. The latter fact is useful, however, since the frequency of appearance of any trait reflects the probability of recombination and this frequency may be used to calculate the gene-centromere distance (Streisinger *et al.*, 1986).

Zebrafish embryos usually live to several days even when they are homozygous for lethal mutations and it is therefore possible to study the effects of lethal mutations on early developmental processes.

Experimental studies

The easy availability of many embryos together with their rapid development and optical clarity have enabled researchers to use a wide variety of experimental methods in their studies of the development of the zebrafish. Here, we present recent examples of such methods, including cellular, molecular, and genetic analyses, that provide some insight into the breadth of studies that have been undertaken. It will often be apparent that methods which were pioneered in other developing organisms have been readily adapted and applied to studies of zebrafish development.

Studies of identified neurons

In experimental embryology, one often wishes to determine the effects of some manipulation on subsequent development. Such studies are greatly facilitated by a detailed understanding of normal development, particularly of identified cells. In the case of the zebrafish, we are fortunate that its embryonic nervous system is extremely simple and so lends itself to such analysis. There are relatively few differentiated neurons, and their positions and connections appear to be almost invariant from one individual to the next.

Within 24hr of fertilisation, a set of interconnecting and early-developing neurons have been formed with large somata and long axons. They are known as 'primary' neurons and fall into three functional groups: first, there are sensory neurons which include touch-sensitive trigeminal neurons of the head and Rohon-Beard neurons of the spinal cord (**8.20**), as well as vibration-sensitive lateral line (**8.21**) and auditory neurons; the second group includes interneurons, such as reticulospinal neurons of the hindbrain (**8.22**), and spinal interneurons; and the last are motoneurons, particularly the primary motoneurons of the spinal cord (**8.23**). These primary neurons interconnect to form an early, functional nervous system that mediates the behaviour of the embryo, including motor response to touch and vibratory stimuli (Kimmel and Westerfield, 1990).

Within the hindbrain and spinal cord, this early primary nervous system is segmentally organized as hindbrain reticulospinal neurons, which are interneurons of the hindbrain that project axons into the spinal cord, show repeating patterns of identified cell types (Metcalfe, 1985). Detailed analysis of zebrafish larvae has shown that many of these cells can be individually identified on the basis of their position, morphology, and axonal projection. There are some 50 reticulospinal neurons on each side of the hindbrain in seven bilateral clusters (**8.24**), and among these 27 different types are recognized, of which 19 are present as single, identified neurons. Analysis of the morphology of individual neurons showed repeating patterns that provided evidence for the notion that the brain is segmentally organized.

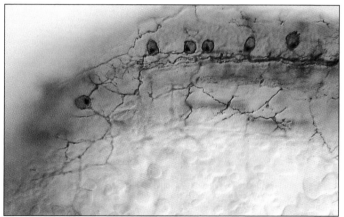

8.20 Rohon-Beard neurons of the spinal cord, labelled with the monoclonal antibody zn-12 (Metcalfe *et al.*, 1990). These neurons are thought to be a transient population of sensory neurons that mediate the early tactile sense of the trunk and tail of the embryo. This function is later served by dorsal root sensory neurons (×360).

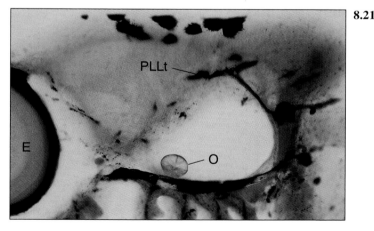

8.21 Lateral view of HRP-labelled lateral-line nerve. This 50μm parasagittal section was near the surface of the left side of the fish. The posterior lateral line nerve can be seen entering the posterior lateral line ganglion, located at the caudal margin of the ear. The central projection enters the brain and bifurcates, with each end terminating in a single column of dorsolateral neuropil (**PLLt**). An otolith (**O**) is visible in the ear cavity. At the far left is a section through the eye (**E**). Several dark pigment cells (melanocytes) are also present in this five-day-old larva. Rostral is to the left, dorsal to the top (×180). (Adapted from Metcalfe *et al.*, 1985.)

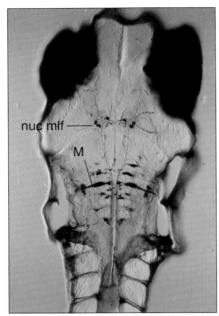

8.22 Dorsal view of HRP-labelled reticulospinal neurons of the midbrain and hindbrain in a five-day-old larva. The rostral cell group is the midbrain nucleus of the medial longitudinal fasciculus (**nuc mlf**); caudally, several hindbrain reticulospinal neurons can be seen. Many of the ventral reticulospinal neurons are not present in this 50μm horizontal section. The largest of the hindbrain neurons are the Mauthner cells (**M**), which have dendrites that terminate laterally in the region of laterally placed vestibulospinal neurons (×120).

8.23 Lateral view of primary motoneurons of the spinal cord, labelled with the monoclonal antibody zn-1. This section passes through the spinal cord at the top and trunk muscle at the bottom. A segmentally repeating pattern of motoneurons and the ventral roots that they form is evident in this 50μm parasagittal section. The primary motoneurons are ventral and darkly stained; other spinal neurons are also labelled. Rostral is to the left and dorsal to the top (×510).

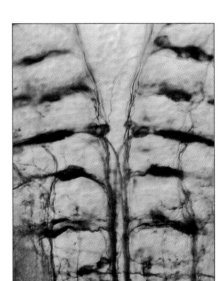

8.24 Hindbrain reticulospinal neurons are segmentally organized. The segmental patterning of the seven bilateral clusters of reticulospinal neurons is evident in this 50μm horizontal section through the hindbrain of a three-day-old larva. Each cluster lies in the centre of a hindbrain segment, or neuromere. The giant Mauthner cells are visible in the fourth hindbrain neuromere (×340). (Photograph of a section prepared by Paul Z. Myers.)

Primary motoneurons of the spinal cord are also present in a segmentally repeating pattern of hemisegments, with each containing three individually identifiable motoneurons, each of which in turn innervates specific non-overlapping regions of the body musculature (**8.25**, **8.26**). The optical clarity and rapid development of the zebrafish embryo enabled Myers, Eisen and Westerfield (1986) to label primary motoneurons in the living embryo with rhodamine and observe that their axons grew out of the spinal cord along stereotyped pathways directly to their appropriate muscle targets. Such analyses have established this as a model system for studies of the determination of cell identity, as well as for mechanisms that guide growing axons to their specific targets.

 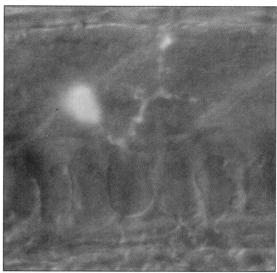

8.25, 8.26 Lateral view of labelled identified primary motoneurons in live zebrafish embryos at 19hr. The first spinal motoneuron to extend its axon out of the spinal cord is called CaP (**8.25**), shown here labelled by intracellular iontophoresis of lucifer yellow. The CaP axon has exited the spinal cord, pioneering the ventral root, and has grown over the lateral surface of the notochord (visible in the centre of the field) and on to the ventral trunk muscle in this segment. The next motoneuron to grow an axon out of the spinal cord, called MiP (**8.26**), has followed the CaP axon out of the spinal cord, then made a cell-specific turn dorsally to innervate the dorsal muscle field of the same segment. A third motoneuron (RoP, not shown) innervates intermediate muscle (×1200). (Photograph courtesy of Judith Eisen.)

Cell lineage analysis

The role of cell lineage in determining cell fate has been investigated in zebrafish (reviewed in Kimmel and Warga, 1988) using techniques applied in similar studies of the leech (Weisblat, this volume). In a typical experiment, a single cell of the embryo is micro-injected with a lineage tracer dye, such as rhodamine-dextran. Often, labelled cells can be followed during development in the live embryo, so allowing the movements and cell division history of the clone of related cells to be recorded. After further development, a labelled clone of cells derived from the injected cell can be identified and the positions and fates of these cells then analysed.

Because of cell re-arrangements, the results of such lineage studies depend critically upon the time of development at which the clone is founded. A clone of cells initiated in the early blastula stay together through several rounds of cell divisions, but, because cells of the deep layers intercalate among the more superficial blastomeres when the blastula begins to thin, clonally related cells become scattered broadly among unlabelled cells as epiboly commences (**8.27**). By the end of gastrulation (10hr), labelled cells are scattered throughout the embryo and by 24hr they are found in a variety of tissues. It is thus clear that cells related by lineages founded in the early blastula do not share a common developmental fate. This conclusion is further supported by labelling equivalent blastomeres of 64-cell stage embryos whose descendants take up widely differing patterns of cell positions and fates.

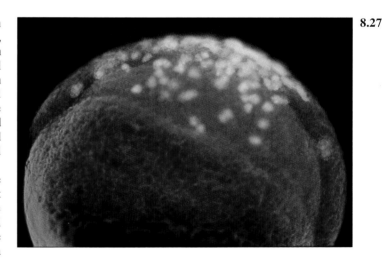

8.27 A labelled clone of cells founded during blastula stages disperses among unlabelled cells during epiboly (×130). (Reprinted with permisssion from Warga and Kimmel, 1990.)

Unlike clones founded in the early blastula, those initiated after late blastula stages become restricted to specific tissue types (e.g. muscle, neural tissue) even though they may be widely distributed throughout the embryo (**8.28**). This important finding is subject to variable interpretations, however, as the position of a cell in the gastrula may also predict cell fate. Indeed, a fate map of the early gastrula (Kimmel *et al.*, 1990) reveals that tissue primordia are distributed in the zebrafish gastrula in a pattern that is similar to the gastrula fate map of other vertebrates such as *Xenopus* (**8.29**). The contributions of cell lineage and position in determining cell fate are not yet understood, nor is the nature of the signals that decide cell fate. These are areas of current research.

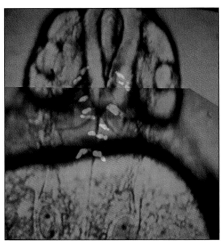

8.28 Dorsal view of a live embryo 24hr after injection of lineage tracer dye. A cell in the late blastula was labelled and generated a clone of cells that was restricted to neural tissue of the floor of the midbrain and hindbrain. Note the tendency to form bilateral cell groups. Composite enhanced video image using a light-amplifying camera (SIT camera, ×100). (From Kimmel and Warga, 1988.)

8.29 The zebrafish fate map (left) at the onset of gastrulation is very similar to that of *Xenopus* (right). In the zebrafish embryo, the shadings show where regions of deep-layer cells give rise to ectoderm (top layer), mesoderm and ectoderm (bottom layer) and it should be noted that there is some overlap here. Although labelling cells at a given latitude and longitude in the gastrula results in predictable cell fates, Kimmel *et al.* (1990) do not give the boudaries of the tissue regions as these are not clearly defined. **AP** and **VP** are the animal and vegetal poles, while **V** and **D** represent the ventral and dorsal poles of the embryos. (Redrawn from Kimmel *et al.*, 1990.)

Mutational analysis of embryonic development

A major goal of developmental biology is to characterize the genetic, molecular, and cellular interactions that give rise to the formed structures of the embryo. One of the most important tools in this endeavour is the availability of mutations that disrupt development. Loss of function often reveals important clues as to the roles of particular gene products in development. Further studies, such as cell transplantation studies between wild-type and mutant strains, which provide genetic mosaics, can be used to test theories of developmental mechanisms. Here, we provide several examples of this experimental approach.

One example utilising the mutational approach in zebrafish is the analysis of the *spadetail* (*spt*) mutant, a recessive lethal mutation which was recovered from wild-type stock. The mutant phenotype is characterized by a lack of trunk somitic mesoderm and an abnormal bump at the end of the tail in the embryo. Kimmel *et al.* (1989) labelled individual blastomeres in the mutant and observed that cells moved abnormally during gastrulation. In wild-type gastrulae, cells that are fated to become trunk muscle normally undergo involution and converge towards the dorsal side of the embryo. In *spadetail* embryos, cells that occupy similar positions in the gastrula fate map fail to converge during gastrulation and come to lie at the vegetal pole in the position of the forming tail. This mass then forms the bump at the tail, and the deficiency in trunk muscle results from the lack of cells moving into the appropriate areas during gastrulation.

The analysis of the mutant raises the question of whether the cell migration deficiency in *spadetail* embryos is due to an absence of guidance information within the gastrula, or if the deficiency is cell-autonomous, that is, residing in an inability of the mutant cells to respond to a normal environment. Ho and Kane (1990) have answered this question by simultaneously transplanting labelled wild-type and mutant cells into wild-type hosts. They showed that wild-type cells migrated normally and formed muscle in the host, whereas the mutant cells failed to converge, even though placed into a wild-type environment (**8.30–8.32**). These experiments established that *spadetail* cells are cell autonomously deficient in their ability to converge during gastrulation.

Another example that combines the use of mutational analysis and cell transplantation involves 'rescue' experiments. The zebrafish mutant *cyclops* (*cyc*) has been shown to be deficient in the ventralmost cells of the spinal cord, called floor plate cells (Hatta *et al.*, 1991). One result of this deficiency is cyclopia, since the two eyes become fused (**8.33**). Transplantation studies revealed an unsuspected 'homeogenetic' cell interaction among the floorplate cells. Wild-type cells transplanted into *cyclops* mutants in the early gastrula floorplate rescued the mutant phenotype by forming normal floorplate cells and also induced neighbouring mutant cells to become floorplate. The wild-type cells were thus able to direct the developmental fate of mutant cells.

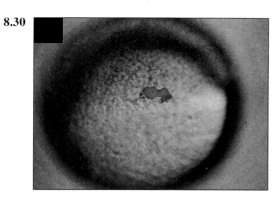

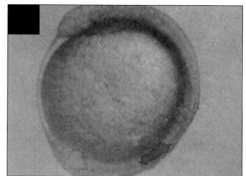

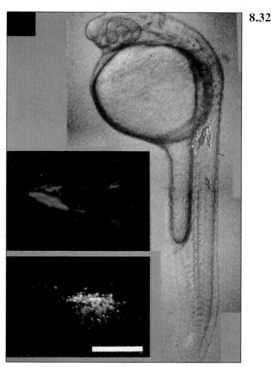

8.30–8.32 The *spadetail* (*spt*) mutation affects the migration of muscle precursor cells during gastrulation in a cell autonomous manner. **8.30** shows mutant (labelled green) and wild-type (red) labelled cells which were simultaneously injected into the lateral marginal zone at the onset of gastrulation in a wild-type host. Fate map analysis shows that trunk muscle derives from this region. **8.31** shows the same embryo after gastrulation. The injected cells segregrated such that the wild-type cells migrated normally into the position where muscle will develop, while the mutant cells moved to the tailbud region. The embryos in **8.30** and **8.31** are 500μm in diametre. **8.32** shows the same embryo at 30hr. Wild-type cells formed muscle, and mutant cells formed mesenchymal cells of the tail region (inset bar: 100μm). (Reprinted with permission from Ho and Kane, 1990.)

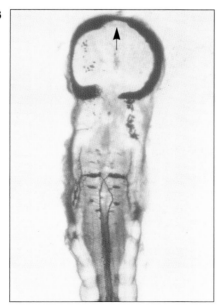

8.33 Horizontal section of a *cyclops* mutant. The eyes are fused at the rostral end (arrow) so that the retina can be seen to be continuous across the front of the embryo. In this mutant embryo, reticulospinal neurons have been labelled with HRP and show abnormal patterning of ventral axon tracts, suggesting that their pattern depends on the floorplate (×60). (Photograph courtesy of K. Hatta.)

Gene expression studies

Another way to identify genes that may play important roles in development, one that complements the analysis of mutant strains, is to look for those that are homologous to developmentally important genes identified in other organisms, whether vertebrates or invertebrates. Control genes that form multigene families and which contain conserved domains, such as homeo-boxes, paired boxes, zinc-fingers, or POU-boxes, appear to be highly conserved across species, and have been the subject of many of the early studies of gene expression in vertebrates, including zebrafish. This strategy, while relatively new, has begun to provide important new insights into vertebrate development.

Pax genes, transcription factors thought to be involved in vertebrate development, provide a good example of the method. The pax family is characterized by the presence of the paired-box, a DNA sequence coding for a DNA binding domain which was originally identified in *Drosophila* developmental control genes. This conserved paired-box sequence was then used to isolate 8 pax genes from the mouse genome, and finally, a mouse pax probe (pax-6) was used to isolate the zebrafish homolog (Puschel *et al.*, 1992). Both the sequence and the expression pattern of this gene are highly conserved between the mouse and zebrafish. The small size and optical clarity of the zebrafish embryo allows analysis of the pax-6 expression pattern in whole mount preparations by *in situ* hybridization, which show that the gene is expressed in the forebrain, including the optic vesicle, and in the hindbrain (**8.34**).

Similarly, homeobox-(Hox)-containing genes, first identified in *Drosophila*, are thought to code for DNA-binding proteins that may act as transcriptional regulators and their restricted pattern of expression in vertebrate embryos suggests a possible role in specifying positional information. To study the regulation of Hox promoters, Westerfield *et al.* (1992) have injected zebrafish eggs with recombinant DNA fragments containing fusions of mouse or human Hox promoters to the β-galactosidase (β-gal) reporter gene. Such DNA injections often result in viable embryos that are mosaically transgenic and the subsequent expression of the gene is detected by the presence of β-gal. Westerfield *et al.* (1992) found that the expression pattern of the Hox promoters was restricted in the same way as it is in mice, and that the specificity of expression in zebrafish depends upon the same regulatory elements within the promoters as in mice. Hox promoters thus appear to be highly conserved through evolution from fish to mammals.

It is now clear that many regulatory genes are highly conserved in both structure and expression pattern across a broad range of animal embryos and the system in which they should be studied is thus largely a matter of convenience. Because the zebrafish is particularly easy to work with, it may well turn out to be a system of choice in understanding how genes regulate developmental processes, perhaps the greatest challenge confronting developmental biology.

8.34

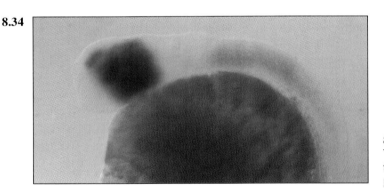

8.34 Pax-6 is expressed in zebrafish during neurulation. These whole 24hr embryos were hybridized with RNA probes that are specific for zebrafish pax-6 (×85). (Reprinted with permission from Puschel *et al.*, 1992.)

The future

The zebrafish presents a unique combination of traits that make it particularly appropriate to work with in solving developmental problems. Indeed, it is hard to think of another animal in which it is as convenient to make mutants, to create transgenic animals or to label cells and study their subsequent behaviour directly in live animals. It is therefore clear that the study of zebrafish will continue to occupy an important role in the analysis of vertebrate development. In addition to the examples discussed above, cell and tissue interactions will prove particularly amenable to analysis by present technology. Here, the availability of mosaic zebrafish, made by cell transplantation between mutant and wild-type embryos, injection of foreign DNA, or other methods, provides a way to manipulate the cellular environment *in vivo* to expose the dependence of one tissue upon another. Over the longer term, studies of gene expression and the role of particular genes in embryogenesis will probably occupy developmental biologists for a long time to come. These are still early days in the experimental analysis of development, and there is an highly complex set of instructions to be deciphered before we fully understand developmental processes and their regulation. We can only hope that many will turn out to be simple sets of interactions and that developmental strategies will be modified and repeated in recognizable form across all animals so that the analysis of development will ultimately provide simple and elegant insights that are generally applicable. Because of its manifest advantages, the zebrafish is likely to make a significant contribution to this quest and complement studies on other animal systems.

References

Creaser, C.W. (1934). The technique for handling the zebrafish (*Brachydanio rerio*) for the production of eggs which are favorable for embryological research and are available at any specified time throughout the year. *Copeia*, **4**, 159–161.

Eisen, J.S., Pike, S.H. and Debu, B. (1989). The growth cones of identified motoneurons in embryonic zebrafish select appropriate pathways in the absence of specific cellular interactions. *Neuron*, **2**, 1097–1104.

Hamilton, F. (1822). *An account of the fishes found in the river Ganges and its branches.* Archibald Constable and Co., Edinburgh and London.

Hatta, K., Kimmel, C.B., Ho, R.K. and Walker, C. (1991). The *cyclops* mutation blocks specification of the floor plate of the zebrafish central nervous system. *Nature*, **350**, 339–341.

Hisoaka, K.K. and Battle, H.I. (1958). The normal development stages of the zebrafish, *Brachydanio rerio* (Hamilton–Buchanan). *J. Morphol.*, **102**, 311–328.

Ho, R.K. and Kane, D.A. (1990). Cell-autonomous action of zebrafish *spt-1* mutation in specific mesodermal precursors. *Nature*, **348**, 728–730.

Kane, D.A. and Kimmel, C.B. (1992). The zebrafish midblastula transition (submitted).

Kimmel, C.B. (1982). Reticulospinal and vestibulospinal neurons in the young larva of a teleost fish, *Brachydanio rerio*. *Prog. Brain Res.*, **57**, 1–23.

Kimmel, C.B., Kane, D.A., Walker, C., Warga, R.M. and Rothman, M.B. (1989). A mutation that changes cell movement and cell fate in the zebrafish embryo. *Nature*, **337**, 358–362.

Kimmel, C.B. and Law, R.D. (1985). Cell lineage of zebrafish blastomeres. I. Cleavage pattern and cytoplasmic bridges between cells. *Dev. Biol.*, **108**, 78–85.

Kimmel, C.B. and Warga, R.M. (1988). Cell lineage and developmental potential of cells in the zebrafish embryo. *Trends Genet.*, **4**, 68–74.

Kimmel, C.B., Warga, R.M. and Schilling, T.F. (1990). Origin and organization of the zebrafish fate map. *Development*, **108**, 581–594.

Kimmel, C.B. and Westerfield, M. (1990). Primary neurons of the zebrafish. In *Signals and Sense* (G.M. Edelman, W.E.Gall, and M.W. Cowan, eds) pp. 561–588, Wiley-Liss, New York.

Metcalfe, W.K. (1985). Sensory neuron growth cones comigrate with posterior lateral line primordial cells in zebrafish. *J. Comp. Neurol.*, **238**, 218–224.

Metcalfe, W.K., Kimmel, C.B. and Schabtach, E. (1985). Anatomy of the posterior lateral line system in young larvae of the zebrafish. *J. Comp. Neurol.*, **233**, 377–389.

Metcalfe, W.K., Myers, P.Z., Trevarrow, B., Bass, M.B. and Kimmel, C.B.(1990). Primary neurons that express the L2/HNK-1 carbohydrate during early development in the zebrafish. *Development*, **110**, 491–504.

Myers, P.Z., Eisen, J.S. and Westerfield, M. (1986). Development and axonal outgrowth of identified motoneurons in the zebrafish. *J. Neurosci.*, **6**, 2278–2289.

Puschel, A.W., Gruss, P. and Westerfield, M. (1992). Sequence and expression pattern of pax-6 are highly conserved between zebrafish and mice. *Development*, **114**, 643–651.

Streisinger, G., Singer, F., Walker, C., Knauber, D. and Dower, N. (1986). Segregation analysis and gene-centromere distances in zebrafish. *Genetics*, **112**, 311–319.

Streisinger, G., Walker, C., Dower, N., Knauber, D. and Singer, F. (1981). Production of clones of homozygous diploid zebra fish (*Brachydanio rerio*). *Nature*, **291**, 293–296.

Stuart, G. W., McMurray, J.V. and Westerfield, M. (1988). Replication, integration, and stable germ-line transmission of foreign sequences injected into early zebrafish embryos. *Development*, **103**, 403–412.

Walker, C. and Streisinger, G. (1983). Induction of mutations by gamma-rays in pregonial germ cells of zebrafish embryos. *Genetics*, **103**, 125–136.

Warga, R.M. and Kimmel, C.B. (1990). Cell movements during epiboly and gastrulation in zebrafish. *Development*, **108**, 569–580.

Westerfield, M. (ed.) (1989). *The Zebrafish Book; A guide for the laboratory use of zebrafish* (Brachydanio rerio). University of Oregon Press, Eugene, Or.

Westerfield, M., Wegner, J., Jegalian, B.G., DeRobertis, E.M. and Puschel, A.W. (1992). Specific activation of mammalian Hox promoters in mosaic transgenic zebrafish. *Genes Dev.*, **6**, 591–598.

9. *Xenopus* and Other Amphibians

Jonathan M.W. Slack

Introduction

Amphibian embryos have been a favourite material for the experimentalist since the late nineteenth century. The eggs are large, usually 1–2mm in diameter, and develop from egg to tadpole outside the mother so remaining accessible to experimentation at all stages. This makes them particularly suitable for microsurgery. In addition, fragments of early embryos and even isolated cells are able to continue development if incubated in simple salt solutions. This is possible because every cell in the embryo contains a supply of yolk platelets which serves as its nutrient supply until the larval blood circulation becomes established. Recent years have seen intense activity involving the combination of the longstanding microsurgical techniques with cell lineage labelling and the techniques of molecular biology.

Although many different species have been used for experimental work in the past, the African clawed toad *Xenopus laevis* (**9.1**) is now the most popular amphibian for the study of early development because of its ease of maintenance, ease of induced spawning, and robustness of the embryos. But other laboratory species such as *Ambystoma mexicanum*, the axolotl (**9.2**), *Rana pipiens*, the American leopard frog, and *Cynops pyrrhogaster*, the Japanese newt, are also used to some extent. Work on one species has obvious advantages of standardization and the rapid transfer of results between laboratories, but there are also problems, principally that different species *are* slightly different and concentration on a single one can lead to unwarranted generalizations.

The two principal taxonomic groups of amphibian are the anura (the frogs and toads) and the urodeles (newts and salamanders). *Xenopus* is not unlike the other anurans formerly used for experimental work except that its eggs are relatively small (1.4mm diameter) and development relatively rapid (1 day to general body plan stage). Urodeles generally have larger eggs, develop more slowly, and have significant differences in their gastrulation movements.

The experimental production of *Xenopus* embryos is very simple. The male and female are both injected with human chorionic gonadotrophin, are put together overnight and the next morning there are embryos. Nowadays, it has become more common to perform *in vitro* fertilizations which generate smaller numbers of embryos, but ones whose time of fertilization is known and whose development shows a high degree of synchrony.

Although an amphibian embryo does become a little bigger over the period covered, this is entirely due to the uptake of water. The embryo has no supply of nutrient from an extra-embryonic yolk mass or a placenta and is therefore actually losing matter throughout development by metabolic activity. All early cell divisions are cleavage divisions in which daughter cells are half the size of the mother cell. These things are true of all holoblastic eggs which develop independently of the mother, but are often forgotten by mammalian or chick embryologists whose organisms undergo growth and development simultaneously.

The stages of *Xenopus* development are described by Nieuwkoop and Faber (1967). According to this series, Stage 8 is the mid-blastula, Stage 10 the early gastrula, Stage 13 the early neurula, and Stage 20 the end of neurulation. Axolotl stages are very similar to this (Bordzilovskaya and Detlaff, 1979) but the series for *Pleurodeles* (Gallien and Durocher, 1957) and *Rana pipiens* (Shumway, 1940) are different and so care is again needed when making interspecies comparisons.

9.1 *Xenopus laevis*, the African clawed toad, adult (×0.4).

9.2 *Ambystoma mexicanum*, the axolotl, adult (×0.4).

Normal development

Fertilization and cleavage

When laid, the amphibian egg has a dark pigmented *animal hemisphere* and a light-coloured *vegetal hemisphere* (**9.3**). It lies within a transparent vitelline membrane inside the jelly coat. After fertilization the membrane lifts from the egg surface and the egg rotates under the influence of gravity so that the less dense animal hemisphere lies uppermost. As in other types of embryo, the two polar bodies appear at the animal pole and contain the unused chromosome sets from the first and second meiotic divisions. The first polar body is produced after maturation and the second after fertilization. The stages of early development are shown for a generalised amphibian embryo in **9.4–9.9**. The sperm enters the animal hemisphere and the side of sperm entry later becomes the ventral side of the embryo. In *Xenopus* and other anurans, the first sperm to enter imposes a block to any others. Urodele eggs are usually fertilized by more than one sperm although only one sperm pronucleus actually fuses with that of the egg. Shortly after fertilization, there is a rotation of the egg cortex relative to the interior which is associated with the transient appearance of an oriented array of microtubules in the vegetal

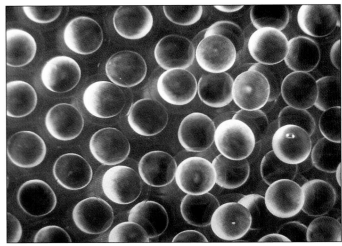

9.3 Fertilized eggs of *Xenopus laevis*. The eggs have rotated and so lie animal pole up within their jelly coats (×7).

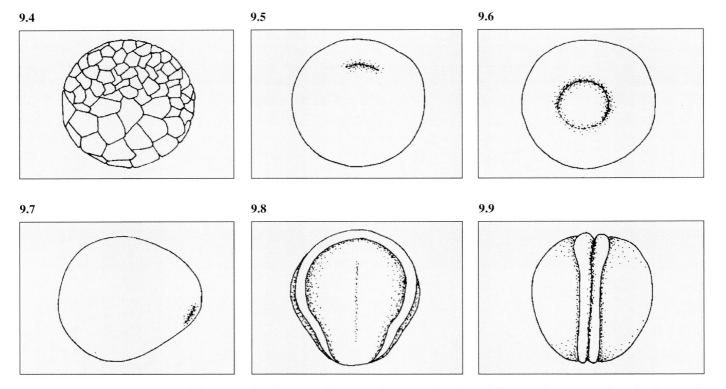

9.4–9.9 External appearance of the early developmental stages of a generalized amphibian embryo: **9.4**, blastula; **9.5**, early gastrula; **9.6**, mid-gastrula; **9.7**, slit-blastopore; **9.8**, open neural plate; **9.9**, closed neural plate, 'lozenge'.

hemisphere. This leads to a reduction in the pigmentation of the animal hemisphere on the prospective dorsal side, opposite the sperm entry point. In some batches of eggs from species other than *Xenopus*, a similar pigmentation change appears as the famous '*grey crescent*' on the dorsal side.

The first cleavage is vertical and usually bisects the egg meridionally, separating it into right and left halves. The second cleavage is at right angles to this and usually separates prospective dorsal from ventral halves. The third cleavage is equatorial, separating animal from vegetal halves. As in other embryos, the large cells resulting from early cleavage divisions are called *blastomeres* (**9.10**). As cleavage takes place a cavity called the *blastocoel* forms in the centre of the animal hemisphere and the embryo is referred to as a *blastula* (**9.11, 9.12**). The outer surface of the blastula consists of the original oocyte plasma membrane which is physically more rigid and binds a number of lectins and antibodies more avidly than the newly formed cleavage membranes. A complete network of close junctions around the exterior cell margins seals the blastocoel from the exterior and renders the penetration of almost all substances highly inefficient. This is why radiochemicals and other substances need to be introduced into the embryo by micro-injection. Internal cells are connected by junctions less 'close' than those at the periphery and are readily disrupted by removal of calcium ions from the medium. Desmosomes are not found until the neurula stages but all the cells of early cleavage stages are connected by gap junctions which allow limited exchange of molecules below about 1000 molecular weight. Cleavage continues rapidly for 12 divisions after which synchrony is lost, the division rate slows, and zygotic gene transcription commences. This time is known as the *mid-blastula transition* or MBT (Newport and Kirschner 1982a, b). There is also a distinct compaction event at which intercellular adhesion increases and the blastula appears as a smooth, instead of knobbly, ball.

9.10 *Xenopus* 8-cell stage (×15).

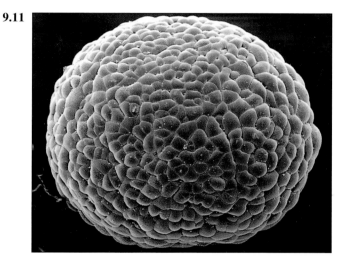

9.11 SEM of *Xenopus* blastula (×50).

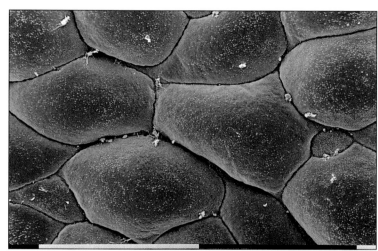

9.12 High-power view of blastula cells, scale bar 0.1mm.

Gastrulation

The next phase of development is one of extensive morphogenetic movements called *gastrulation* (**9.13–9.16**). The motor for gastrulation is mainly provided by a belt of tissue around the equator of the embryo called the *marginal zone*, of which the surface layer ends up as the lining of the *archenteron* cavity, while the deep part is the prospective mesoderm. Although some autonomous expansion of the animal hemisphere and some dorsal convergence of the marginal zone have already occurred, the conventional start of gastrulation is marked by the appearance of a pigmented depression in the dorsal vegetal quadrant (**9.20, 9.21**). This is the *dorsal lip of the blastopore*. The blastopore becomes elongated laterally and soon becomes a complete circle. When it is circular the part referred to as the dorsal lip is the dorsal segment of the complete circle and that described as the ventral lip is the ventral part of the complete circle. The blastopore is associated with the formation of bottle cells and is the locus of invagination of the superficial marginal zone which results in the creation of the archenteron. The invagination occurs all round the blastopore but is much more extensive on the dorsal side where it proceeds until the leading edge of tissue is well past the animal pole, with the archenteron roof being closely apposed to the overlying ectoderm. In the lateral and ventral parts of the blastopore there is only a small invagination. By

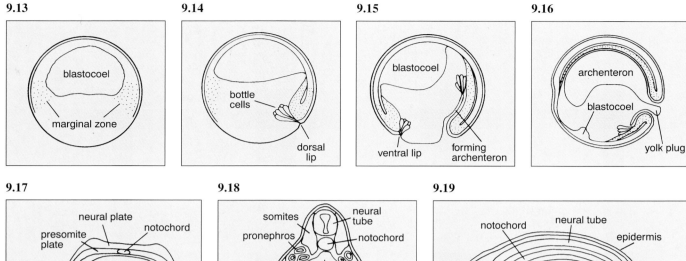

9.13–9.19 Outline of early *Xenopus* development: **9.13**, blastula, medial section; **9.14**, early gastrula, medial section; **9.15**, mid-gastrula, medial; **9.16**, late gastrula, medial; **9.17**, open neural plate, transverse; **9.18**, tailbud, transverse; **9.19**, tailbud, sagittal.

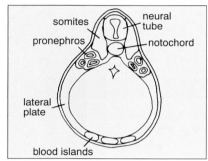

9.20 SEM of early *Xenopus* gastrula (vegetal view, ×50).

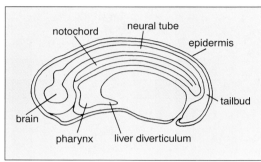

9.21 Axolotl gastrula (vegetal view, ×32).

the end of gastrulation the entire archenteron roof and floor have been formed from the dorsal invagination except for a small crescent of archenteron floor at the posterior end which is formed from the lateroventral invagination.

Simultaneously with the formation of the archenteron, the deep marginal zone 'involutes'. This zone can be imagined as a cylinder around the equator of the embryo and involution consists of the cylinder turning inward at the lower edge and then extending towards the animal pole. The dorsal involution is most extensive while the ventral involution is slight and the process is accompanied by a general shift of tissue in a dorsal direction. At the same time as the marginal zone is involuting, the animal cap tissue spreads vegetally so that the blastopore lip becomes a smaller and smaller ring around the vegetal pole and eventually narrows to a small slit.

Extensive studies have been made by the group of Keller who consider that it is not the bottle cells but the movements of the deep marginal zone cells that provide the motive force for the invagination and involution processes. These can be considered as made up of two components: *convergence* and *extension*. Convergence represents the packing together of cells to shrink the circumference of the total marginal zone and occurs equally around this zone. Extension refers to an active process of cell intercalation leading to elongation in the animal vegetal direction, and this occurs only in the dorsal region (Keller *et al.*, 1985; Keller and Danilchik, 1988).

In the axolotl and other urodeles, the gastrulation movements are somewhat different. The archenteron is formed entirely by the dorsal invagination, while in the lateral and ventral regions of the blastopore lip there is a separation of marginal zone from vegetal tissue and the future mesoderm extends towards the animal pole as a cell sheet with a free edge. In *Xenopus*, the archenteron roof is at all times lined with endoderm while in urodeles the forming notochord is the archenteron roof and is the part which is common to the dorsal lip derived invagination and the whole blastopore derived involution of mesoderm (see **9.24**). There is no clear distinction between surface and deep marginal zone, both contributing substantially to the mesoderm. In the newt, *Pleurodeles,* the injection of antibodies to fibronectin can inhibit gastrulation and so the tissue movements are thought to depend on adhesion between the cells at the leading edge and fibronectin deposited on the blastocoelic surface of the animal hemisphere (Boucaut *et al.,* 1984).

By the end of gastrulation, the former animal cap ectoderm has covered the whole external surface of the embryo, the yolky vegetal tissues have become a mass of endoderm in the interior, and the former marginal zone has become a layer of mesoderm extending from the slit-shaped blastopore that reaches the anterior end on the dorsal side but only a short distance on the ventral side. In other words, the three classic germ layers, ectoderm, mesoderm and endoderm, have achieved their final trilaminar arrangement. The archenteron has become the principal cavity at the expense of the blastocoel, and the embryo has rotated so that the dorsal side is uppermost. It now has a true anteroposterior (or craniocaudal) axis which runs from the leading edge of the mesoderm to the blastopore.

Neurulation and later development

The next stage of development is called the *neurula* (**9.8, 9.17** and **9.22–9.24**), and here the ectoderm on the dorsal side becomes the central nervous system. The *neural plate* becomes visible as a keyhole-shaped region delimited by raised neural folds and covering much of the dorsal surface of the embryo. Quite rapidly the folds rise and move together to form the *neural tube* which becomes covered by the ectoderm from beyond the folds, now known as epidermis. Tissue from the folds which comes to lie dorsal to the neural tube is the *neural crest*. The internal arrangement of tissues is shown in **9.24** and **9.25**. In the mesodermal layer, the posterior part of the dorsal midline segregates as a distinct *notochord*, and the mesoderm on either side begins to become segmented in anteroposterior sequence to form paired *somites*. In *Xenopus* and other anurans, the entire archenteron is lined by endoderm from its formation, but in urodeles the archenteron roof initially consists of the notochord and presomite plate, and the endoderm later sends up folds which meet beneath the notochord.

By the end of neurulation the rudiments for all the major structures of the body are in their definitive positions (**9.18, 9.19**). Although by this stage the cells still appear to be yolk-filled bags with little visible histological differentiation, most of the major tissues are now known to commence their terminal differentiation at the end of gastrulation. The central nervous system is formed from the neuro-epithelium of the neural tube. The mesoderm of the trunk region develops into several structures in dorsoventral sequence: notochord, myotomes, pro-nephros and mesonephros, and ventral blood islands (**9.26**). Anteriorly, the prechordal mesoderm forms part of the jaw muscles and branchial arches. The heart, limbs and

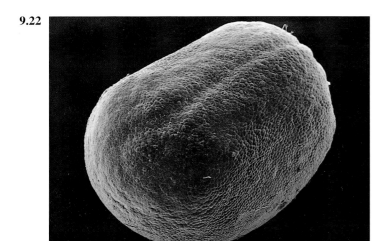

9.22 SEM of *Xenopus* neurula (dorsoanterior view, ×40).

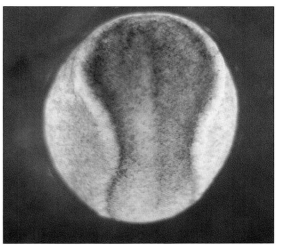

9.23 Axolotl neurula (dorsal view, ×25).

future haemopoietic system also exist as rudiments in the mesodermal mantle. The endoderm becomes the epithelial components of the pharynx, stomach, liver, lungs (if any), intestine and rectum, and the tail arises from the tailbud. Later stages of *Xenopus* show the elongation of the body, the pigmentation of the retina and appearance of pigment cells on the trunk, and the formation of a transparent tailfin (**9.27, 9.28**). In the axolotl, pigmentation develops later and there is no cement gland (**9.29**).

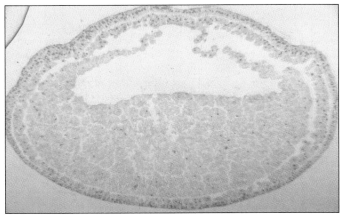

9.24 Axolotl, transverse section of open neural plate stage. The endoderm and mesoderm are well separated in this picture, and it can be seen that they come together with the forming notochord in the dorsal midline (×40).

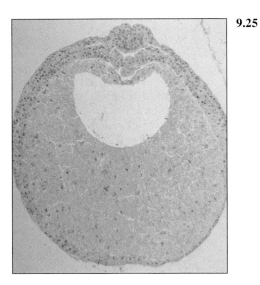

9.25 Axolotl, transverse section of lozenge stage. Now the neural tube has closed, the notochord has rounded up, the somites are forming, and the endoderm has formed a continuous layer below the notochord (×40).

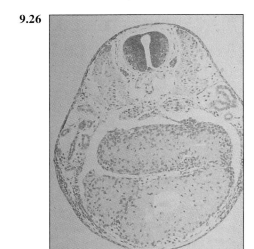

9.26 Axolotl, transverse section of prelarval stage. Histological differentiation is now advanced and it may be seen that the mesoderm consists of notochord, somites, kidney tubules, lateral plate and blood islands in dorsal to ventral sequence (×40).

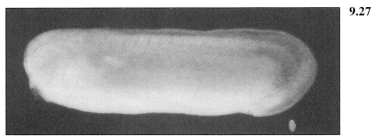

9.27 *Xenopus* early tailbud stage (×20).

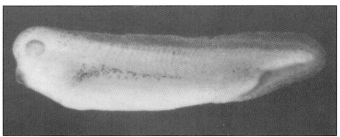

9.28 *Xenopus* late tailbud stage (×20).

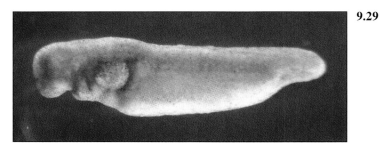

9.29 Axolotl tailbud stage (×15).

Techniques for studying Normal development
Modern micro-anatomical methods

Traditional embryological experiments have relied on the histological differentiation of cells or on characteristically shaped multicellular arrangements to identify structures. In recent years, a search has been made for early molecular and immunological markers which can be used to identify cells at a time nearer to that of the determinative events, or even be the molecular basis of the determinative events themselves.

For *Xenopus* there is now available a fairly comprehensive set of antibodies and other reagents for identifying the main tissue subdivisions. **9.30–9.36** show immunofluorescence

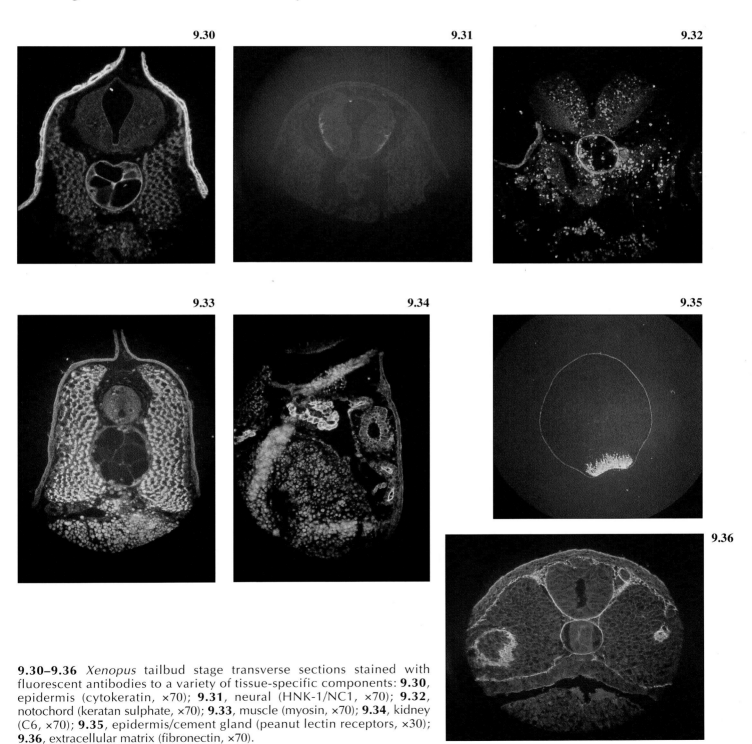

9.30–9.36 *Xenopus* tailbud stage transverse sections stained with fluorescent antibodies to a variety of tissue-specific components: **9.30**, epidermis (cytokeratin, ×70); **9.31**, neural (HNK-1/NC1, ×70); **9.32**, notochord (keratan sulphate, ×70); **9.33**, muscle (myosin, ×70); **9.34**, kidney (C6, ×70); **9.35**, epidermis/cement gland (peanut lectin receptors, ×30); **9.36**, extracellular matrix (fibronectin, ×70).

photographs using a panel of re-agents which react specifically with epidermis (Dale, Smith and Slack, 1985), neural tissue (Tucker *et al.*, 1988), notochord (Smith and Watt, 1985), muscle (Kintner and Brockes, 1984) and kidney (Slack, 1985a). Lectins can be useful, as shown for example by the specificity of peanut lectin for the outer layer of the epidermis and the cement gland (Slack, 1985b). The material between cells is critical for correct performance of morphogenetic movements and components such as fibronectin tend to surround all the tissue masses.

The use of tissue-specific antibodies has been increased by the recently introduced technique of wholemount immunostaining. For amphibian embryos, the key step in the devising of effective protocols was the realization that yolk is not intrinsically opaque but can be rendered transparent by immersion in a solvent of sufficiently high refractive index (Dent *et al.*, 1989). Tetrahydronaphthalene has been introduced for this purpose, which is a very effective clearing agent and does not dissolve out the most commonly used histochemical reaction products too quickly. In **9.37** and **9.38** two examples are shown of wholemount staining of muscle, one cleared and the other uncleared. Uncleared specimens are attractive and can be informative for fairly superficial structures but to see deep structures clearing is obviously necessary.

Wholemount staining can also be used to visualize and identify structures by histochemical staining. **9.39** shows a lozenge stage *Xenopus* stained for alkaline phosphatase. The notochord is particularly rich in this enzyme and shows up clearly. In **9.40** the peroxidase activity of haemoglobin is used to show the location of the ventral blood islands. In this specimen the notochord is also stained light blue with alcian blue, a dye which preferentially stains sulphated polysaccharides.

9.37 *Xenopus* tailbud stage stained with an anti-muscle antibody and cleared in tetrahydronaphthalene (×15).

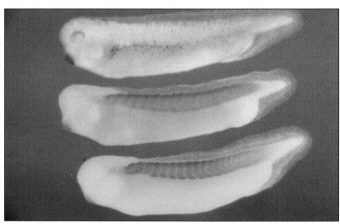

9.38 Similar specimens to **9.41** but uncleared and showing the use of contrasting detection methods (×12).

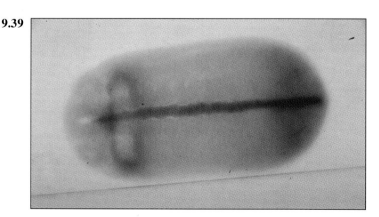

9.39 Visualization of the notochord by histochemical stain for alkaline phosphatase in a lozenge stage *Xenopus* embryo (×35).

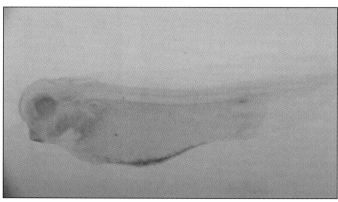

9.40 Simultaneous visualization of the notochord by staining with alcian blue, and of the blood islands by histochemical stain for peroxidase. *Xenopus* pre-larval stage (×15).

Fate mapping

Like many animal embryos, amphibians have a continuous topographic projection from the fertilized egg to later stages. In other words, it is possible at any stage to construct a fate map of the embryo which shows what each volume element will become and where it will move in the subsequent stages. The map cannot be exact however because there is some short range mixing of cells during gastrulation.

Classic fate maps for amphibians were constructed by staining parts of the embryo with vital dyes and locating the stained patch at later stages (Vogt, 1929; Pasteels, 1942; **9.41**). Despite their honourable history, vital dyes do have limitations. It is difficult to mark accurately regions within the embryo, and the dyes have a tendency to spread and fade and so cannot be regarded as cell-autonomous labels. For these reasons high molecular weight lineage labels are now preferred for fate mapping and for cell labelling as an adjunct to other experiments. Two of the most popular are horseradish peroxidase (HRP) (Jacobson and Hirose, 1978) and fluorescein dextran amine (FDA) (Gimlich and Braun, 1985). The labels can be used for fate mapping in two ways: either by injection into a single identified blastomere at an early cleavage stage, or by injection of a fertilized egg and the use of the resulting uniformly labelled embryo as a donor for grafts. **9.42** shows some *Xenopus* neurulae which were either uninjected, injected into the zygote and so filled all over, or injected into one blastomere of the 2-cell-stage, and hence half-labelled. **9.43** shows a clone of cells in the neural plate, arising from the injection of a single blastula cell in the axolotl. This reveals both the overall morphogenetic movement of extreme anteroposterior elongation, and also the local cell mixing which brings about a salt-and-pepper appearance in the labelled region. **9.44–9.47** show the different behaviour of dorsal and ventral marginal zone sectors labelled by orthotopic grafting of early gastrulae with grafts from FDA-filled donors. Both sectors were about 45° at the time of grafting but the dorsal region contracts in circumference while the ventral region greatly expands during gastrulation. All the lineage label studies have shown that there is more local cell mixing than was previously suspected, particularly in the lateroventral region. The prospective regions are not therefore sharply delineated as in classic fate maps but rather they are fuzzy patches which overlap with their neighbours to a greater or lesser extent (Dale and Slack, 1987a; **9.48**).

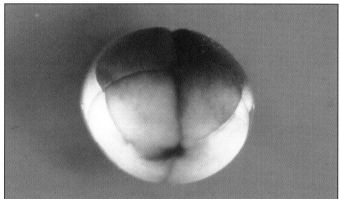

9.41 Application of a patch of the vital dye, Nile Blue, to the dorsal side of an 8-cell *Xenopus* embryo (×35).

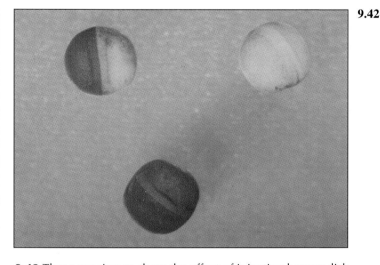

9.42 These specimens show the effect of injecting horseradish peroxidase enzyme into the zygote (bottom) or one blastomere of the 2-cell-stage (top left). The *Xenopus* embryos were left to develop until neural plate closure and then fixed and stained for peroxidase (×15).

9.43 Clone of HRP labelled cells in the neural plate of an axolotl, following injection at the 128-cell-stage. Notice how the clone has elongated as a result of morphogenetic movements (×30).

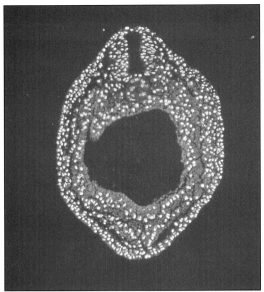
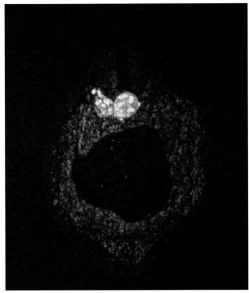

9.44, 9.45 Normal fate of the dorsal marginal zone in the axolotl. A graft was made of the dorsal lip region from a labelled to an unlabelled gastrula. The host was allowed to develop to tailbud stage, then fixed and sectioned. **9.44** shows cell nuclei stained with a nuclear stain and **9.45** shows yellow fluorescence from graft derived cells. These are distributed along the whole anteroposterior axis in the notochord and somites (×30).

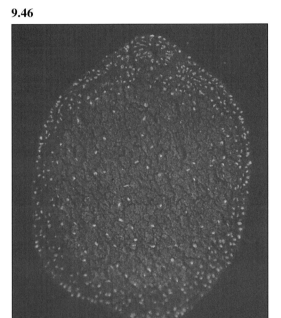
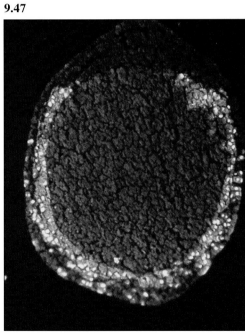

9.46, 9.47 Normal fate of the ventral marginal zone. A similarly sized graft of ventral marginal tissue is confined to the posterior half of the body, but expands to fill most of the embryo circumference (×35).

Although the fate map has a statistical character it remains an essential baseline for interpretation of embryological experiments. Unless we know what a group of cells will do in normal development we cannot very well claim to have changed their behaviour by experimental treatment. The important features are as follows:

- The epidermis and neural plate arise from the animal hemisphere, the neural plate from the dorsal and the epidermis from the ventral region.
- The mesoderm arises from a belt around the equator but, in *Xenopus*, not from the surface layer of cells carrying the oocyte-derived membrane.
- The dorsal side of the blastula contributes mainly to the anterior of the later body and the ventral side of the blastula to the posterior; however the dorsal marginal zone also populates the dorsal midline axial structures along the entire body length.
- There is a rough mapping of mesodermal tissue types in the fate map from dorsal to ventral in the order: notochord, heart, myotomes, pronephros, lateral plate, blood islands. However, the region forming muscle is rather extensive, and about 60% of muscle normally derives from the ventral half of the blastula.
- The yolk mass part of the endoderm comes from the yolk-rich vegetal hemisphere, but the lining of the pharynx from the dorsal part of the marginal zone.

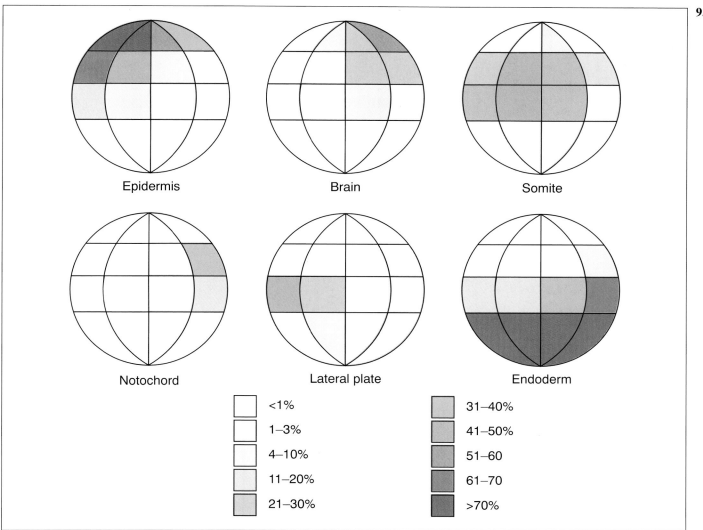

9.48 Fate map for the 32-cell-stage *Xenopus* embryo compiled from many specimens injected with FDA. Because cleavage planes are not identical in different individuals and because there is some cell mingling during gastrulation, the prospective regions overlap to some extent.

Methods for changing embryonic pattern

Although *Xenopus* has no genetics, there are two well-known treatments that can alter the overall pattern in just as spectacular way as can the maternal-effect mutations in *Drosophila*. These are irradiation with ultraviolet light, and treatment with lithium ion.

Ultraviolet irradiation

If fertilized eggs are irradiated from the vegetal side, the resulting embryos develop with a deficiency in the structures normally formed from the dorsal side of the blastula (Malacinski, Benford and Chung, 1975). As we have seen above, the dorsal part of the blastula contributes most cells to the anterior part of the post-gastrulation body, and also populates the dorsal midline structures along the entire body length. Hence a mild UV effect involves a reduction of the head, a more severe one the absence of the head, and a most severe one, the absence of both the head and the dorsal midline structures in the trunk. The effects of UV are dose-dependent, and the limit form resulting from the highest doses is a radially

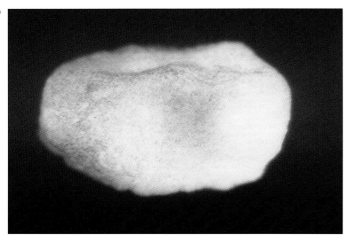

9.49 Hyperventral *Xenopus* embryo resulting from irradiation of the vegetal hemisphere with UV light prior to cortical rotation (×40).

symmetrical embryo which contains a normal proportion of mesoderm, but mainly in the form of blood islands, normally the tissue characteristic of the ventroposterior part of the mesodermal mantle (**9.49**; Cooke and Smith, 1987). Since the radiation only penetrates a short distance it has relatively little effect on cellular viability and the limit forms can survive for many days.

Originally, the UV effect was thought to be due to the destruction of a dorsal or axial determinant, but it was later found by Scharf and Gerhart (1980) that embryos that had been subjected to UV irradiation just after fertilization could still form a normal pattern if they were tipped on their sides before the time of subcortical rotation. So the majority view now is that the UV effect involves damage to a component required for the dorsoventral polarization, perhaps the subcortical microtubule array, rather than destruction of a cytoplasmic determinant for the dorsal region. However, Elinson and Pasceri (1989) showed that eggs which had been UV irradiated as *oocytes* underwent the normal subcortical rotation and could not be rescued by tipping. The pattern defects from irradiation of oocytes are similar to those from irradiation of eggs and this result has given the dorsal determinant theory a new lease of life.

Lithium

Kao, Masui and Elinson (1986) showed that treatment of cleavage stages with high concentrations of LiCl could have drastic effects on pattern. In principle the change is the converse of the UV effect, embryos behaving as though they contained a higher than normal proportion of dorsal tissue. The limit form here is a radially symmetrical hyperdorsal (or hyperanterior) form in which a continuous band of pigmented retina and of cement gland extends around the body and there is a central core of notochord with little muscle (Cooke and Smith 1988; **9.50**). The maximum sensitivity of whole embryos is found at the 32–64 cell stage, but this may simply reflect the period of maximum exposure of new cleavage membrane at the surface and hence penetrability of the embryos. Although external concentrations of the order of 0.3M are used, whole embryos are not very permeable and the intracellular concentrations measured have been only a few millimolar. Embryos which have been ventralized by early UV irradiation can later be rescued back towards a normal pattern by Li treatment, and this is true not only of cases irradiated as

eggs, but also of cases irradiated as oocytes, which as we have seen above are refractory to rescue by tipping. This suggests that Li is mimicking some event which occurs at a multicellular stage and is normally a consequence of the subcortical rotation. If LiCl is injected into ventrovegetal blastomeres of normal embryos at the 32-cell-stage, it is possible to induce twinning, implying that a second centre of dorsal character has been formed on the ventral side.

Curiously enough, Li had been known to have patternaltering effects on amphibian embryos for decades preceding the recent burst of activity. But it had always been applied to much later stages, usually gastrulae, and the results were somewhat less interesting, consisting simply of reductions of the head and, in extreme cases, suppression of the trunk notochord as well (Lehmann, 1937; Bäckstrom, 1954). At least the head reductions are probably secondary to an inhibition of gastrulation movements which mean that the leading edge of the dorsal invagination reaches less far than usual, and the induction of anterior neural structures cannot occur. The suppression of the notochord is less easy to explain as an indirect effect. It occurs without cell death, the dorsal midline cells becoming muscle rather than notochord, and does therefore seem to indicate a ventralization of the dorsal midline occurring during gastrulation. This is more or less the opposite effect to that shown on the cleavage stages and, according to Yamaguchi and Shinagawa (1989), the switchover in effects occurs at the mid-blastula transition.

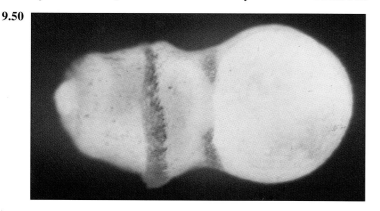

9.50 Hyperdorsal *Xenopus* embryo resulting from treatment with lithium during cleavage (×40).

Differentiation of explanted tissues

The explantation of small regions followed by culture in buffered salt solutions is a method introduced by Holtfreter in the 1930s (Holtfreter, 1938a, b). The cells of the explant are sustained by their own internal yolk supply and so will survive, divide and differentiate in a very simple medium. Almost all explant cultures give rise to more than one histological cell type, suggesting that regional differences exist within the explant, either inherited from the intact embryo or generated internally after isolation. Fortunately, several cell types arise in characteristic locations in the vertebrate body plan and so the regional identity acquired by an explant can often be known from the cell types which it contains or from the arrangement of structures which form within it.

Many isolation studies have been performed over the years using a variety of species, (for one example, see Dale and Slack, 1987b). In the early blastula, explants from most parts of the animal hemisphere will develop into solid balls consisting entirely of epidermis (**9.51**). The outer layer is somewhat different from the rest in that the cells are more regularly arranged and certain epidermal markers normally only expressed on the outer cell layer are found. However, in favourable cases it can be shown that no cells die, no cells are lost, and every cell in the explant expresses epidermal markers. On the dorsal side, many animal hemisphere explants from near the pigment boundary will also form some neural tissue, especially at the later stages (MBT-gastrula), and most explants from this position taken at any stage will contain cement glands. Explants from the vegetal hemisphere coming from the regions nearer the pigment boundary tend to form dense-staining yolky cells which express endodermal markers, while those from the vegetal pole region remain as clumps of large light-staining yolky cells of the type normally destined to end up in the gut lumen.

Explants from the equatorial region (the marginal zone) will usually produce some mesodermal tissues together with endoderm, all surrounded by a well-organized two-cell layer epidermis. There is a sharp difference between explants from the dorsal marginal zone (DMZ) which produce clumps of notochord, muscle and neuro-epithelium and those from ventro-lateral marginal zone (V+LMZ) which usually produce mesenchyme and erythrocytes (**9.52, 9.53**). This difference is also apparent in the shape changes shown by these explants at the time that intact control embryos are gastrulating. The DMZ explants will elongate in an attempt to reproduce the convergent extension they undergo *in situ*, while the VMZs remain fairly spherical. It is also possible to obtain differentiation of disaggregated blastula cells, even in monoclonal cultures, if they are plated on plastic coated with the extracellular matrix proteins fibronectin and laminin (Godsave and Slack, 1989). **9.54** shows a patch of muscle cells from such a culture.

If this type of information is put together in one diagram, we get a specification map as shown in **9.55** and **9.56**. We know from the single cell culture experiments that there are regions containing mixtures of differently specified cells, so it should not be thought that at this stage the embryo is necessarily a mosaic of regions between which sharp lines can be drawn. As to the origin of the regional differences, these are certainly partly derived by passive partition of the egg

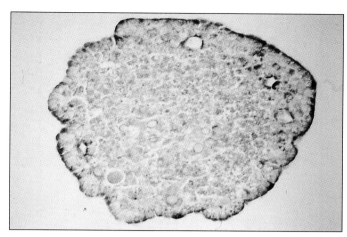

9.51 An explant of tissue from the animal pole of a blastula (an 'animal cap') differentiates into a solid ball of epidermis (×80).

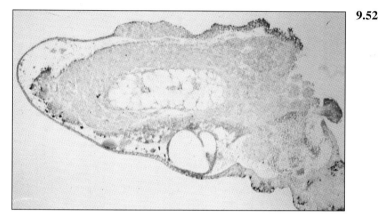

9.52 Dorsal marginal explants differentiate into notochord, muscle and neural tissue (×80).

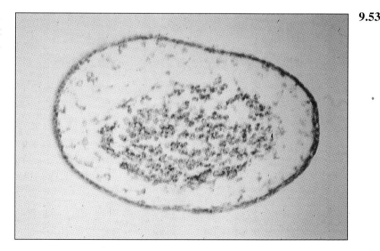

9.53 Ventral marginal explants form loose mesenchyme and blood cells (×80).

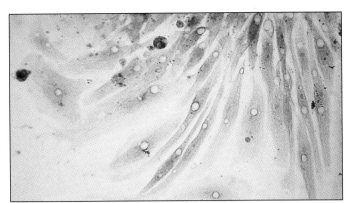

9.54 Differentiation of muscle cells from isolated blastula animal cap cells treated with fibroblast growth factor (×100).

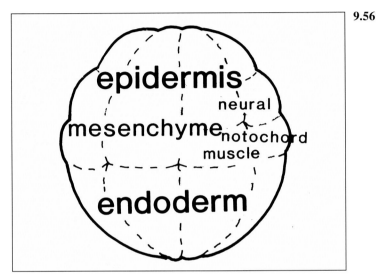

9.55, 9.56 A simplified version of the fate map presented in **9.48** is shown in **9.55**. **9.56** is a specification map, showing the self-differentiation of isolated tissue fragments removed from early blastulae.

cytoplasm. After fertilization, the subcortical rotation is responsible in some way for generating a small dorsovegetal and a large ventrovegetal region with axial and non-axial mesoderm inducing properties respectively. At multicellular stages, the marginal zone appears to consist of a small organizer zone and an extended ventral mesodermal zone. The animal hemisphere is mainly specified to become epidermis, with a rather small neural/cement gland zone on the dorsal side.

Alteration of pattern by grafting

If specification maps are constructed at different stages, it is clear that a more and more complex pattern is being set up as development proceeds. This increase in complexity arises from a sequence of inductive interactions in each of which a competent region of tissue becomes subdivided into several differently specified zones in response to a graded signal emitted by an organizer region. In amphibian embryos this sequence of events has been deduced from grafting and combination experiments. In a graft a small region of tissue is transplanted from one individual embryo to another. Nowadays the graft is usually labelled with a lineage label such as FDA introduced into the donor embryo by injection of the zygote. If transplanted to the same position (orthotopic) there should be no effect on pattern and this is one type of fate mapping experiment. If the graft is implanted in a different position (heterotopic) then we can ask whether its pathway of development is altered by its new position, or conversely whether the surrounding tissues of the host have their pathway of development altered. If the latter is the case then the graft itself must be a signalling centre.

As an example of this type of experiment let us consider the most famous graft in the whole of embryology: the organizer graft originally performed by Spemann and Mangold (1924). We have already seen that the dorsal marginal zone of the blastula is specified to form notochord and somites. If this region is grafted into the ventral marginal zone of a late blastula or early gastrula (**9.57**), then it dorsalizes its surroundings and forms a secondary axis in which the notochord is predominantly formed from the graft and the somites mainly from the host ventral mesoderm. During gastrulation the secondary axis becomes apposed to the ventral epidermis and there induces a secondary neural plate. **9.58** illustrates a double embryo resulting from an organizer graft while **9.59** shows a case in which the graft was labelled with horseradish peroxidase (HRP) and it may be seen that it is predominantly the notochord of the secondary embryo that is labelled. In **9.60–9.62** the host embryo was labelled in two ventral blastomeres which do not normally contribute to the neural tube and it may be seen that they contribute heavily to the induced neural tube.

Although the neural inducing activity of the organizer graft was long known as *primary* embryonic induction, it is not really primary at all and we now know that certain regional differences are set up in the blastula by a process called *mesoderm induction* (see Smith, 1989). Comparison of the fate map and the specification map (**9.55, 9.56**) shows that a part of the animal hemisphere, roughly corresponding to the second tier of cells at the 32-cell-stage, contributes heavily to the mesoderm in normal development, but, if isolated at the early blastula stage, it forms only epidermis. This suggests that it needs a signal from the vegetal region to cause it to become mesodermal and indeed when combinations are made of

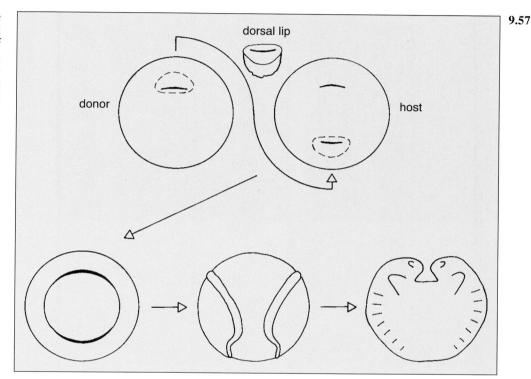

9.57 In the organizer graft a piece of dorsal marginal tissue is grafted into the ventral marginal zone of a host gastrula. As described in the text, the graft forms a new mesodermal axis itself, dorsalizes the surrounding mesoderm, and induces the neighbouring ectoderm to form a second neural plate.

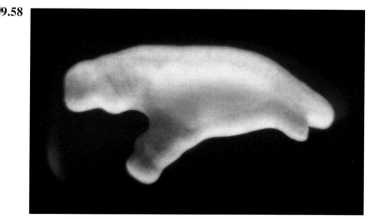

9.58 Double embryo of the ribbed newt, *Pleurodeles*, resulting from an organizer graft (×15).

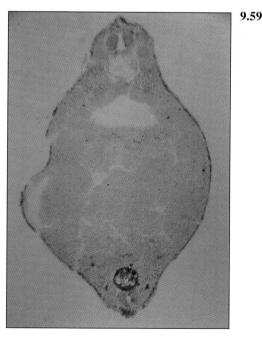

9.59 Transverse section through a *Xenopus* embryo which received an organizer graft labelled with HRP. It may be seen that the graft forms the notochord of the secondary axis (×32).

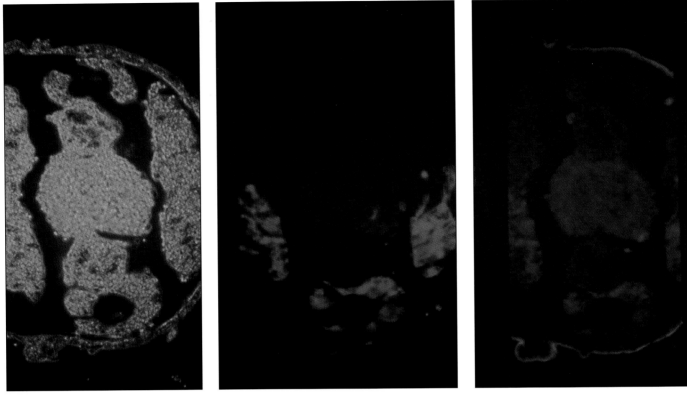

9.60–9.62 Neural induction by an organizer graft is shown in this specimen where the host was injected with RDA (red) into the two ventral cells of the third tier at the 32-cell-stage. These do not normally contribute to the neural tube (see **9.48**) but in the double embryo they have contributed heavily to the secondary neural tube. **9.60** shows a dark field view; **9.61**, a RDA label; and **9.62** a peanut lectin label of the epidermis (see also **9.35**) (×70).

animal and vegetal tissue from blastulae, the animal component then forms a substantial amount of mesoderm. This is the first inductive process shown in **9.63**. The signal comes in two flavours, a *dorsal type* signal on the dorsal side and a *ventral type* signal around the remainder of the circumference. The dorsal-type signal induces the organizer, and the ventral signal induces a belt of ventral-type mesoderm around the remainder of the circumference. Recent work has shown that cytokines of the activin group mimic the dorsal signal and those of the fibroblast growth factor group mimic the ventral signal (Smith, 1989), although whether they are actually doing these jobs *in vivo* remains unclear.

During gastrulation the organizer dorsalizes the surrounding mesoderm so that the mesoderm closest to it becomes somite, that further away becomes kidney, and that furthest away remains in the ventral ground state of mesothelium and blood islands. It induces the neighbouring ectoderm to become neural plate, partly in the plane of the tissue (i.e. from the dorsal lip towards the animal pole) and partly from below after it has invaginated and become the roof of the archenteron. During invagination it somehow generates a series of anteroposterior levels in the dorsal midline which makes up the core of the body plan and which serves as the source for a number of additional inductive signals which are necessary for the correct formation of the ecto- and endodermal derivatives of the head, trunk and tail. A summary of our current understanding of this sequence is provided in **9.63**, although this will undoubtedly prove to be an oversimplification as the molecular biology details are filled in.

Future role of the organism

The death of amphibian experimental embryology has been announced many times but remarkably *Xenopus* and the other species refuse to lie down and accept that their day is done! The main ground for impending redundancy is usually that amphibians have no genetics. It is true that they have no genetics, or at least we are unlikely to learn much of their genetics because of rather long generation times and rather large genome sizes. But is genetics really indispensable for understanding development? It was not necessary for earlier work in which metabolic pathways were elucidated, or in which the mechanism of the nerve action potential was understood, or for the isolation of hormones and growth factors. Although genetics was essential for the understanding of *Drosophila* development, it may be that the increased

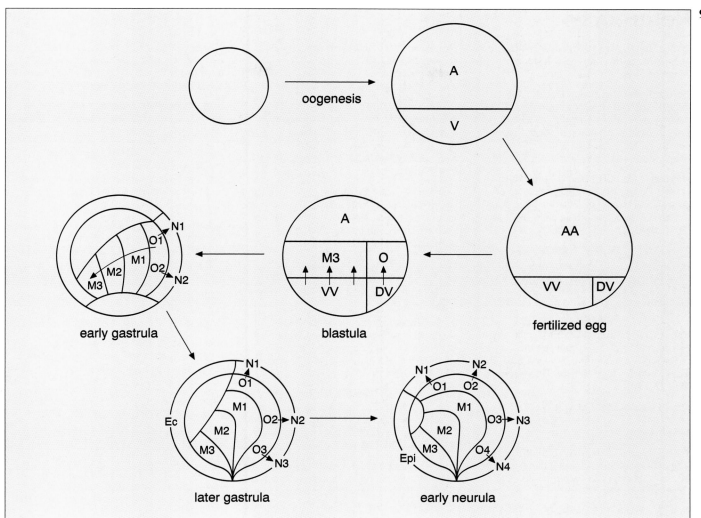

9.63 Overall model of the inductive interactions responsible for forming the body plan of the amphibian embryo. During oogenesis it is thought that the cytoplasm of animal and vegetal hemispheres becomes different. After fertilization the cortical rotation establishes a 'DV' territory in the vegetal hemisphere. In the blastula, mesoderm induction occurs. The DV region emits an activin-like signal which induces the organizer and the VV region emits an FGF-like signal which induces the ventral-type mesoderm. During gastrulation, the organizer tissue generates a sequence of anteroposterior territories in the mesoderm (O1–O4), dorsalizes the surrounding mesoderm, and induces the neural plate (N1–N4) from the ectoderm.

genome size of vertebrates, and the associated redundancy of function, means that informative embryo-lethal phenotypes will be the exception rather than the rule.

In *Drosophila*, genetics has been important both for the identification of developmentally important genes and for the investigation of gene function. In vertebrates, we already know some of the developmentally significant genes, either by homology cloning from *Drosophila* (e.g. homeobox genes) or from biochemical work (e.g. extracellular matrix components and growth factors). Also, the investigation of gene function does not necessarily require genetic techniques. In *Xenopus*, methods have been devised for overexpressing individual genes by injection of synthetic mRNA (Krieg and Melton, 1984; Moon and Christian, 1989). Maternally expressed genes can be inactivated by injecting antisense mRNA into the oocyte, and it may be possible to inactivate zygotic genes by extracellular application of antisense oligodeoxynucleotides, as has proved successful in other systems.

So amphibian embryos are likely to remain as mainline research organisms at least until the outline molecular understanding of primary body plan formation has been attained. Once the molecules have been identified and their functions established in amphibians, it may be possible to confirm that they are playing a similar role in mammals by observation of expression patterns or application of the more restricted set of manipulations which is possible with them. For the later stages of embryonic development in which the internal parts of individual organs become specified, the advantage of amphibians is much less, because useful *in vitro* isolation and combination experiments can already be carried out using mammalian material.

References

Bäckstrom, S. (1954). Morphogenetic effects of lithium on the embryonic development of Xenopus. *Arkiv for Zoologi*, **6**, 527–536.

Bordzilovskaya, N.P. and Detlaff, T.A. (1979). Table of stages of normal development of axolotl embryos and the prognostication of timing of successive developmental stages at various temperatures. *Axolotl Newsletter*, **7**, 2–22.

Boucaut, J.C., Darribère, T., Boulekbache, H. and Thiery, J.P. (1984). Prevention of gastrulation but not neurulation by antibodies to fibronectin in amphibian embryos. *Nature*, ,**307**, 364–367.

Cooke, J. and Smith, E.J. (1988). The restrictive effect of early exposure to lithium upon body pattern in Xenopus development studied by quantitative anatomy and immunofluorescence. *Development*, **102**, 85–99.

Cooke, J. and Smith, J.C. (1987). The mid-blastula cell cycle transition and the character of mesoderm in UV induced non-axial Xenopus development. *Development*, **99**, 197–210.

Dale, L. and Slack, J.M.W. (1987a). Fate map for the 32 cell stage of *Xenopus laevis*. *Development*, **99**, 527–551.

Dale, L. and Slack, J.M.W. (1987b). Regional specification within the mesoderm of early embryos of *Xenopus laevis*. *Development*, **100** 279–295.

Dale, L., Smith, J.C. and Slack, J.M.W. (1985). Mesoderm induction in *Xenopus laevis*: a quantitative study using a cell lineage label and tissue-specific antibodies. *J. Embryol. Exp. Morph.*, **89**, 289–313.

Dent, J.A., Polson, A.G. and Klymkowsky, M.W. (1989). A whole-mount immunocytochemical analysis of the expression of the intermediate filament protein vimentin in Xenopus. *Development*, **105**, 61–74.

Elinson, R.P. and Pasceri, P. (1989). Two UV sensitive targets in dorsoanterior specification of frog embryos. *Development*, **106**, 511–518.

Gallien, L. and Durocher, M. (1957). Table chronologique du dévelopment chez *Pleurodeles waltlii* Michah. *Bull. Biol. Fr. Belg.*, **91**, 97–114.

Gimlich. R.L. and Braun, J. (1985). Improved fluorescent compounds for tracing cell lineage. *Dev. Biol.*, **109**, 509–514.

Godsave, S.F. and Slack, J.M.W. (1989). Clonal analysis of mesoderm induction in *Xenopus laevis*. *Dev. Biol.*, **134**, 486–490.

Holtfreter, J. (1938a). Differenzierungspotenzen isolierter Teile der Urodelengastrula. *Wilhelm Roux' Archiv fur Entwicklungsmechanik der Organismen.* **138**, 522–656.

Holtfreter, J. (1938b). Differenzierungspotenzen isolierter Teile der Anurengastrula. *Wilhelm Roux' Archiv fur Entwicklungsmechanik der Organismen.* **138**, 657–738.

Jacobson, M. and Hirose, G. (1978). Origin of the retina from both sides of the embryonic brain: A contribution to the problem of crossing over at the optic chiasma. *Science*, **202**, 637–639.

Kao, K.R., Masui, Y. and Elinson, R.P. (1986). Lithium induced respecification of pattern in *Xenopus laevis* embryos. *Nature*, **322**, 371–373.

Keller, R. and Danilchik, M. (1988). Regional expression, pattern and timing of convergence and extension during gastrulation of *Xenopus laevis*. *Development*, **103**, 193–209.

Keller, R.E., Danilchik, M., Gimlich, R. and Shih, J. (1985). The function and mechanism of convergent extension during gastrulation of *Xenopus laevis*. *J. Embryol. Exp. Morph.*, **89**, Suppl., 185–209.

Kintner, C.R. and Brockes, J.P. (1984). Monoclonal antibodies identify cells derived from dedifferentiating muscle in newt limb regeneration. *Nature*, **308**, 67–69.

Krieg, P.A. and Melton, D.A. (1984). Functional messenger RNAs are produced by Sp6 in vitro transcription of cloned cDNAs. *Nucleic Acid Res.*, **12**, 7057–7070.

Lehmann, F.E. (1937). Mesodermisierung des praesumptiven Chordamaterials durch Einwirkung von Lithiumchlorid auf die Gastrula von Triturus alpestris. *Wilhelm Roux' Archiv fur Entwicklungsmechanik der Organismen.*, **136**, 112–146.

Malacinski, G.M., Benford, H. and Chung, H.M. (1975). Association of an ultraviolet irradiation sensitive cytoplasmic localization with the future dorsal side of the amphibian egg. *J. Exp. Zool.*, **191**, 97–110.

Moon, R.T. and Christian, J.L. (1989). Microinjection and expression of synthetic mRNAs in Xenopus embryos. *Technique*, **1**, 76–89.

Newport, J. and Kirschner, M. (1982a). A major developmental transition in early Xenopus embryos: I. Characterization and timing of cellular changes at the midblastula stage. *Cell*, **30**, 675–686.

Newport, J. and Kirschner, M. (1982b). A major developmental transition in early Xenopus embryos II. Control of the onset of transcription. *Cell*, **30**, 687–696.

Nieuwkoop, P.D. and Faber J. (1967). *Normal Table of* Xenopus laevis. N.Holland, Amsterdam.

Pasteels, J. (1942). New observations concerning the maps of presumptive areas of the young amphibian gastrula (*Ambystoma* and *Discoglossus*). *J. Exp. Zool.*, **89**, 255–281

Scharf, S.R. and Gerhart, J.C. (1980). Determination of the dorsoventral axis in eggs of *Xenopus laevis*: Complete rescue of UV impaired eggs by oblique orientation before first cleavage. *Dev. Biol.*, **79**, 181–198

Shumway, W. (1940). Stages in the normal development of Rana pipiens. *I. External Form. Anat. Rec.*, **78**, 139–147.

Slack, J.M.W. (1985). Peanut lectin receptors in the early amphibian embryo: regional markers for the study of embryonic induction. *Cell*, **41**, 237–247.

Slack, J.M.W., Cleine, J.H. and Smith, J.C. (1985). Regional specificity of glycoconjugates in Xenopus and axolotl embryos. *J. Embryol. Exp. Morph.*, **89**, Suppl., 137–153.

Smith, J.C. (1989). Mesoderm induction and mesoderm inducing factors in early amphibian development. *Development*, **105**, 665–677.

Smith, J.C. and Watt, F.M. (1985). Biochemical specificity of the Xenopus notochord. *Differentiation*, **29**, 109–115.

Spemann, H. and Mangold, H. (1924). Uber Induktion von Embryonenanlagen durch Implantation artfremder Organisatoren. *Arch. fur microscopische Anatomie und Entwicklungsmechanik*, **100**, 599–638

Tucker, G.C., Delarue, M., Zada, S., Boucaut, J.C. and Thiery, J.P. (1988). Expression of the HNK-1/NC-1 epitope in early vertebrate neurogenesis. *Cell Tissue Res.*, **251**, 455–465.

Vogt, W. (1929). Gestaltungsanalyse am Amphibienkeim mit örtlicher Vitalfärbung. II Teil, Gastrulation und Mesodermbildung bei Urodelen und Anuren. *Wilhelm Roux' Archiv fur Entwicklungsmechanik der Organismen*, **120**, 384–706

Yamaguchi, Y. and Shinagawa, A. (1989). Marked alteration at midblastula transition in the effect of lithium on formation of the larval body pattern of *Xenopus laevis*. *Develop. Growth and Differ.*, **31**, 531–541.

10. The Chick

Claudio D. Stern

Introduction

The avian embryo has a long and distinguished history as a subject of embryological study. The Ancient Egyptians appear to have been the first to investigate its development in a systematic way; they incubated hens' eggs for varying periods of time and opened them to examine the state of development of the embryo. It was also a study of avian embryos that led William Harvey, in the 17th century, to the discovery of the circulation of the blood, and observations of the beating heart of the chick embryo that occupied René Descartes in part of his *Discours de la Méthode*. Notwithstanding this, the avian embryo is still one of the organisms of choice for modern developmental biology because it is easily obtained, large and relatively translucent, it allows delicate microsurgical manipulations to be performed easily, and because its development is relatively well understood (for reviews, see Lillie, 1952; Patten, 1971).

Birds are amniotes, like mammals, whose development they closely resemble. The main differences between the early stages of mammalian and avian development are implantation and placentation, which are characteristics of mammals. The avian embryo does not have a placenta and is a self-contained developing system, whilst the mammalian embryo depends on the mother for its nutrition, exchange occurring across the placenta.

To the modern developmental biologist, the avian embryo offers a very accessible system in which molecular studies can be combined with 'classical' embryology. Excellent staging systems are also available for the chick embryo (Eyal-Giladi and Kochav, 1976, for early stages and Hamburger and Hamilton, 1951, for later stages). Transplantation, cell labelling, immunocytochemistry and chemical treatments can now be combined with *in situ* hybridization and Northern analysis to examine changes in the patterns of gene expression resulting from experimental manipulations with well-studied effects. Moreover, with the advent of techniques for producing transgenic birds, some of which have recently become available, the developmental effects of targeted mutations can be studied in cellular as well as molecular detail.

The egg

The avian egg consists of the yolk (mainly phospholipids), which is the main source of nutrients for the embryo, enveloped by a translucent, non-cellular vitelline membrane. This is surrounded by albumen (egg white) and the whole contained within a calcareous shell, from which it is separated by two egg membranes. At the blunt end of the egg, between the two membranes, is an air sac. All of these components are secreted by the mother. The embryo itself lies initially on the surface of the yolk, just under the vitelline membrane. The viscous albumen surrounding the vitelline membrane is connected with the inner membrane by two glycoprotein threads called the chalazae, which allow the yolk to rotate so that the embryo always faces the top of the egg. The structure of the egg is designed to conserve water, to allow gaseous exchange and to prevent micro-organisms from coming into contact with the embryo, as well as to provide nutrition and protection to the embryo. It also provides a uniquely complex environment for the embryo to develop. For example, the pH of the yolk is slightly acid, while that of the albumen is quite alkaline. During early stages of development, the edges of the single-cell thick embryo attach to the inner (yolk) face of the vitelline membrane, on which it expands, and is poised in a pH gradient that may amount to as much as 3pH units across a single cell.

In addition to nutritive components and glycosaminoglycans to increase its viscosity, the albumen contains the enzyme lysozyme, a bacteriostatic agent. For this reason, egg albumen is often used in embryo cultures to prevent the growth of micro-organisms as well as to provide a source of nutrients.

Chick and quail embryos hatch 19–21 days after laying if incubated at 38°C; quail embryos develop slightly faster (19–20 days) than chicks (20–21 days).

Normal development

Development in utero

After fertilization, which occurs internally in the mother, a hen's egg spends about 5hr in the oviduct, and then moves to the uterus, where cleavage begins. This happens after about 5+hr after laying of the previous egg and the cleaving embryo remains in the uterus some 20hr before it is laid. This is a major disadvantage of working with avian embryos, because the earliest stages of development are not easily accessible.

Cell division (cleavage) of the fertilized egg occurs in a planar way, with the daughter cells staying in the plane between the vitelline membrane and the yolk (Eyal-Giladi and Kochav, 1976; Burley and Vadehra, 1989; Tullett, 1991). Unlike mammalian embryos, which have holoblastic cleavage, cleavage in avian embryos is meroblastic. This means that the membranes separating new cells arise as open cleavage planes, stretching into the yolk. As a result, the cells forming in the centre of the cleaving blastodisc are smaller than those around its periphery, while those around the edge contain more yolk. Because of this quirk, the morula and blastula stages, as defined in lower vertebrates, are difficult to distinguish in avian embryos.

As the egg descends along the mother's reproductive system, it rotates while the albumen and shell components are secreted around it. During this rotation, the plane of the cleaving blastodisc remains at an angle of about 45° to the radius of the earth. This angle and the rotation are believed to be important in determining the polarity of the embryo. The edge of the blastodisc facing downwards will form the caudal (posterior) part of the embryo, under the influence of gravity.

Formation of the embryonic axis and origin of the mesoderm

At the time of laying, the chick blastoderm consists of a disc of cells, some 2mm in diameter, comprising an inner, translucent *area pellucida* and an outer, *area opaca* (**10.1**), with this latter region only contributing to extra-embryonic structures. The first cell layer to be present as such is the epiblast, which is continuous over both *areae opaca* and *pellucida*. It is a one-cell thick epithelium which soon becomes pseudostratified and columnar, the apices of the cells facing the albumen. From this arises a second layer of cells, the hypoblast. At this time, the hypoblast is no more than several unconnected islands of 5–20 cells. By about 6hr of incubation, however, more cells are added to it and it becomes a loose but continuous epithelium. These further hypoblast cells arise from two regions: the posterior (future caudal) part of the marginal zone, which is a region of epiblast at the outer edge of the area pellucida, and a flap of yolky cells which lies deep to the posterior marginal zone, and is called the posterior germ wall margin (**10.2**; see Stern, 1990).

The embryo now consists of two layers: the epiblast proper, facing the albumen, from which will arise all of the embryonic tissues, and the hypoblast, facing the yolk, which will only give rise to extra-embryonic tissues (mainly the yolk sac stalk), although it may also contain some primordial germ cells. As the hypoblast continues to spread as a layer from posterior to anterior parts of the blastodisc (**10.2**), cells appear between the previous two germ layers. These are the first cells of the mesoderm and, as more of these accumulate, they coalesce in the first axial structure of the embryo, the primitive streak, which makes its appearance at the posterior margin of the *area pellucida* at about 10hr of incubation (**10.3**). Later, as the streak elongates along the antero-posterior axis of the embryo, more mesodermal cells are recruited into the primitive streak by migration from the epiblast (**10.4**). All the mesoderm eventually migrates out of the streak to give rise to four mesodermal components: the lateral plates, intermediate mesoderm, paraxial mesoderm and the axial notochord (**10.5, 10.6**). The lateral plates will give rise to the circulatory system, the lining of the coelom, the limb skeleton, and most of the remaining mesodermal organs. The intermediate mesoderm is a cord separating the lateral plate from the paraxial mesoderm, which gives rise to the transitory embryonic kidneys (pronephros and mesonephros; see **10.10**). The paraxial mesoderm becomes subdivided into somites (see below), which give rise to the skeletal (voluntary) musculature, to the dermis and to the skeleton of the trunk. The notochord forms a rod in the midline of the embryo and mainly contributes to the intervertebral discs.

Before formation of the lateral plate, the mesoderm is packed densely at the primitive streak; as the lateral plate forms, most of it migrates away from the axis of the streak. The left and right halves of the lateral plate later become separated from each other by the regression (shortening) of the primitive streak that occurs after the end of gastrulation. The notochord is laid down as a rod of mesoderm by the cranial tip of the primitive streak (Hensen's node), and elongates as the primitive streak regresses (**10.5, 10.6**) (see Selleck and Stern, 1991).

Some primitive streak cells insert into the hypoblast, displacing it towards the edges of the *area pellucida*. These primitive-streak-derived endodermal cells form the definitive endoderm, which will give rise to the lining of the gut. The elongation of the primitive streak and the expansion of the blastodisc, together with further recruitment of cells derived from the posterior marginal zone (this contribution now forming the junctional endoblast, which is also extra-embryonic) confine the original hypoblast to a crescent shaped region underlying the anterior portion of the *area pellucida*; this region is known as the germinal crescent because it contains, transiently, the primordial germ cells which will later migrate into the gonads.

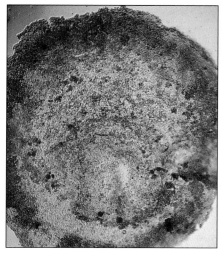

10.1 A chick embryo at the time of laying. The embryo is a disc of some 2–3mm diameter. It is stained with carmine as a whole mount and viewed from its ventral surface. The posterior marginal zone can be distinguished as a more translucent arc towards the lower part of the photograph. (×17. Eyal-Giladi and Kochav stage X).

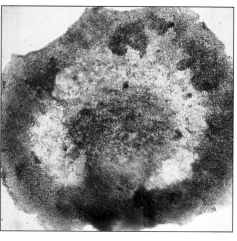

10.2 By about 5 hours' incubation the hypoblast can be seen as a darker sheet of cells, emerging from the posterior margin and starting to cover the surface of the blastoderm. (×17. Eyal-Giladi and Kochav stage XII).

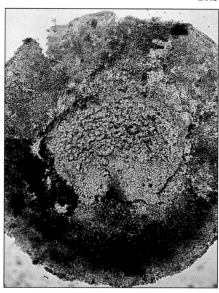

10.3 At about 8–10 hours' incubation the primitive streak appears. It can be seen as a thick triangle of cells which have accumulated between the other two layers (epiblast and hypoblast) at the posterior end of the blastoderm (×12.5. Hamburger and Hamilton, stage 2).

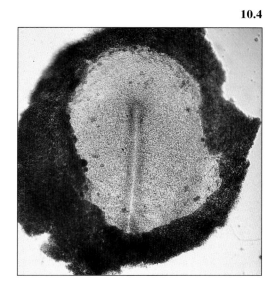

10.4 By some 15 hours' incubation, the primitive streak has developed a groove and has elongated to occupy about ²/₃ of the diameter of the blastoderm. Anteriorly, the streak is terminated by a condensation of cells, Hensen's node. (×10; Hamburger and Hamilton, stage 4.)

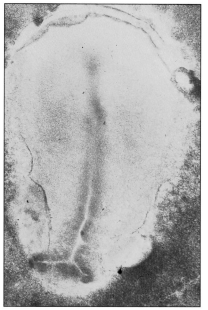

10.5 After 18 hours' incubation, the head process (notochordal rod underlying the head) has emerged from Hensen's node. The primitive streak now begins to shorten (regression). (×8.5; Hamburger and Hamilton, stage 5.)

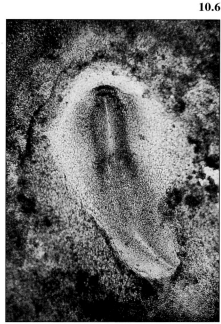

10.6 By about 24 hours' incubation, the primitive streak has shortened to about half of its original length, a fold (head fold) involving ectoderm and endoderm appears at the anterior tip of the head process, and the neural plates begin to elevate on either side of the embryonic axis. Very shortly afterwards, the first somite will appear (×5; Hamburger and Hamilton, stage 7-.)

Segmentation of the mesoderm

The segmental organization of vertebrate embryos is most obvious in the pattern of somites (**10.7–10.11**), from which derive the vertebrae and ribs of the axial skeleton, the dermis of the trunk and all the voluntary musculature of the adult. The metameric pattern of somites determines the segmental arrangement of other structures in the embryo, such as the peripheral nervous system (see below).

In the chick embryo, some 55 pairs of somites form, each somite being, at first, an epithelial sphere, or rosette (**10.9**), budding off the rostral (anterior) end of each of the paired segmental plates of paraxial mesoderm (**10.10**) that appear about 1.5 days after the egg is laid. Each pair of somites takes about 1.5hr to form. Some 6–8hr after its initial appearance, each somite splits up (**10.11**) into the dermomyotome dorso-laterally, which retains some epithelial characteristics and gives rise to the dermis of the trunk and to skeletal muscle, and the sclerotome ventro-medially, which is a loose mesenchyme that gives rise, along with the notochord, to the axial skeleton. Recent results (Primmett *et al.*, 1989) suggest that the segmental pattern of somites is controlled in part at least by cell autonomous properties that depend on the cell division cycle (see below).

The subdivision of the sclerotome into rostral and caudal halves is maintained by mixing restrictions between the two kinds of cells (for review see Keynes and Stern, 1988). Rostral cells are able to mix with rostral cells, and caudal with caudal, but when rostral and caudal cells meet they generate a boundary that separates them and is probably important in the morphogenesis of the vertebral column. Traditionally, it is thought that the rostral half of one sclerotome joins the caudal half of the preceding sclerotome to give rise to one vertebra, a model which accounts for the observation that the vertebrae are out of phase with the corresponding myotome (axial muscle blocks). Recent experimental work (Stern and Fraser, in preparation) tracing the derivatives of individual somites has however cast some doubt on this interpretation; it suggests that each somite gives rise to one vertebra, but that sclerotome cells shift anteriorly by a half segment by migrating along the notochord.

10.7

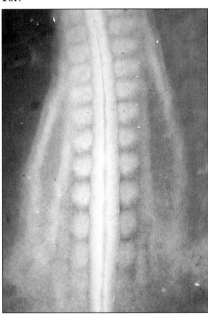

10.8

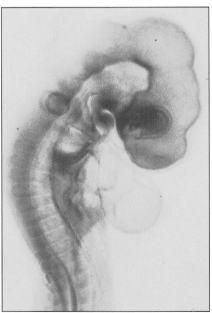

10.9

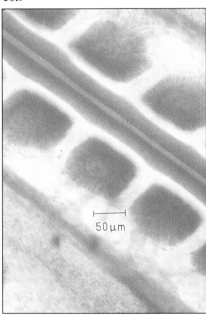

10.7 An embryo at 2.5 days' incubation, showing somites at the level of the wing bud. Photographed *in ovo*, using reflected light after injection of Indian ink under the embryo to improve contrast. Myelomeres (indentations in the neural tube) can be seen next to each somite. (×50; Hamburger and Hamilton, stage 15.)

10.8 Head of embryo at 3 days' incubation. Whole mount, stained with carmine. The eye and otic vesicle (ear) can be distinguished, as can the primitive subdivisions of the brain. (×20; Hamburger and Hamilton, stage 16.)

10.9 Newly-formed somites towards the posterior end of an embryo at 2 days' incubation. Whole mount, stained with Fast Green. The rosette shape of the somites can be seen: the cells surround a central lumen. (Hamburger and Hamilton, stage 11.)

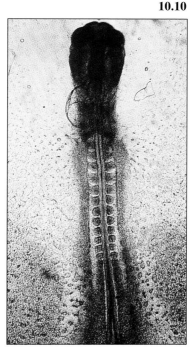

10.10 Lower power view of an embryo at 2 days' incubation. The somites can be seen flanking the embryonic axis. At the posterior end of the embryo (lower part of photograph), the segmental plates of presomitic mesoderm can be seen. Lateral to the rows of somites, the condensed mesoderm represents the presumptive intermediate mesoderm rods, which will give rise to the embryonic kidney (pro- and mesonephros). In extraembryonic regions, blood islands begin to appear. Whole mount, stained with carmine. (×13; Hamburger and Hamilton, stage 11.)

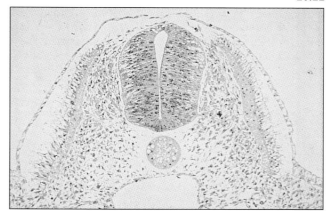

10.11 Transverse section through a well developed somite in a 3-day embryo, at the level of the wing bud. The neural tube and notochord are seen in the midline. On either side, the somite has split up into the dermomyotome dorso-laterally and the mesenchymal sclerotome ventro-medially. Plastic section, stained with toluidine blue. (×330; Hamburger and Hamilton, stage 16.)

Early development and segmentation of the peripheral nervous system

After a hen's egg has been incubated for about 24hr, the epiblast in the midline of the embryo overlying the developing notochord thickens to become the neural plate (**10.6**). This region elongates towards the tail end of the embryo as the primitive streak continues to regress and as it does so, it elevates to form the neural folds, which then roll up into a tube, the neural tube (**10.7, 10.9** and **10.11**). The whole process progresses roughly in an anterior-to-posterior direction so that the neural plate is still elevating in the tail region while it has already closed in cranial regions (for review, see Schoenwolf and Smith, 1990).

The rims of the neural plate that meet during tube closure (the dorsal part of the neural tube) give rise to the neural crest, a special, multipotent population of migratory cells that colonises many regions of the embryo and give rise to melanocytes, to the adrenal medulla, to the sensory component of the peripheral nervous system, to the autonomic ganglia (sympathetic and parasympathetic), to glial (Schwann) cells that accompany peripheral nerves and to skeletal elements of the cranial vault and face (for review see Le Douarin, 1983). The motor nerves do not arise from the neural crest; instead, axons emerge from the ventral portion of the neural tube, from where they extend outwards to find their peripheral targets.

The neural-crest cells and motor axons that emerge from the neural tube display segmental organization (for review see Keynes and Stern, 1988). The earliest motor axons can be seen to exit from the neural tube in the trunk at about three days' of incubation (**10.12**), and about one day later many axons can be seen to fasciculate together to form the spinal nerves (**10.13–10.15**). When looked at in the whole mount, using staining techniques that reveal axons, such as zinc iodide/osmium tetroxide (**10.12**) or an antibody against neurofilaments

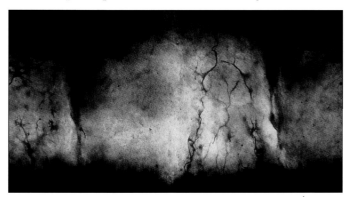

10.12 The earliest motor axons emerging from the neural tube (lower part of photograph), confined to the anterior (right in photograph) half of each sclerotome. Growth cones can be seen at the tips of these 'pioneer' axons. Whole mount of embryo stained with zinc iodide/osmium tetroxide. (×500. 3 days' incubation, Hamburger and Hamilton, stage 17).

(**10.13–10.15**), the relationship between the growing motor axons and the adjacent somite mesoderm can be observed: motor nerves are confined to the anterior (rostral) halves of the somites (**10.12–10.14**). The neural crest cells are also confined to the rostral halves of the somites during their migration. They can be visualized using an antibody, HNK-1 (**10.16**).

Recent evidence suggests that the segmental outgrowth of motor nerves and migration of neural crest cells are determined at least in part by inhibitory cues present in the caudal (posterior) half of the sclerotome. First, rostrocaudal rotation of the neural tube by 180° does not alter the segmental pattern of motor nerves or crest cells, whilst rotation of the somitic mesoderm reverses the pattern, suggesting that the segmentation is controlled by differences between the cells in the rostral and caudal halves of the somite, rather than by intrinsic segmentation of the neural tube (see Keynes and Stern, 1988).

Second, the caudal halves of the sclerotomes express glycoprotein receptors to peanut agglutinin (PNA-receptors; Stern *et al.*, 1986; **10.17**) and it has now been shown that PNA-binding fraction of membrane glycoproteins from the sclerotome causes the collapse of growth cones of cultured dorsal root ganglion axons (Davies *et al.*, 1990).

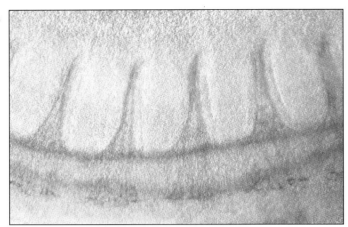

10.13 Embryo at about 4 days' incubation (Hamburger and Hamilton, stage 20), showing forming nerve roots. Motor axons are now fasciculated within the anterior half of each somite. Whole mount, stained with an antibody against neurofilaments, and revealed with a second antibody coupled to horseradish peroxidase and the diaminobenzidine reaction. In the lower part of the figure the neural tube can be seen, where the cell bodies are revealed by the antibody as a continuous bar of staining. The fan-like arrangement of neurons still lower down in the photograph represents the sensory neurons, whose cell bodies are located in the dorsal root ganglia, within the anterior half of each sclerotome (×170).

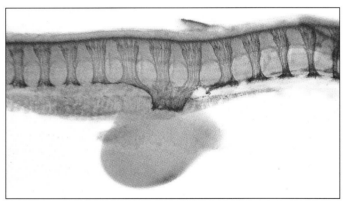

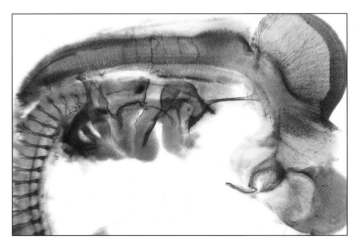

10.14 Side view of whole mount of embryo at 4 days' incubation (Hamburger and Hamilton, stage 21), at the level of the wing bud, stained with antibody against neurofilaments. At this level, axons collect to form the brachial plexus, which will innervate the wing (×70).

10.15 Head of whole mounted embryo (Hamburger and Hamilton, stage 21), stained with antibody against neurofilaments. Axons in the head and branchial arches are starting to form complex patterns (×40).

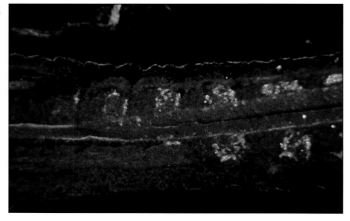

10.16 Neural crest cell migration through the sclerotome. Frozen, coronal section through Hamburger and Hamilton stage 12 embryo (head to right), stained with monoclonal antibody HNK-1, which recognizes neural crest cells, revealed with a fluorescein-coupled second antibody. In this embryo, the last few somites on the right (lower part of photograph) had been excised and replaced with a graft of posterior segmental plate. Migration of crest cells into the grafted segmental plate is inhibited (×100).

10.17 Somites in a whole-mounted Hamburger and Hamilton stage 16 embryo, stained with peanut agglutinin coupled to peroxidase and revealed by diaminobenzidine. Peanut lectin recognizes cells in the posterior half of the sclerotome. The lectin receptors appear to be involved in inhibiting the growth of motor axons and perhaps the migration of neural crest cells in the posterior sclerotome (×115).

Segmentation of the central nervous system

Despite the finding (see above) that the somites are responsible for the segmental organization of the peripheral nervous system, the central nervous system (neural tube) itself also displays segmental organization. In the trunk, evidence for such segmentation (myelomeres) is limited for the present to morphological signs: the developing spinal cord shows indentations in line with each neighbouring somite (see **10.7, 10.9**).

In the head, however, the evidence for intrinsic segmentation is stronger. Even in unfixed specimens (**10.18, 10.19**), the hindbrain (rhombencephalon) can be seen to be subdivided into a series of metameric units called rhombomeres (**10.19–10.21**; see Lumsden and Keynes, 1989). Seven or eight such segments can be distinguished, displaying an alternating pattern of neuronal differentiation (**10.20, 10.21**) and other properties. These segments ('rhombomeres') are separated from each other by lineage restrictions that prevent cells from one rhombomere from crossing to the adjacent one (Fraser *et al.*, 1990). The diencephalon is similarly subdivided into four segments ('diencephalic neuromeres'), again with an alternating pattern of neuronal differentiation and cell surface properties, and lineage restrictions maintaining the boundaries between adjacent neuromeres (Figdor and Stern, in preparation).

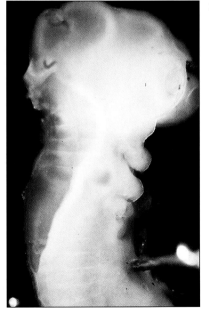

10.18 Appearance of embryo of about 3 days' incubation after injection of Indian ink under the embryo to improve contrast, seen in its egg through a window cut into the shell (×8.5).

10.19 View of hindbrain region of same embryo as in **10.18**. The dorsal aspect of the neural tube has been slit open, revealing the subdivisions of the hindbrain (rhombomeres). Unfixed embryo, reflected light (×40).

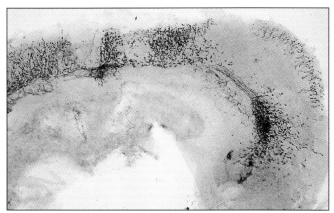

10.20 The hindbrain region at Hamburger and Hamilton stage 12 (2 days' incubation), seen by immunoperoxidase using an antibody to neurofilaments. Bands of axons develop first in even-numbered rhombomeres (2, 4 and 6; ×85).

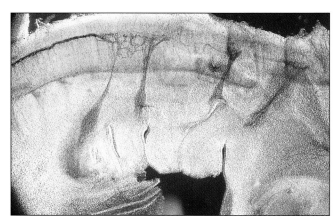

10.21 Whole mount of embryo at Hamburger and Hamilton stage 17, seen by immunoperoxidase with nuerofilament antibody. Nerves supplying the branchial arches can be seen to emerge from rhombomeres 2, 4, 6 and 7 (cranial nerves V, VII/VIII, IX and X, respectively). The otic vesicle (ear), which lies opposite rhombomeres 5 and 6, can be seen intervening between the roots of VII/VIII and IX (×67).

Experimental manipulations
Regulation and induction

Higher vertebrate embryos are capable of extensive regulation (**10.22–10.24**). If a pre-primitive-streak chick blastoderm is cut into several portions, most of the portions are able to produce a complete, albeit somewhat smaller, embryo (**10.22**). Furthermore, if pieces derived from several embryos are combined, a fairly normal embryo develops which receives contributions from all the donor fragments. These findings indicate that, at this stage of development, groups of cells are not committed to a pre-ordained programme of development. Development at this stage must, therefore, transcend the cell autonomous interpretation of a genetic programme; cell interactions are thus important in determining the pattern of gene expression in individual cells.

There are, in theory, two possible interpretations for these phenomena: the first is that cells in the embryo are pluripotent, and become specified to their ultimate fates by cell interactions, so that, when a fragment is separated from the embryo, cells in the fragment interact to give the correct proportions of different cell types. The second interpretation of the phenomenon of regulation requires the embryo to consist of a fairly random mix of cells already specified to their ultimate fates, which later sort to their correct destinations. According to this second interpretation, isolation of a fragment of embryo leads to sorting of the cell types to newly-defined sites, but the cells do not change fates. We do not yet know which of these two interpretations is correct, although the first is generally favoured.

The hypoblast layer appears to be involved in the specification of embryonic polarity. If it is rotated by 180° about its anteroposterior axis, the primitive streak now forms from the opposite (anterior) pole of the blastodisc. The nature of the interactions between hypoblast and presumptive primitive streak cells in the epiblast is not understood, although it has been suggested (Mitrani *et al.*, 1990; see also Stern, 1992) that the hypoblast induces the formation of mesoderm from the epiblast, as the vegetal pole cells induce mesoderm from the animal cap in amphibian embryos. It has not yet however been possible to demonstrate mesoderm induction in isolated fragments of avian embryos, as has been done in amphibians.

The cells of Hensen's node, which lie at the rostral (anterior) tip of the primitive streak, have rather special properties. If a node is grafted into the lateral region of a host embryo, a second nervous system is formed (**10.23**; see also **10.36**), just as a graft of the dorsal lip of the blastopore generates a second nervous system in amphibians. This is neural induction. It can be demonstrated, using chick/quail chimaeras or other methods of cell marking (see below), that the supernumerary nervous system that forms is derived from cells of the host epiblast rather than of the grafted node. Some signal from Hensen's node can therefore change the fate of the overlying epiblast cells to make them become nervous system, rather than epidermis.

10.22

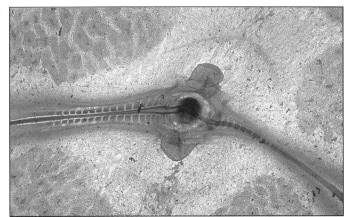

10.23

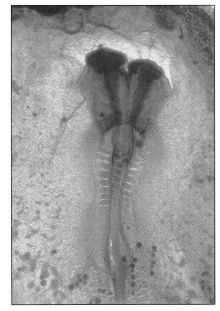

10.24

10.22–10.24 Three different arrangements of twins generated by manipulations of the early chick blastoderm. In **10.22** (×8.5), a pre-primitive streak stage blastoderm was completely transected along its midline. Two completely separate embryos have developed. Whole mount, stained with carmine. In **10.23** (×13), the anterior tip of the primitive streak of a stage 3 embryo was grafted into the anterior part of a stage 3 host blastoderm. A second axis developed, but the heads of the two embryos have merged. In the axis on the right no notochord developed, and a single, medial row of somites has formed. Whole mount, stained with Fast Green. In **10.24** (×110), the primitive streak of a Hamburger and Hamilton stage 2 embryo was cut along its length. Two parallel axes developed, which share one row of somites and a single, wide neural tube in the trunk. In the head region, however, the two axes separate. Whole mount, stained with Fast Green.

Other transplantation experiments: development of the limbs

The avian embryo has also been used extensively for studies on limb development (**10.25–10.30**; for review, see Tabin, 1991). The limb buds begin to form at about 3 days' incubation after laying, when the ectoderm opposite somites 15–21 (wing bud; **10.7**, **10.11** and **10.14**) and 27–33 (leg bud) begins to elevate from the surface of the body, enclosing a mass of loosely packed lateral plate mesoderm. In time, the cells of this mesoderm will condense to form the primordia of the limb skeleton (**10.28**). The muscles (**10.26**) and dermis (**10.25**) arise from somite-derived cells which migrate into the limb buds.

Grafting experiments have shown that the posterior margin of the chick limb bud contains special cells, able to re-specify the pattern of the whole limb. This region has been called the zone of polarizing activity (ZPA). If such a region is grafted into the anterior margin of a host limb bud, a mirror-image pattern of bone and muscle elements is formed (**10.30**). It is believed that the ZPA produces a diffusible 'morphogen' (thought by some to be retinoic acid), the local concentration of which at different positions in the limb determines the type of bone and muscle elements that will develop at those positions (Tickle, 1991).

The distal part of the developing limb bud is also special. It consists of a thickened ridge of surface ectoderm (the apical ectodermal ridge, or AER) which covers a region of special mesoderm, called the progress zone. The AER is required for continued growth and development of the limb, while the progress zone appears to be involved in controlling the proximo-distal sequence of skeletal elements.

10.25 Wing of chick embryo at stage 35 (approximately 9 days), viewed with reflected light. Signs of early feather bud formation can be seen in the skin. (×25. Courtesy of Drs J.H. Lewis and P. Martin.)

10.26 Wing of chick embryo at stage 35, viewed through cross-polarizing filters, which reveals the developing muscles by their birefringence. (×25. Courtesy of Drs J.H. Lewis and P. Martin.)

10.27 Wing of chick embryo at stage 35, after staining with silver to reveal the nervous system. (×25. Courtesy of Drs J.H. Lewis and P. Martin.)

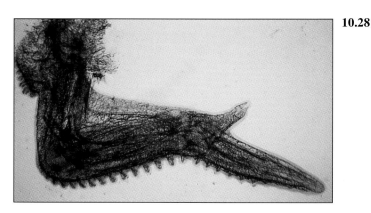

10.28 Wing of chick embryo at stage 35, after injection of Indian ink into the circulation to reveal the pattern of blood vessels. (×25. Courtesy of Drs J.H. Lewis and P. Martin.)

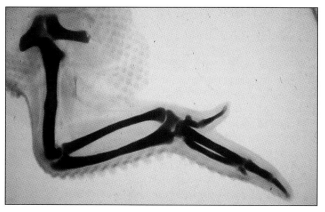

10.29 Wing of chick embryo at stage 35, after staining with Alcian Green to reveal skeletal elements. (×25. Courtesy of Drs J.H. Lewis and P. Martin.)

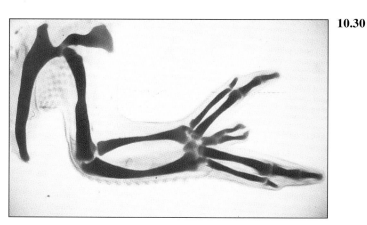

10.30 Wing of chick embryo with duplication of digits generated by grafting a zone of polarizing activity from a donor embryo to the anterior margin of the limb bud at stage 21. Whole mount, stained with Alcian Green. (×25. Courtesy of Dr C. Tickle.)

Fate mapping and cell lineage analysis

Two new techniques for identifying the descendants of specific groups of cells have recently begun to be used extensively. One of these involves the injection into a single cell of the fluorescent tracer Lysine-Rhodamine-Dextran (see Fraser *et al.*, 1990). This tracer has many advantages: it is non-toxic, it is not transferred between cells other than during cell division, it is intensely fluorescent and it can be fixed using aldehyde fixatives for subsequent histological analysis. The fluorescent derivatives of the labelled cell can be seen for two days, or up to about 10 cell divisions later (**10.31**).

Another technique involves the use of the carbocyanine dyes DiI and DiO (see Stern, 1990; Selleck and Stern, 1991). These are lipophilic and water insoluble, and therefore insert themselves into cell membranes (**10.32**). They are also non-toxic, cell autonomous and intensely fluorescent, but, because they are soluble in organic solvents, they cannot be subjected to conventional histological procedures. Nevertheless, the labelling can be stabilised by a photo-oxidation process, in which the fluorescence emitted is used to oxidize the substrate diaminobenzidine, which then produces a brown-black product, insoluble in organic or aqueous solvents, and which allows histological processing (see Stern, 1990; Selleck and Stern, 1991).

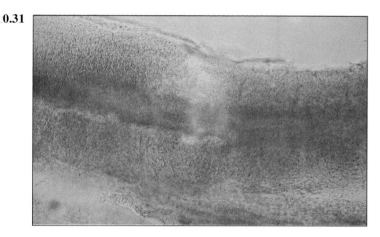

10.31 Clone of fluorescent cells derived from a single, ventral neural tube cell that had been injected with Lysinated Rhodamine Dextran at Hamburger and Hamilton stage 11 (two days earlier). Double exposure, showing the clone by fluorescence (red) and the whole mounted embryo by transmitted light. (×170. Reproduced from Stern *et al.* (1991), with permission from the Company of Biologists Ltd.)

10.32 Cells derived from a region in the anterior part of the primitive streak. A small group of cells in this region of a Hamburger and Hamilton stage 4 embryo was labelled with the carbocyanine dye, DiI. After 1.5 days' further incubation, the labelled descendants are found restricted to the lateral half of the somitic mesoderm. Double exposure: fluorescence (red) to reveal DiI and transmitted light. (×50; reproduced from Selleck and Stern (1991) with permission from the Company of Biologists Ltd.)

Teratology

Like other vertebrate embryos, avian embryos lend themselves to teratological studies. Specific agents can be added to chick embryos during culture by the technique of New (1955), injected under the blastoderm *in ovo*, or even just dropped onto the embryo in the egg. They can also be applied locally to particular regions of the embryo, by, for instance, using carrier ion exchange beads (Tickle *et al.*, 1982). However, the immunologist Peter Medawar once pronounced:

> 'The classification and investigation of abnormal embryos... are the subject matter of "teratology", a word that suggests pretensions to the stature of a science (a designation not really deserved)..., but teratology has not — as had at one time been hoped — thrown a flood of light upon developmental processes, and it has not helped us very notably in the interpretation of normal development. Teratology is more deeply in debt to embryology than the other way around.'

Although perhaps a little extreme, there is some truth in this statement. There are very few instances where teratology as an approach has helped to elucidate a developmental mechanism, at least in avian embryos. There are some exceptions, however, and three examples have already been mentioned above *en passant*. First, when an isolated chick epiblast is treated with an appropriate dose of activin, a whole embryonic axis can develop (Mitrani *et al.*, 1990; see also Stern, 1992). Second, if a bead, soaked in retinoic acid, is implanted into the anterior margin of a chick wing bud, an extra set of digits is induced to develop, mimicking the action of the ZPA (Tickle *et al.*, 1982; see **10.30**). Third, embryos subjected to brief heat shock or to drugs that affect the cell cycle cause specific and predictable effects on somite segmentation (Primmett *et al.*, 1989; **10.33, 10.34**).

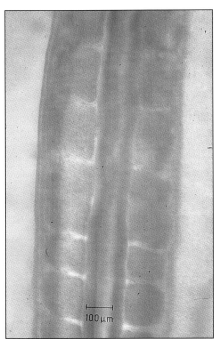

10.33 Treatment of an embryo with drugs that can synchronize cells by interference with the cell cycle can cause localized segmental defects in the somite pattern. Here, an embryo that had received a brief pulse of 5-fluorouracil 24hr earlier: a pair of somites on each side of the embryo have fused together. Whole mount, stained with Fast Green (×70).

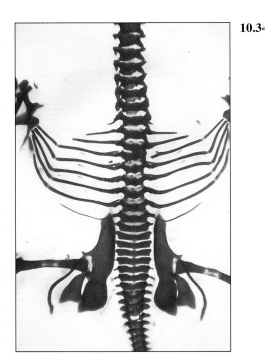

10.34 Embryos treated with methods that affect the cell cycle (see **10.33**) generate periodic segmental anomalies. This embryo that had received a brief heat shock, after culture for a few days, displays three sites with vertebral column anomalies: a fusion of neural arches in the upper thoracic region, an ectopic lumbar rib and a more caudal neural arch defect. Whole mount, stained with Alcian Blue (×125).

Region- and cell-type-specific markers

Recent research has identified a number of molecular markers that are expressed specifically in different regions of the embryo. These are extremely useful when conducting studies of regional patterning. In particular, the avian homologues of many homeobox-containing genes first identified in mouse embryos have now been cloned, and in some instances antibodies are available that recognize the protein product of these genes. **10.35**, for example, shows the expression pattern of the homeobox gene *Chick-En* (homologous to the mouse *engrailed-2* gene), which is localized in the posterior half of the midbrain and anterior part of the first rhombomere. **10.36** shows the expression of another homeobox gene, homologous with the mouse gene *Hox3.3* and the *Xenopus* gene *XlHbox1*. At early stages of development, this is expressed in the entire nervous system posterior to the occipital region, and in the mesoderm of the trunk but excluding the occipital region. In this figure, an ectopic nervous system has been induced by the graft of Hensen's node from a donor embryo. The induced axis

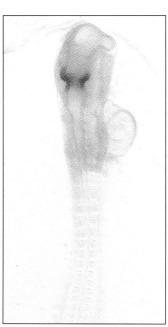

10.35 Whole mount of a chick embryo at Hamburger and Hamilton stage 11, stained with a monoclonal antibody against the engrailed-2 gene product and revealed by immunoperoxidase. The gene is expressed in the posterior midbrain and anterior part of rhombomere 1. (×170. Reproduced from Storey *et al.* (1992) with permission from the Company of Biologists Ltd.)

10.36 Whole mount of chick embryo at Hamburger and Hamilton stage 12, stained with polyclonal antiserum against the *XlHbox1/Hox3.3* gene product and visualized with a second antibody coupled to alkaline phosphatase. The antibody recognizes the neural tube posterior to the hindbrain and somitic mesoderm posterior to the occipital region. When at Hamburger and Hamilton stage 4, a Hensen's node from a donor embryo of the same stage had been grafted into this embryo. The grafted node induced a secondary neural axis, seen on the upper right of the photograph. This ectopic axis also expresses the homeoprotein, indicating that it has undergone regionalization. (×170. Reproduced from Storey *et al.* (1992) with permission from the Company of Biologists Ltd.)

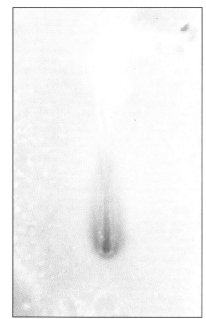

also expresses the homeoprotein in its posterior part, showing that regionalization has occurred.

Other antibodies have been generated which recognize specific cell types and many such antibodies are now available. For example, immunoreactivity with monoclonal antibody against the L5 epitope (Roberts *et al.*, 1991) is expressed at early stages of development, specifically in the neural tube and structures derived from it (**10.37**). Monoclonal antibody Not1 (Yamada *et al.*, 1991) is specific for the notochord and HNK-1 recognizes migrating neural crest cells in somite-stage embryos (see above) (**10.38**).

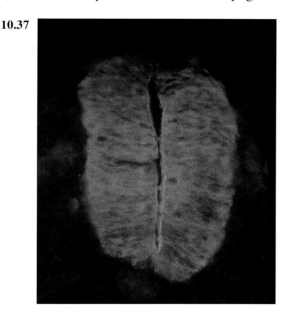

10.37 Transverse section through a chick embryo at Hamburger and Hamilton stage 12, stained with monoclonal antibody against the carbohydrate epitope L5 (see Roberts *et al.*, 1991), revealed with a fluorescein-conjugated second antibody. The L5 antigen is expressed in the developing nervous system (×600).

10.38 Coronal section through a chick embryo at Hamburger and Hamilton stage 12, stained by double-label indirect immunofluorescence with monoclonal antibody HNK-1 (fluorescein, green; see **10.16**), which recognizes neural crest cells migrating through the anterior half of the sclerotome, and monoclonal antibody Not1 (Yamada *et al.*, 1991) which recognizes the notochord (rhodamine, red; ×250).

HNK-1 as a useful marker at different stages of development

We have already mentioned the monoclonal antibody HNK-1 as a marker for migrating neural crest cells. At other stages of development, however, it transiently recognizes different cell types. **10.39–10.44** show some examples of the immunoreactivity patterns seen during very early development (Canning and Stern, 1988). Initially, the antibody recognizes cells of the forming hypoblast (**10.39, 10.41**), but when the primitive streak appears (**10.41**), the antibody recognizes all the cells contained in it. Before the appearance of the streak, HNK-1 reveals a mixture of positive and negative cells in the epiblast (**10.42**). It has recently been shown (Stern and Canning, 1990) in two experiments that the positive cells in the epiblast are the precursors of the HNK-1-positive cells in the primitive streak. First, when HNK-1 is coupled directly to colloidal gold and this is used to map the fates of the early HNK-1-positive cell population (**10.43**), it is found that all the HNK-1-gold-labelled cells contribute to the streak and not to the epidermis. Second, when the HNK-1- positive cells are ablated with the antibody and complement, no primitive streak develops; these treated embryos can however be rescued with a graft of quail HNK-1-positive cells, in which case the grafted quail cells contribute to the mesoderm (**10.44**).

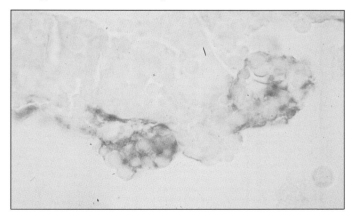

10.39 At about the time of laying (Eyal-Giladi and Kochav stage X), when the hypoblast consists only of isolated islands of cells, monoclonal antibody HNK-1 recognizes cells that ingress from the epiblast to form the hypoblast islands. Frozen section, indirect immunoperoxidase (×1000).

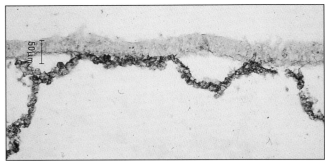

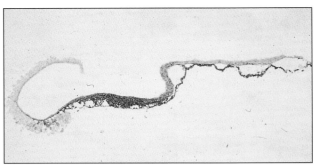

10.40 When the hypoblast is fully formed (Eyal-Giladi and Kochav stage XIII), the antibody recognizes the entire sheet of hypoblast cells, although some of these cells are derived from the epiblast by ingression and others from the posterior margin by migration.

10.41 By the time the primitive streak makes its appearance (Hamburger and Hamilton stage 2), the antibody recognizes the primitive steak cells themselves. The photograph is of a sagittal section through a stage 3 embryo, posterior towards left, stained by indirect immunoperoxidase with the HNK-1 antibody (×20).

10.42 Just before primitive streak formation (Eyal-Giladi and Kochav stage XII-XIV), the epiblast contains an apparently random mixture of HNK-1-positive and -negative cells, seen here in a whole mounted embryo seen from the epiblast side, by indirect immunoperoxidase.

10.43 The HNK-1-positive cells in the epiblast can also be visualized (and their descendants followed) if the antibody is directly coupled to colloidal gold. After incubation of an embryo in this reagent, the HNK-1-positive cells endocytose the complex and can be followed by virtue of the gold particles in the cytoplasm. This section through an epiblast at Eyal-Giladi and Kochav stage XIII labelled in this way shows 4 marked cells. Wax section, unstained; gold label intensified with silver.

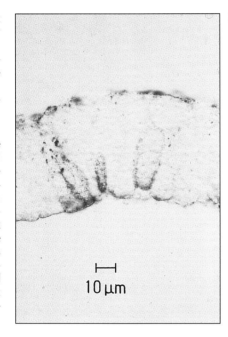

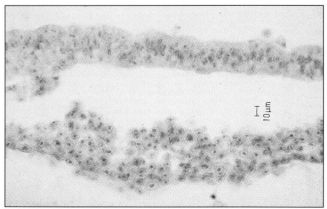

10.44 When HNK-1-positive cells in the chick epiblast are ablated with the antibody and complement, no primitive streak forms. It is however possible to rescue such embryos by a graft of quail HNK-1-positive cells. A primitive streak now develops, and the mesoderm contained within it is derived from the quail donor. The photograph shows a section through such a rescued embryo. Quail cells can be recognized from chick host cells by the condensed heterochromatin associated with the nucleolus, here visualized using a haematoxylin technique following acid hydrolysis.

Quail-chick chimaeras

This powerful technique, introduced in the early 1970s by Nicole Le Douarin in France (reviewed by Le Douarin, 1983), has been extensively used to map the descendants of particular groups of cells in the avian embryo. It was this technique that allowed the pathways of migration of the neural crest cells to be mapped accurately in the avian embryo, and the results have since been confirmed by other techniques in mammalian and amphibian embryos. Quail and chick embryos develop more or less synchronously (the quail very slightly faster at later stages of development) and cells from one species seem able to behave normally (i.e. according to their position) when grafted into an embryo of the other species. Moreover, cells from the two species can be recognized from each other by the mass of condensed heterochromatin associated with the quail nucleolus (**10.44**), and the marker does not become diluted with cell division.

In vitro development

It is possible to culture avian embryos out of the egg from late uterine stages until the formation of axial structures (head primordium, beating heart, somites, neural tube, notochord, etc.), using the technique described by New (1955). Although it is still difficult to access the very earliest stages of development (cleavage) for experimental manipulation, a technique for 'complete embryo culture' has recently been developed (Perry, 1988a) which allows access to the 1-cell stage and further incubation to the hatching chick stage. Using this technique, it is now possible to inject DNA constructs into the single-cell embryo and to look at their expression at much later stages of development.

Developmental mutants

Unlike *Drosophila* and the mouse, avian mutants have not yet been extensively used. This is not because they do not exist (a few hundred avian mutants have already been identified, see Cooke and Buckley, 1987) but mainly because avian species are more difficult and cumbersome to keep in an animal house than mice, and unlike *Drosophila* cannot be reared in yoghurt bottles. With the advent of recombinant DNA techniques and reverse genetics, however, it is now becoming more desirable that mutants be studied and even generated by targeted mutagenesis. Methods for doing this are now beginning to become available.

In the last two years or so, methods for generating transgenic birds have begun to be devised (for example, see Perry, 1988b; Petitte *et al.*, 1990; Shuman, 1991). Although none of these methods is yet easy to perform, there are currently rumours that a much more straightforward method, using avian sperm as a carrier for DNA, is under test in several laboratories and is producing very encouraging results.

The future

The avian embryo has been used as an experimental subject for embryological studies for more than 2000 years. Hens' eggs were initially used to observe different stages of development and the circulation of the blood. When experimental embryology was introduced in the late 1800s, the chick quickly became one of the two most popular subjects for this approach, together with amphibians. But there was never an avian developmental genetics comparable with those offered by the fly and the mouse. Now, with the advent of reverse genetics and recombinant DNA techniques, the avian embryo is once again becoming one of the main organisms of choice. The ease with which experimental manipulations can be done, together with the multiplicity of cellular and region-specific markers now available, will become a powerful combination with molecular biology to address questions of gene action and function during development. When the techniques for routine production of transgenic birds and for targeted mutagenesis become established, no doubt in the very near future, birds will be one of the most versatile models for modern developmental studies.

References

Burley, R.W. and Vadehra, D.V. (1989). *The Avian Egg; Chemistry and Biology*. John Wiley and Sons, New York.

Canning, D.R. and Stern, C.D. (1988). Changes in the expression of the carbohydrate epitope HNK-1 associated with mesoderm induction in the chick embryo. *Development*, **104**, 643–656.

Cooke, F. and Buckley, P.A. (1987). *Avian Genetics*. Academic Press, London.

Davies, J.A. Cook, G.M.W. Stern, C.D. and Keynes, R.J. (1990). Isolation from chick somites of a glycoprotein fraction that causes collapse of dorsal root ganglion growth cones. *Neuron*, **4**, 11–20.

Eyal-Giladi, H. and Kochav, S. (1976). From cleavage to primitive streak formation: A complementary normal table and a new look at the first stages of the development of the chick. *Devl. Biol.*, **49**, 321–337.

Fraser, S. Keynes, R. and Lumsden, A. (1990). Segmentation in the chick embryo hindbrain is defined by cell lineage restrictions. *Nature*, **344**, 431–435.

Hamburger, V. and Hamilton, H.L. (1951). A series of normal stages in the development of the chick. *J. Morph.*, **88**, 49–92.

Keynes, R.J. and Stern, C.D. (1988). Mechanisms of vertebrate segmentation. *Development*, **103**, 413–429.

Le Douarin, N.M. (1983). *The Neural Crest*. Cambridge University Press, Cambridge.

Lillie, F.R. (1952). *Development of the Chick: An Introduction to Embryology*. Holt, New York.

Lumsden, A.G.S. and Keynes, R.J. (1989). Segmental patterns of neuronal development in the chick hindbrain. *Nature*, **337**, 424–427.

Mitrani, E. Ziv, T. Thomsen, G. Shimoni, Y. Melton, D.A. and Bril, A. (1990). Activin can induce the formation of axial structures and is expressed in the hypoblast of the chick. *Cell*, **63**, 495–501.

New, D.A.T. (1955). A new technique for the cultivation of the chick embryo *in vitro*. *J. Embryol. Exp. Morph.*, **3**, 326–331.

Patten, B.M. (1971). *The Early Embryology of the Chick*. McGraw-Hill, New York.

Perry, M.M. (1988a). A complete culture system for the chick embryo. *Nature*, **331**, 70–72.

Perry, M.M. (1988b). Update on gene-transfer in chicks. *World's Poultry Sci. J.*, **44**, 224–226.

Petitte, J.N. Clark, M.E. Liu, G. Verrinder Gibbins, A.M.V. and Etches, R.J. (1990). Production of somatic and germline chimeras in the chicken by transfer of early blastodermal cells. *Development*, **108**, 185–190.

Primmett, D.R.N. Norris, W.E. Carlson, G.J. Keynes, R.J. and Stern, C.D. (1989). Periodic segmental anomalies induced by heat-shock in the chick embryo are associated with the cell cycle. *Development*, **105**, 119–130.

Roberts, C. Platt, N. Streit, A. Schachner, M. and Stern, C.D. (1991). The L5 epitope: an early maker for neural induction in the chick embryo and its involvement in inductive interactions. *Development*, **112**, 959–970.

Schoenwolf, G.C. and Smith, J.L. (1990). Mechanisms of neurulation: traditional viewpoint and recent advances. *Development*, **109**, 243–270.

Selleck, M.A.J. and Stern, C.D. (1991). Fate mapping and cell lineage analysis of Hensen's node in the chick embryo. *Development*, **112**, 615–626.

Shuman, R.M. (1991). Production of transgenic birds. *Experientia*, **47**, 897–905.

Stern, C.D. (1990). The marginal zone and its contribution to the hypoblast and primitive streak of the chick embryo. *Development*, **109**, 667–682.

Stern, C.D. (1992). Mesoderm induction and formation of the embryonic axis in amniotes. *Trends Genet*, **8**, 158–163..

Stern, C.D. and Canning, D.R. (1990). Origin of cells giving rise to mesoderm and endoderm in chick embryo. *Nature*, **343**, 273–275.

Stern, C.D. Jaques, K.F. Lim, T.M. Fraser, S.E. and Keynes, R.J. (1991). Segmental lineage restrictions in the neural tube of the trunk of the chick embryo depend on the adjacent somites. *Development*, **113**, 239–244.

Stern, C.D. Sisodiya, S.M. and Keynes, R.J. (1986). Interactions between neurites and somite cells: inhibition and stimulation of nerve growth in the chick embryo. *J. Embryol. Exp. Morph.*, **91**, 209–226.

Storey, K.G. Crossley, J.M. De Robertis, E.M. Norris, W.E. and Stern, C.D. (1992). Neural induction and regionalisation in the chick embryo. *Development*, **114**, 729–741.

Tabin, C.J. (1991). Retinoids, homeoboxes and growth factors: toward molecular models for limb development. *Cell*, **66**, 199–217.

Tickle, C. (1991). Retinoic acid and chick limb bud development. *Development*, Suppl. **1**, 113–121.

Tickle, C. Alberts, B.M. Lee, J. and Wolpert, L. (1982). Local application of retinoic acid to the limb bud mimics the action of the polarising region. *Nature*, **296**, 564–565.

Yamada, T. Placzek, M. Tanaka, H. Dodd, J. and Jessell, T. M. (1991). Control of cell pattern in the developing nervous-system: polarizing activity of the floor plate and notochord. *Cell*, **64**, 635–647.

11. The Mouse

Jonathan B.L. Bard and *Matthew H. Kaufman*

Introduction

For many years, mammalian development attracted relatively little attention because the embryos were thought to be inconvenient to handle and hard to study. As the result of a series of technical advances over the last twenty or so years, however, the situation has changed dramatically and, although the mouse embryo will never be the easiest to work with, it is perhaps the developmental system to which the greatest attention is now being paid. There are four main reasons for this newfound enthusiasm. First, the procedures for the manipulation of pre-implantation embryos and the culture of early postimplantation embryos are well established; second, the techniques required for the molecular-biological study of mouse development are indistinguishable from those for any other system; third, mouse embryogenesis is the best and most accessible model for both normal and abnormal human development; and, finally, the relative ease with which transgenic mice can be made is allowing the production of animals with almost tailor-made mutations that can not only be used to investigate gene function, but can also model a wide range of human disorders. Together, these reasons have provided the basis for a great deal of funding, and the results of the work are demonstrating the power of the mouse as a developmental system.

Mice have one further advantage over all other vertebrates in common use except the zebra fish and that is their use in genetic studies: because they are small, cheap and easy to breed, it has been possible to collect and make large numbers of mutants (see Lyon and Searle, 1989). Apart from the random-bred varieties, there is now a wide variety of inbred strains (approximately 300) that have well-understood markers for purposes that are both genetic (e.g. coat-colour) and biochemical (e.g. isozymal variants that are both autosomal and X-linked). These inbred strains are useful for experimental purposes and reproduceability as they are genetically, and hence immunologically, identical, while F1 hybrids between two inbred strains are also commonly used because of their increased vigour and phenotypic features.

The net result of these advances has been that, over the last few years, there has been a flood of papers on mouse development. Such has been the pace of progress that this chapter can do little more than discuss the major features of mouse embryogenesis and point to some of the main areas of research and thus hope to give the reader a sense of current work. It will start with a summary of normal mouse development, and go on to consider ways in which pre-implantation and postimplantation embryos are being used, with the making of polyploid embryos and the development of the kidney being chosen as case studies. We will then touch on mouse developmental genetics and the making of transgenic mice, and end by briefly discussing the future for studies of mouse development.

Normal development

Introduction

This section gives a description of the normal development of the mouse embryo based on the standard texts of Theiler (1989) and Kaufman (1992a), but includes some data on the mechanisms underpinning particular aspects of embryogenesis. It should be pointed out that mouse development differs in several ways from that of most mammals (e.g. it turns; for details, see Balinsky, 1981).

The stages of mouse development are from Theiler (1989) and are abbreviated to 'TS', while the ages given are days after conception abbreviated to 'dpc'. Rat stages are, from implantation onwards, about 1–2 days later than those of the mouse.

Maturation of the egg

Maturation starts within the ovary when ovarian follicles are stimulated to undergo folliculogenesis by the pituitary gonadotrophin follicle-stimulating hormone (FSH). This induces them to progress from an immature state to form antral-stage follicles (**11.1**). About 12hr later, they are exposed to another pituitary gonadotrophin, luteinising hormone (LH) whose roles include initiating oocyte maturation. This allows the egg to move from the resting (or dictyate/dictyotene) state, an early stage of prophase of the first meiotic division, to complete the first meiotic division. The first polar body is then extruded and the egg continues to differentiate until it 'blocks' at the metaphase stage of the second meiotic division shortly before ovulation (also stimulated by LH. See Edwards and Gates, 1959). The egg requires a further stimulus to complete the second meiotic division and this is usually provided by the fertilizing spermatozoon. The cells within the antral follicles from which the eggs were ovulated are then stimulated by LH, so that this unit becomes converted into the *corpus luteum*. This secretes progesterone which maintains the pregnancy until its critical role is taken over by the hormone producing tissues of the placenta. Should fertilization not occur, the corpus luteum regresses, and a new oestrus cycle begins.

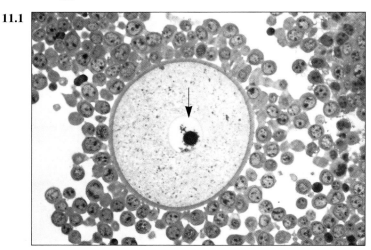

11.1 Germinal vesicle stage oocyte surrounded by follicle cells (termed the *corona radiata*). Within the clear nucleus (arrow) is a darkly staining nucleolus (× 600).

The immediate effect of fertilization: Theiler stages 1–2, 0–24hr

Once the egg has been activated by the fertilizing sperm, usually in the ampullary region of the oviduct, various events occur in rapid succession. The universal activating stimulus (which has been observed in all species studied) is believed to be the release of intracellular-bound calcium ions, some of which are associated with membrane phospholipids. This release seems to be related to the change that occurs in the resting potential of the vitelline membrane which is termed the 'fast (or vitelline) block to polyspermy', which temporarily inhibits the fertilization of the egg by supplementary (perivitelline) spermatozoa. Shortly afterwards, the cortical granules that are located just beneath the vitelline membrane release their contents into the perivitelline space. These consist of proteolytic enzymes which alter the biochemical and physical properties of the *zona pellucida* (the acellular glycoprotein membrane surrounding the egg and produced by the follicle cells). This second release, sometimes termed the 'zona reaction', constitutes the 'slow block to polyspermy', since it has the effect of inhibiting supernumerary spermatozoa from attaching to the surface of the *zona*, while those sperm that are in the process of penetrating the *zona* fail to progress further (Kaufman, 1983).

It is of considerable importance that mating only occurs when the female mouse is in oestrus, as the efficiency of the various 'blocks' decreases with increased post-ovulatory ageing of the egg. Similarly, the second meiotic spindle tends to lose its subcortical attachment and drifts towards the centre of the egg. Under these circumstances, second polar-body extrusion may be inhibited. In addition, ultrastructural evidence of deterioration of the components of the spindle apparatus has also been described, and it is likely that this is associated with other evidence of age-related cytoplasmic deterioration.

Post-fertilization events: Theiler stages 2–3, 1–2dpc

The most obvious event that occurs in association with the completion of the second meiotic division is the extrusion of the second polar body, and this is usually seen within a few hours of activation (**11.2–11.13**). Shortly afterwards, the sperm nucleus decondenses to form the male pronucleus, while the product of the second meiotic division that remains within the egg differentiates into the female pronucleus (Austin, 1965). The two pronuclei are first evident at about 4–5hr after

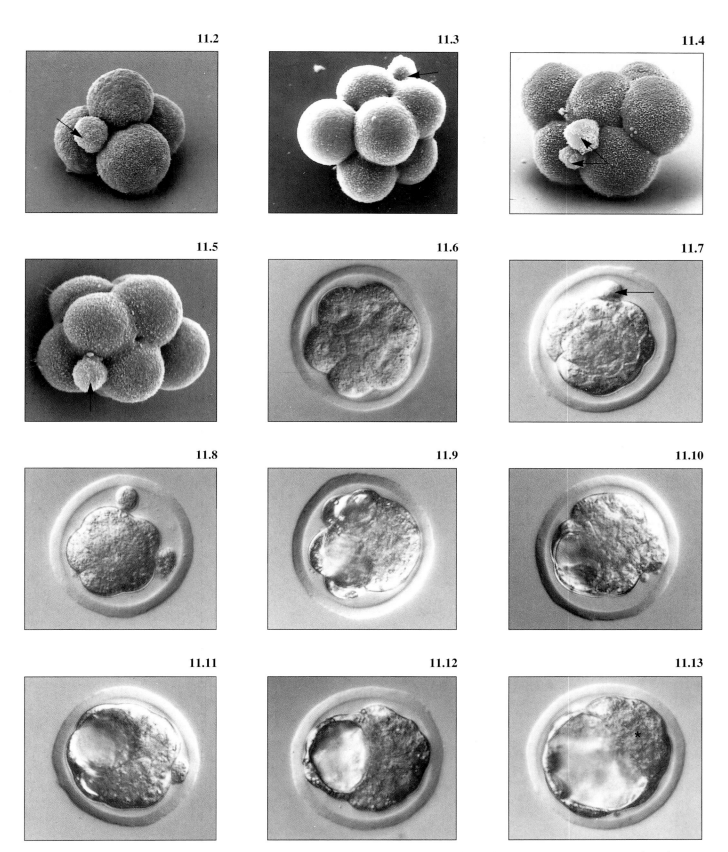

11.2–11.13 Scanning electron micrographs of zona-denuded (**11.2–11.5**) and interference-contrast micrographs of zona-intact mouse embryos (**11.6–11.13**) showing the 4-cell-stage embryo (**11.2**), the early (**11.3–11.5**) and later (**11.6–11.8**) stages of compaction and the formation of the blastocyst (**11.9–11.13**). Note that the second polar body (arrow), which occasionally divides (see **11.4**), is often recognizable up to the blastocyst stage. The trophectoderm and inner cell mass components (asterisk) of the blastocyst are only clearly distinguishable in the most advanced embryo illustrated (**11.2–11.5** ×900, **11.6–11.13** ×600, from Kaufman, 1992a, with permission).

activation, and, despite the fact that both have a haploid chromosome constitution, the male pronucleus is always about twice the volume of its female counterpart. Both increase in volume throughout the first cell cycle, and achieve their maximum volume just before the first cleavage division. Shortly after the pronuclei are first observed, the DNA within them is doubled (during the S-phase of the first cell cycle).

As the fertilized egg is about to enter the first cleavage division, the outlines of the two pronuclei gradually disappear, and this coincides with the condensation of their maternally- and paternally-derived chromosomes. These initially associate (during prometaphase) on the equator of the first cleavage spindle; the paired chromatids then become apparent and these segregate (during anaphase) to the two poles of the spindle apparatus. All the events associated with the first cleavage division, from the time that the pronuclear outlines disappear to cytoplasmic cleavage (cytokinesis) with the formation of the 2-cell stage embryo, are completed in about 2hr. The total duration of the first cell cycle, between fertilization and the first cleavage division, is normally about 16–18hr. After about a further 36hr and two approximately synchronous divisions, the 8-cell stage is reached; subsequent divisions become increasingly asynchronous. By this stage, the embryo has usually reached the utero-tubal junction and will shortly pass into the lumen of the uterus. Reproduction in the mouse is extremely efficient and most of the fertilized eggs successfully develop to birth.

The pre-implantation stage: Theiler stages 4–5, 3–4dpc

Perhaps the most important event to occur in this period is the process of *compaction* when the cleavage-stage embryo condenses and becomes a morula with the macroscopically observable individuality of the blastomeres being lost. A proportion of the component cells now becomes surrounded by a second population of cells and, a little later, fluid starts to accumulate within the morula. The former group of cells becomes the inner cell mass (ICM) and will eventually give rise to the embryo proper, while the outer shell of cells forms the trophectoderm or trophoblast cells which mainly form the extra-embryonic tissues such as the embryonic component of the placenta (**11.2–11.13**; Gardner and Papaioannou, 1975).

The mechanism for this is both simple and elegant: at the 8- and 16-cell stages, individual blastomeres have probabilities of about 45% and 20%, respectively, of giving morphologically dissimilar products (for review see Pedersen, 1986), the process being termed *polarization* (Johnson and Ziomek, 1981). Cells that will end up on the inside of the morula (ICM) tend to be smaller, have fewer surface microvilli and tend to produce extracellular glycoproteins that make them stickier than their complementary division products which will form the outer shell of trophoblast cells. Once this division has occurred, the different fates of the two division products are determined.

The critical fact that determines when compaction/polarization occurs is not cell number but developmental age, that is the number of DNA replications that have taken place since fertilization. Thus, isolated blastomeres will behave according to this developmental clock, while the individual cells within aggregates of cleavage-stage embryos will each behave according to their inbuilt clocks (Smith and McLaren, 1977).

As compaction occurs, junctional complexes (desmosomes) form between the cells. Slightly later, fluid accumulation takes place between the cells of the morula and, due to the water-tight seals (tight junctions) that form between the trophectoderm cells, fluid of this accumulates to form the blastocoel and the embryo reaches the blastocyst stage (Ducibella, 1977). The presence of the water-tight seals isolates the blastocoelic fluid from the outside of the embryo and allows it to become a special microenvironment. It is this which probably stimulates the ICM cells exposed to it to form the primary embryonic endoderm, while the remaining ICM cells maintain their pluripotential properties.

Implantation: Theiler stages 6–8, 4.5–6dpc

Embryos will only develop once they have implanted themselves in the wall of the uterus, an event that normally occurs after the 64-cell stage when about one third of the embryo's cells comprise the ICM, with the remainder forming the trophectoderm. For implantation to take place, the *zona pellucida* has to be lost since it is a prerequisite for the normal decidual response that the trophectoderm makes direct contact with the endometrial cells lining the uterus. The *zona* gradually thins and splits open due to a combination of factors, the most important of which is believed to be its exposure to proteolytic enzymes (uterolysins). This splitting is the consequence of the gradual expansion of the blastocyst due to the principally unidirectional flow of water into the blastocoelic cavity, and due to the weakness caused by the pathway created by the fertilizing sperm (TS6).

Once the blastocyst is zona-free, it is capable of stimulating the endometrium to undergo a series of changes associated with the decidual response (TS7). It is only the cells that constitute the polar region of the trophectoderm that play an active role in implantation, and these cells will shortly give rise to the ectoplacental cone (Gardner and Papaioannou, 1975). When they migrate away from the central part of this region, most are destined to form the trophoblast giant cells that will eventually be located between the embryonic and maternal components of the placenta. At or shortly after implantation (approximately 4.5dpc), the cells on the blastocoelic surface of the ICM delaminate to form the primary embryonic endoderm, and these cells subsequently migrate and will eventually cover the entire blastocoelic surface of the mural trophectoderm. The blastocoelic cavity is then converted into the primary yolk sac cavity (for lineage relationships, see **11.14**).

By about 5dpc, the cells of the ICM rapidly proliferate to form the egg cylinder, and within this a pro-amniotic canal soon becomes apparent. This conceptus is soon divisible into an embryonic pole from which the embryo will develop and which occupies the distal half of the egg cylinder, and the abembryonic pole. The proximal part of the latter forms in close proximity to the ectoplacental cone, while the distal part will principally give rise to the yolk sac and allantois (Snell and Stevens, 1966).

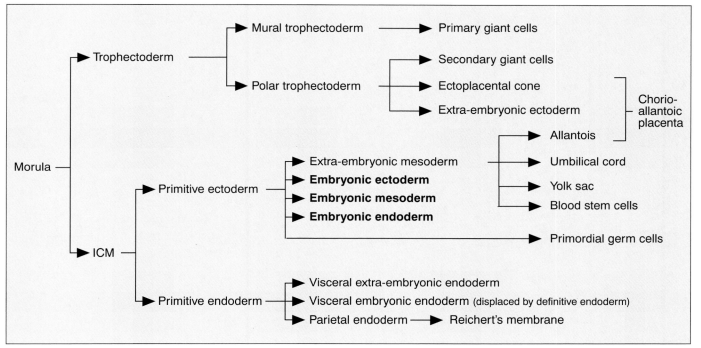

11.14 A diagram of the cell lineage relationships in the early mouse embryo showing the origins of the embryonic and extra-embryonic tissues (redrawn and modified from Rossant and Papaioannou, 1977).

Gastrulation: Theiler stages 9–10, 6.5–7dpc

At this early stage of development (**11.15**), the embryo is cup-shaped and bilayered with a subjacent hypoblast (a layer of endoderm) and a superior ectodermal epiblast, with this latter sheet of cells eventually forming the neural and surface ectoderm, together with the other tissues of the embryo (for more details, see later and Lawson *et al.*, 1991). Gastrulation within the epiblast layer generates the intervening layer of mesoderm and this is first apparent at about 6.5dpc when a proportion of the epiblast cells migrate into and through the primitive groove (**11.16**). These mesoderm cells then migrate laterally, separating the overlying ectoderm from the underlying endoderm in all areas except the buccopharyngeal and cloacal membranes which will subsequently be located at the stomodaeum and procto-

daeum (the two ends of the primitive gut tube).

The primitive groove is first clearly seen at about 7dpc at the so-called primitive-streak stage, and provides the first evidence of an embryonic axis. Hensen's node is located at the rostral end of the primitive streak, and, while this is not as well differentiated and obvious in the mouse as in the chick embryo, it is believed to play a similar role in controlling the subsequent morphogenesis of the embryo. Some of the cells of the node form a notochordal process which is initially associated with the axial endoderm but later separates from it and these tissues, together with the nascent mesoderm, express the *Brachyury (T)* gene which may be involved in the specification of the mesoderm (Rashbass *et al.*, 1991).

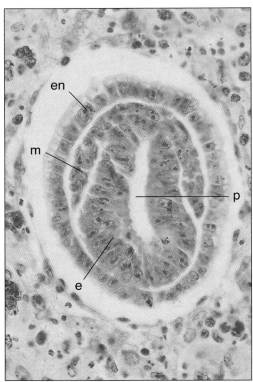

11.15 An advanced egg-cylinder/early-primitive-streak stage embryo (TS10) showing the three primary germ layers with early evidence of mesoderm formation (×305). (Key: **e**, ectoderm; **en**, endoderm; **m**, mesoderm; **p**, pro-amniotic canal.)

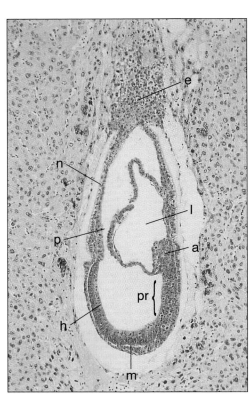

11.16 Primitive-streak-stage embryo (TS10/11, ×90). (Key: **a**, allantois; **e**, ectoplacental cone; **h**, neuro-epithelium of headfold region; **l**, posterior amniotic fold; **m**, early evidence of intra-embryonic mesoderm; **n**, extra-embryonic endoderm (of future visceral yolk sac); **p**, pro-amniotic canal; **pr**, primitive-streak region.)

Neurulation and early organogenesis: Theiler stages 11–12, 7.5–8dpc

Over the period between about 7.5–8.5dpc the neural folds form on either side of the embryonic axis, becoming elevated and apposed along the dorsal midline. This process of primary neurulation soon forms the primitive neural tube which subsequently differentiates into the brain and possibly all but the most caudal part of the spinal cord. The initial site of neurulation is in the future occipital/cervical region, and this process gradually extends rostrally and caudally. Similarly, the cardiogenic plate is first seen within the intra-embryonic coelom (**11.17**) and starts to differentiate into the heart. While this happens, the intra-embryonic mesoderm forms three reasonably well-defined columns of tissue, termed the paraxial, intermediate-plate and lateral-plate mesoderm. The paraxial mesoderm subsequently differentiates and forms somites which give rise to the dermomyotome and sclerotome; the intermediate-plate mesoderm also segments and gives rise to the nephrogenic cord (to be involved in the differentiation of components of the urogenital system), while the lateral-plate mesoderm does not segment, but splits into two layers, the splanchnic and somatic components. These are involved in the formation of the wall of the gut tube and in the differentiation of the lungs, and in the formation of the lateral and ventral parts of the body wall respectively. (For further details on the differentiation of these tissues, see Hamilton and Mossman, 1972; Moore, 1982).

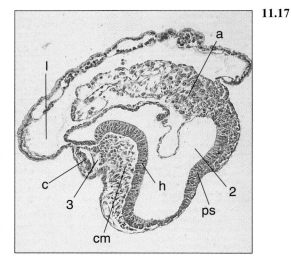

11.17 Early-headfold-stage embryo with 1–2 pairs of somites (TS11/12, ×95). (Key: **1**, exocoelomic cavity; **2**, amniotic cavity; **3**, intra-embryonic-coelomic cavity (future region of pericardial cavity); **c**, cardiogenic plate; **cm**, cephalic-mesenchyme tissue; **h**, neuroepithelium of headfold region; **ps**, primitive streak.)

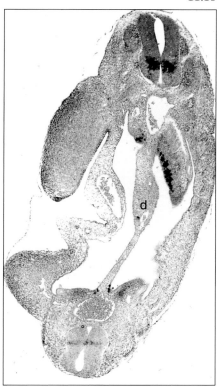 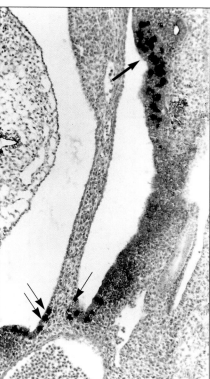 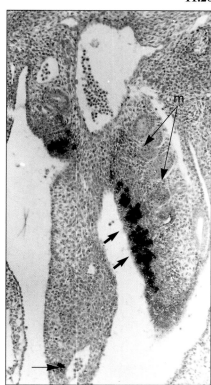

11.18–11.20 The migration of the primordial germ cells (marked by their alkaline-phosphatase activity which shows black in this preparation). Some of the germ cells are migrating along the dorsal mesentery (**d**) of the hind gut (**11.19**, arrows), while others have reached the gonadal (medial) part of the urogenital ridge (**11.20**, thick arrows) whose lateral part is occupied by the mesonephros (**m**). Note that the primitive motor neurons located in the ventral horn of the spinal cord also have a high level of alkaline phophatase activity. (Sections from a 10.25dpc embryo, 30–35 somites, TS16, **11.18** ×50, 11.19 and 11.20 ×125.)

Red blood cell formation (erythropoiesis) initially occurs in association with the blood islands which form within the mesodermal component of the visceral yolk sac during the primitive-streak stage. This is also the first location where the primordial germ cells (pgc) are seen and can be recognized by their characteristic morphology and alkaline-phosphatase activity (**11.18–11.20**). The pgc become incorporated into the base of the allantois and migrate via the wall of the hindgut and its dorsal mesentery to the urogenital ridge (Chiquoine, 1954; Ozdenski, 1967). Once in that location, they differentiate and play a key role in gonad differentiation.

Turning: Theiler stage 13, 8.5dpc

When they have about 8–14 pairs of somites, embryos of the mouse and other related species undergo the process of 'turning' during which the configuration of the embryonic axis changes. By this process, it converts from being inside-out in relation to most other mammalian species to adopting the more familiar 'fetal' position (as seen in these other species). At the same time, the embryo rolls into its extra-embryonic membranes and becomes completely surrounded by its amnion and yolk sac. The process of 'turning' is described in detail elsewhere (Kaufman, 1990) and illustrated in **11.21–11.26** which also show how the extra-embryonic membranes come to surround the embryo.

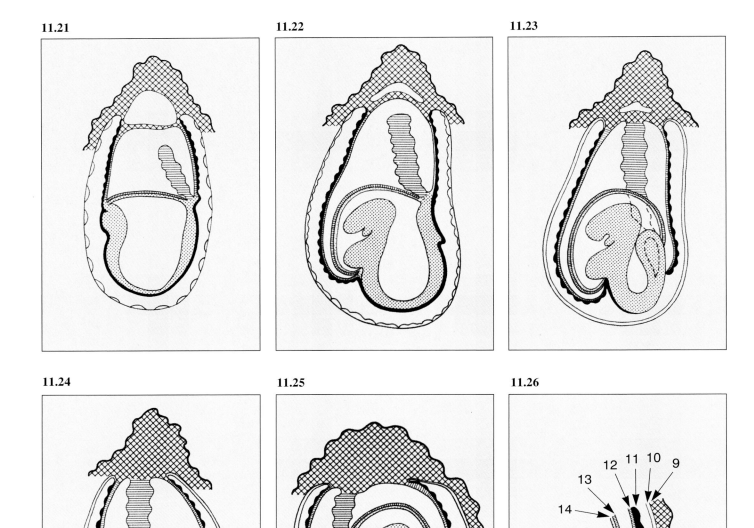

11.21–11.26 A simplified sequence of diagrams illustrating the turning that the embryo undergoes and the way in which the extra-embryonic membranes develop; **11.21**, the pre-somite headfold stage (7.5–8dpc, TS11/12); **11.22**, 8–10 pairs of somites (8.5dpc, TS13); **11.23**, 10–12 pairs of somites (8.75dpc, TS13/14); **11.24**, 12–14 pairs of somites (9dpc, early TS14); **11.25**, 15–20 pairs of somites (9.25dpc, late TS14); and **11.26**, the embryonic layers, extra-embryonic tissues and cavities illustrated in **11.25**. (Key: **1**, embryonic endoderm; **2**, embryonic ectoderm and mesoderm; **3**, amniotic cavity; **4**, amnion; **5**, exocoelomic cavity; **6**, yolk sac; **7**, yolk cavity; **8**, ectoplacental cone and trophectoderm derivatives; **9**, Reichert's membrane; **10**; parietal (extra-embryonic) endoderm; **11**, visceral (extra-embryonic) endoderm; **12,** extra-embryonic mesodermal component of yolk sac; **13**, mesodermal component of amnion; **14**, ectodermal component of amnion. From Kaufman, 1992a, with permission.)

Filling in the body plan: Theiler stages 14–21, 8.5–13dpc

Once the embryo has 'turned' and adopted the characteristic 'fetal' position, it develops rapidly and, over the period 8.5–13.5dpc, the principal features of each of the various body systems become established (**11.27–11.30** and **11.31–11.34**). The heart and vascular system is the first of these to function and the initial contractions are observed at about 8.5dpc when the embryo has 8–10 pairs of somites (TS13). Shortly before this, the extra-embryonic and embryonic vascular systems amalgamate (they have previously developed independently), and this allows the primitive nucleated red blood cells that develop in the visceral yolk sac to enter the embryonic circulation. In embryos with 13–20 pairs of somites (TS14), the rostral neuropore closes to form not only the rostral extremity of the neural tube, but also the optic vesicles which are outgrowths from the primitive forebrain. At about this time, the anterior neural tube starts to segment and form the transitory rhombomeres, an event associated with the expression of homeobox genes (see Murphy and Hill, 1991). A

11.27–11.30 Scanning electron micrographs of representative embryos over the period 8–11.5dpc: **11.27**, 5–6 pairs of somites and early evidence of neural-fold apposition (8dpc, TS12, ×60); **11.28,** 10–12 pairs of somites (8.5dpc, TS13, ×60); **11.29,** 20–25 pairs of somites (note that the forelimb bud (arrow) is starting to differentiate and that, while the rostral neuropore has closed, the caudal neuropore has yet to do so; 9.5dpc, TS15, ×50); and **11.30,** 11.5dpc (TS19, ×15). The distal part of the limb buds is now paddle-shaped and the tail has elongated and narrowed.

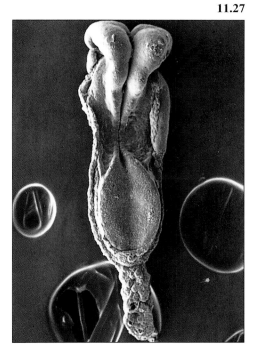
11.27

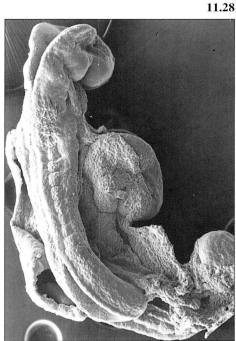
11.28

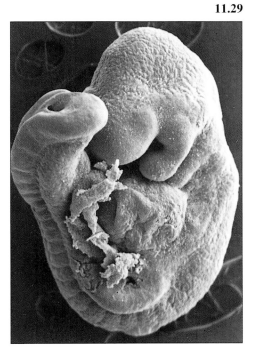
11.29

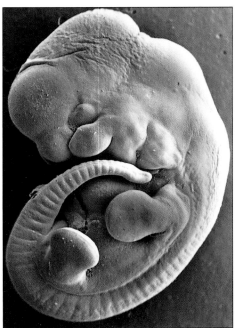
11.30

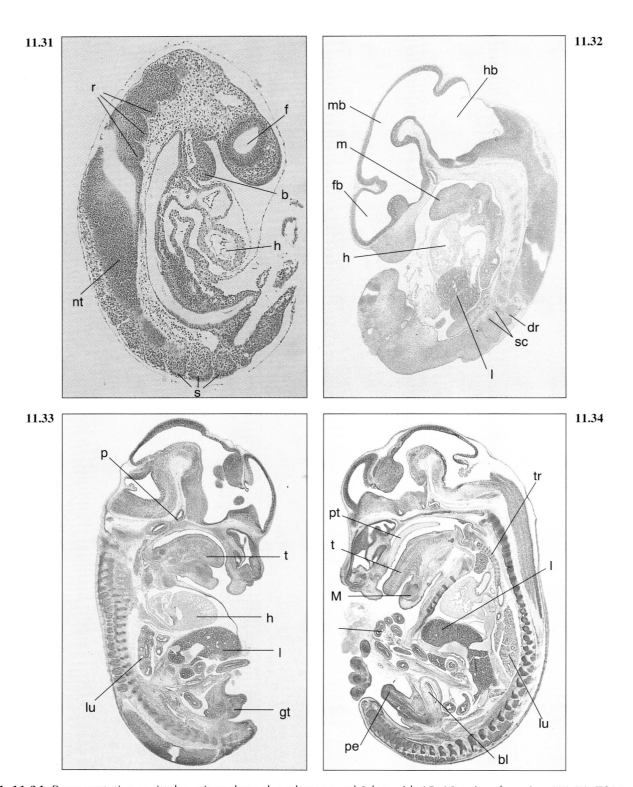

11.31–11.34 Representative sagittal sections through embryos aged 9dpc with 15–18 pairs of somites (**11.31**, TS14, ×53), 11–11.5dpc (**11.32**, TS18/19, ×19), 12.5–13dpc (**11.33**, TS21, ×12), and 14.5dpc (**11.34**, TS22/23, ×9). These show the increase in complexity that takes place over this period. (Key: **b**, first branchial arch (mandibular component); **bl**, bladder; **dr**, dorsal root ganglion; **f**, forebrain vesicle (third ventricle); **fb**, forebrain vesicle (telencephalic vesicle); **gt**, genital tubercle; **h**, heart (ventricular region); **hb**, hindbrain (fourth ventricle); **l**, liver rudiment; **lu**, lung; **m**, mandibular component of first branchial arch (with tongue rudiment); **md**, loops of midgut (within physiological umbilical hernia); **nt**, neural tube; **p**, pituitary gland; **pe**, penis; **pt**, palate; **r**, rhombomeres; **s**, somites; **sc**, condensations of sclerotome material; **t**, tongue rudiment; **tr**, trachea (with cartilage rings).)

day later (TS16; 30–34 pairs of somites) there are four pharyngeal arches in the cephalic region and the lens placode, the otic vesicle and the nasal pits are all beginning to form. At about this time, the caudal neuropore closes.

Within the embryo, the mesenchyme of the *septum transversum* is first seen at about 9dpc and, over the next day or so, the hepatic/biliary primordia diverticulum start to form within it. Sinusoids, principally from the vitelline venous system, also proliferate and the liver is soon the largest of the organs, taking over from the yolk sac as the source of red blood cells whose stem cell precursors are probably of yolk sac origin. Over the period 10–12dpc, the lungs, salivary glands and teeth appear and, a day later, the gonads start to differentiate.

The neural crest cells that will form the sensory cells of the dorsal root ganglia are first seen at about 9.5dpc and the ganglia themselves are clearly apparent by 10.5dpc in the trunk region and throughout the neural axis by 11–11.5dpc.

The cranial ganglia also differentiate over a similar time course and are also distinguishable by 11.5dpc. Segmental spinal nerves are not easily seen before 10.5–11dpc, and then only in silver-stained material.

The later development of the gut is achieved in an unexpected way. The primitive midgut expands into the proximal part of the umbilical cord which forms the physiological umbilical hernia. This hernia disappears with the return of the midgut to the peritoneal cavity at about 16.5dpc, an event facilitated by the regression of the mesonephroi and their replacement by the metanephroi, the definitive kidney, which is relatively smaller than the mesonephros. The most obvious developmental features over this period, however, and those most useful in staging embryos, are the limb buds. The forelimb buds first appear at about 9dpc (TS15; 20–25 pairs of somites) with the hindlimb buds a day later and their progress over the next 6 days is easy to follow.

Consolidation: Theiler stages 22–26, 14–18dpc

The period between 13.5dpc and birth is one of consolidation when the fine structural detail of each system is laid down as it grows and becomes functional (Kaufman, 1992a), and assessment of developmental age can now be made by analysing ossification centres. These appear in a defined sequence from about 14–14.5dpc and extend to involve all components of the skeletal system, many of which are still present in the form of cartilage primordia for some considerable time after birth (**11.35**). The changes taking place as the embryo develops during the last week of embryogenesis (for growth details see **11.36**) closely resemble those in man, for comparison, and the interested reader should consult textbooks of human embryology for detailed information on these later events (e.g. Hamilton and Mossman, 1972; Moore, 1982).

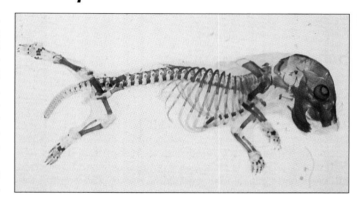

11.35 An intact and cleared 18.5dpc embryo (TS26) stained with alizarin and alcian blue. The cartilage is stained blue, while the ossified tissue is stained red.

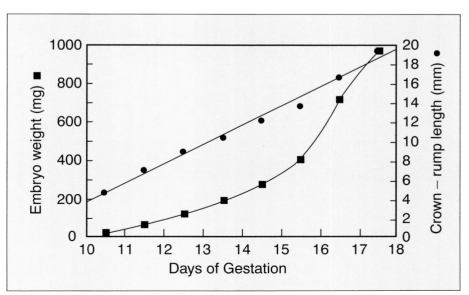

11.36 A graph showing how the mouse embryo grows during the second half of its gestation.

Experimental manipulations and *in vitro* development

A: The pre-implantation embryo

Most experimental work has used pre-implantation embryos as they can be maintained *in vitro* and obtained in relatively large numbers by flushing them out of the female reproductive tract (7–15 embryos per animal following spontaneous ovulation and 20–40 after superovulation). These embryos are used for making chimeras, parthenogenetic activation, manipulating the number of nuclei in the egg and, most recently, for the making of transgenic mice (see later). As such pre-implantation embryos only do well in culture up to the blastocyst stage, but then fail to progress for reasons that remain unclear, they have to be grown in pseudopregnant recipients. These mice are first impregnated by vasectomized males and the embryos are then either transferred to the oviduct or into the uterine lumen. It is usually best if the recipient is at a developmental stage some 24hr earlier than the embryos being transferred.

There is an extensive literature on the techniques required for the various manipulations and, for details on these, the reader is referred particularly to the books of Hogan, Constantini and Lacy (1986), Robertson (1987) and Rossant and Pedersen (1986).

Chimeras

Pre-implantation embryos are used to make both aggregation and ICM injection chimeras for cell lineage studies. For the former, either 8-cell stage embryos or morulae are aggregated together and the developmental potential of the resultant embryos investigated, while, in the latter case, an ICM from the blastocyst of a 'donor' embryo is transferred into the blastocoelic cavity of a 'host' blastocyst. In both approaches, high rates of efficiency in chimera production are often achieved, although the second procedure is technically the more difficult, and requires considerable expertise. In addition, chimeras can be made using ES cells.

Embryonic stem cells

It is now possible to make cell lines from the pluripotential ICM cells which will give rise to the three germ layers (Evans and Kaufman, 1981). These are *embryonic stem* (ES) cells (**11.37, 11.38**) and, although they differentiate *in vitro* to give a wide variety of cell types, they can be maintained in their pluripotential state if cultured with a glycoprotein called LIF (Smith *et al.*, 1988). If these pluripotential cells are transferred into particular sites in adult mice, they rapidly develop to form a teratocarcinoma. If, however, they are transferred to the blastocoelic cavity of the normal blastocyst, they incorporate into the ICM and the resultant chimeric mouse shows no evidence of malignancy. The significance of these cells is that they can be prepared from any mouse strain and maintain their original normal genetic constitution; they can thus incorporate themselves into the germ line of chimeric individuals and so generate further generations (Bradley *et al.*, 1984). For this

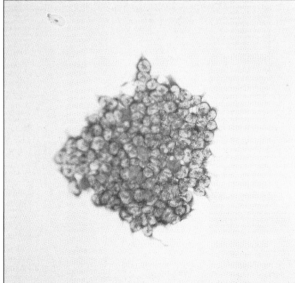
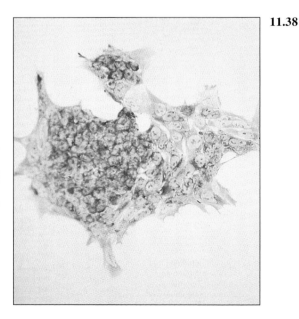

11.37, 11.38 Embryonic stem (ES) cells cultured in LIF and maintaining their undifferentiated state are shown in **11.37** (×60). When the LIF is removed, the cells start to differentiate (**11.38**, ×60). (Micrographs courtesy of Austin Smith.)

reason, they are a much more powerful model system than embryo carcinoma (EC) cells which are tumour-derived and invariably appear to have abnormal genetic constitutions. It is now possible to generate mutations in ES cells and use these mutated cells to make transgenic mice; this important aspect of ES cells will be discussed later. It is worth mentioning that it is now also possible to maintain primordial germ cells in culture (Resnick et al., 1992).

Parthenogenetic development

Unfertilized eggs may readily be activated to develop parthenogenetically to the early limb-bud stage. The most efficient and commonly used technique involves the short exposure of unfertilized eggs, briefly aged *in vivo*, to a dilute solution of ethanol (7%) in either PBS or tissue culture medium (Kaufman, 1982, 1983). By modifying the post-ovulatory age of the egg at the time of activation, and/or the culture conditions during or shortly after activation, parthenogenones may be induced to develop along one of various pathways, depending on whether the extrusion of the second polar body is allowed or suppressed, whether a single or two haploid pronuclei form, or whether the egg cleaves prematurely so that each blastomere contains a single haploid pronucleus (one of these blastomeres represents the second polar body).

Triploids and tetraploids

Mouse embryos with a range of abnormal chromosome complements can be made fairly easily (**11.39**) and are useful in analysing imprinting phenomena. They are also important for analysing the influence of ploidy on development (Kaufman 1991a).

Triploids are sometimes formed spontaneously by the LT/Sv strain as up to 50% of their eggs are ovulated as primary rather than secondary oocytes (O'Neill and Kaufman, 1987). A single diploid female pronucleus forms so that, following fertilization, there are three sets of chromosomes, two from the female and one from the male. Such digynic triploids arise more commonly, however, when, following completion of the second meiotic division, the second polar body fails to be excluded. This may be achieved by ageing the egg either *in vivo* or *in vitro* before fertilization. These triploids survive to the forelimb-bud stage (approximately 10dpc, 20–25 somite stage), but they tend to develop cephalic neural tube defects and heart abnormalities (Kaufman and Speirs, 1987). The more usual way of making digynic triploids, however, is by exposing very recently ovulated eggs to cytochalasin B or D which inhibits second polar-body extrusion, but does not inhibit fertilization or the normal sequence of events that follows fertilization (Niemierko and Opas, 1978). Alternatively, nuclear micromanipulation may be used to insert a 'donor' female pronucleus into a genetically normal (i.e. diploid) 'recipient' fertilized egg. When such embryos reach the forelimb-bud stage they are significantly smaller than controls of the same developmental age due to an ill-understood imprinting effect.

The other class of triploids, the diandrics, can be formed *in vivo* in two distinct ways, by fertilizing an egg with either two (normal) haploid sperm or rarely with a single (abnormal) 'diploid' sperm. Diandric triploids can be made using micromanipulation to transfer a male pronucleus to a genetically normal diploid recipient egg. Such triploids again develop to the forelimb-bud stage where they have 20–25 pairs of somites and mostly seem morphologically normal, although again they are significantly smaller than control diploids of the same developmental age (Kaufman, Lee and Speirs, 1989; for comparisons with humans on this point, see Szulman and Surti, 1978a,b).

There are also, in principle, two types of tetraploid embryos: those that are genetically homozygous, arising from the duplication of a normal diploid embryo (and are thus XXXX or XXYY), and those that are heterozygous, deriving from trispermic fertilization. The latter are difficult to make, but the former can be made *in vitro* by electrofusing the two blastomeres of two-cell-stage embryos (Kaufman and Webb, 1990) or, but with less consistent results, by inhibiting the first or second cleavages with agents such as cytochalasin B or D (Snow, 1973). Once the tetraploids are replaced in appropriate recipients, many develop quite well for up to about 10 days. They then start to die off, so that there are relatively very few left by day 16 of gestation, at which stage they are developmentally equivalent to normal embryos of about 14dpc (Kaufman, 1991b).

These embryos do not develop as normally as diandric triploids (although they are capable of much more advanced development) and, while about half have a normal postcranial axis, the remainder show postcranial vertebral axis abnormalities almost invariably associated with a gross deficiency of the anterior abdominal wall (omphalocoele), while most male tetraploids (XXYY) develop aortic-arch arterial abnormalities (**11.40, 11.41**; Kaufman, 1992b). There is also one particularly interesting feature associated with these embryos, they all have abnormal craniofacial development (**11.42, 11.43**) which seems to derive from incomplete differentiation of the forebrain and its derivatives (i.e. the telencephalic vesicles, the pituitary gland and the eyes). Associated with these macroscopic abnormalities is an interesting though not entirely unexpected cellular feature: tetraploid cells are about twice the size of diploid cells, and, as the embryos are of almost normal size, they have only about half as many cells. It is therefore possible to interpret the head abnormalities as deriving from there being too few forebrain cells to accommodate the map instructions for their differentiation (Henery, Bard and Kaufman, 1992).

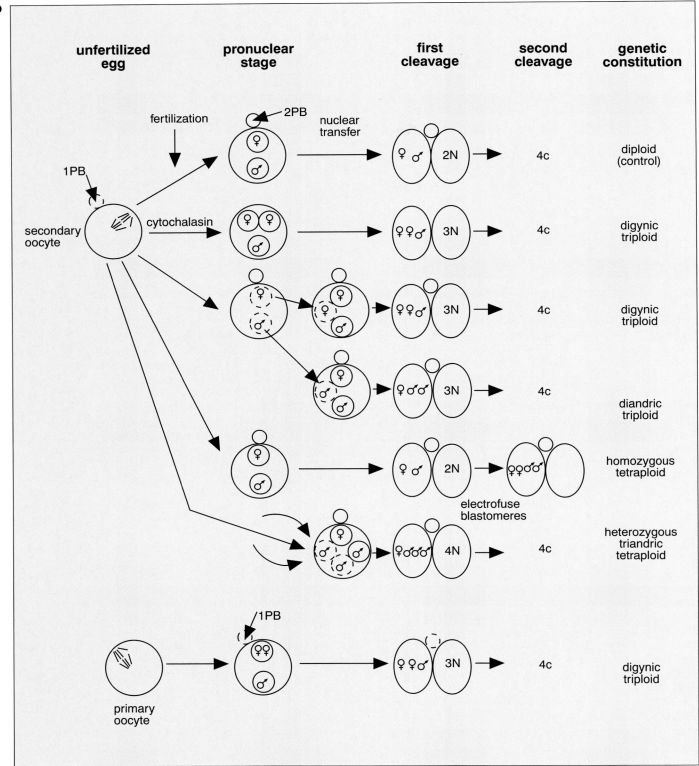

11.39 Diagram illustrating some spontaneously occurring and experimentally induced developmental pathways that give rise to embryos with an abnormal genetic constitution. (Key: **1PB**; first polar body; **2PB**, second polar body: **2N**, **3N**, **4N**, diploid, triploid and tetraploid; **4c**, 4-cell stage.)

11.40–11.43 Normal (**11.40**, **11.42**) and tetraploid (**11.41**, **11.43**) embryos at about 14–14.5 days. The intact tetraploid embryo has abnormal craniofacial features, absent or minute eyes, and a narrowing of the face and forehead as compared to the control (postcranial features are normal in this tetraploid). The representative sections are through the diencephalon (**d**, principally thalamus and hypothalamus) and telencephalic (**t**, cerebral) vesicles (which are derivatives of the primitive forebrain). In the tetraploid (**11.43**), the morphology of this region is grossly distorted, often due to the incomplete separation (**is**) of the two cerebral hemispheres, while that of the medulla oblongata region of the hindbrain (**h**) is normal (NB the sections are at slightly different levels; **11.40** and **11.41** ×6.5, 11.42 and **11.43** ×7.2).

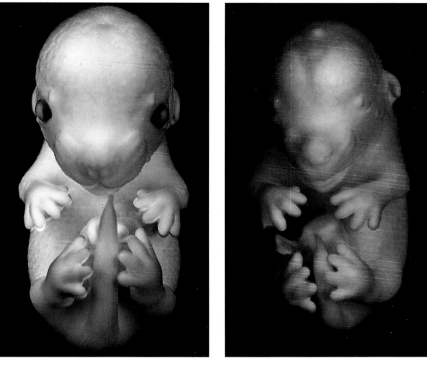

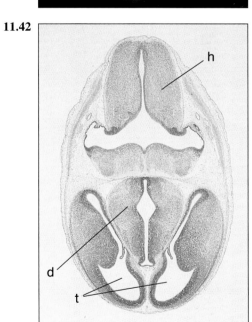

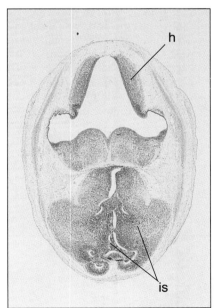

B: The postimplantation embryo

Whole-embryo culture

Postimplantation embryos have mainly been used to elucidate lineage relationships, and to culture whole embryos and organ rudiments. It is not, however, easy to isolate and handle the early postimplantation stages of mouse development as they cannot be re-implanted into mice, but have to be cultured, something that can only be done for up to 2–3 days. Although time-consuming and expensive to do on any scale, such culturing renders mammalian embryos accessible to experimental and surgical procedures. The technique is mainly used for maternal-free studies, for examining the effects of drugs, yolk-sac function, injections of metabolites and antibodies, and for embryos that have been experimentally manipulated (e.g. Gressens *et al.*, 1993). Rat embryos are the easiest to culture, but mice can also be used, although their behaviour in culture is less consistent.

The most usual stage for such work is the headfold (pre-somite) stage (7.5dpc) and, for the succeeding 24–48hr, growth is much as *in vivo*, i.e. to about the early limb-bud stage. The easiest technique for maintaining these early postimplantation embryos is to maximize oxygen uptake by using a rotating bottle culture system with freshly centrifuged rat serum to which may be added defined media (Cockroft, 1990). The technique seems to work because the yolk sac provides a large surface for nutritional and respiratory exchange. Limb-bud-stage embryos can be maintained *in vitro* for up to 48hr, but require sophisticated culture techniques (New, 1978). It is also possible to undertake very restricted operations on later mouse embryos within the uterus and even on embryos removed from the uterus but attached to their placentas (for details, see the relevant chapters in Copp and Cockroft, 1990).

Lineage studies

Elucidating the fates of the early cells, and hence the origins of each tissue, is harder in the mouse, which develops *in utero*, than in other embryos. Various cell-marking techniques have therefore been adapted to follow mouse lineages (see Beddington and Lawson, 1990). For pre-implantation work, chimeras are made with marked cells, while, for postimplantation embryos, marked grafts are inserted into embryos, cells are injected with a marker such as horseradish peroxidase or cells are infected with retroviruses which integrate into the genome and express a reporter gene.

Pre-implantation chimeras can be formed between a normal mouse embryo and cells from any source that can integrate into the embryo (Bradley, 1987), and donor cells have included other genetically normal or abnormal mouse strains, ES cells and rat cells. It is only necessary that the second cells carry some biochemical, chromosomal or immunological marker (for experimental details, see Beddington and Lawson, 1990). While these markers show the genetic origins of any tissue in the chimeras, it is hard to work out the immediate lineage of the tissue as one cannot know where in the inner cell mass the marked cells were placed.

One can thus only know the geographic origin of marked cells in postimplantation embryos where the point of insertion can be closely controlled, and here the most important work is the recent study of Lawson *et al.* (1991). They injected individual epiblast cells with horseradish peroxidase prior to gastrulation, cultured the embryos for a day or so, and correlated the eventual destiny of the cell with its initial position. A summary of their results is given in **11.44** which illustrates how the epiblast surface is transformed into later embryonic tissues. In addition to providing this fate map of the pre-gastrulation embryo, Lawson *et al.* also showed that there were no clonal restrictions on cell fates at this stage, nor could they find evidence of sorting out as the cells move through the primitive streak, since the descendants of marked cells do not cohere. It therefore seems likely that the details of cell fates are not completely determined at the pre-gastrulation stage.

For later embryos, [^{3}H]thymidine-labelled cells have been used to study neural-crest-cell migration (Morriss-Kay and Tan, 1987), while transplants of somites from a transgenic mouse carrying a reporter gene have enabled Beddington and Martin (1989) to follow the fate of somite cells (**11.45, 11.46**). An interesting variation on this approach is the use of a retrovirus carrying a reporter gene to infect the embryo under conditions which ensure that only a few cells will take up the virus. In due course, the infected cells will divide and each cell in the resulting clone will display the reporter gene. If an early embryo is used, the type of cell infected cannot be predicted. With later embryos and the judicious placing of the virus, however, it is possible to control the process more closely. In this way, it has been possible to follow the lineages of the differentiating brain cells in the rat (**11.47, 11.48**; Price, 1987).

11.44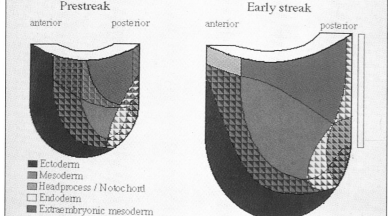

11.44 Fate maps of the epiblast of the prestreak and early streak stages showing the original sites of the key tissues present in the late-streak and neural-plate stages respectively. The bar shows the approximate extent of the primitive streak. (Photograph courtesy of Kirstie Lawson; from Lawson *et al.*, 1991.)

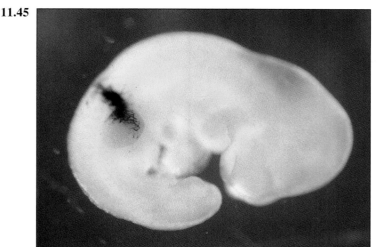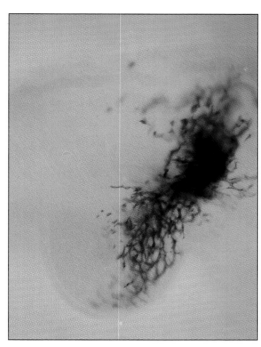

11.45, 11.46 A mouse that has had a somite replaced by one from a transgenic mouse carrying a β-galactosidase reporter gene (**11.45**, ×25). As the somite breaks up, its cells migrate into the limb (**11.46**, ×100). (Micrographs courtesy of Rosa Beddington).

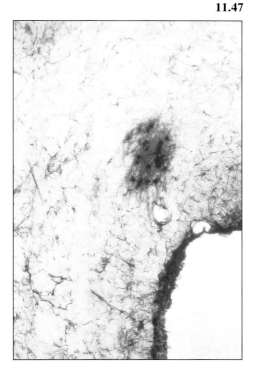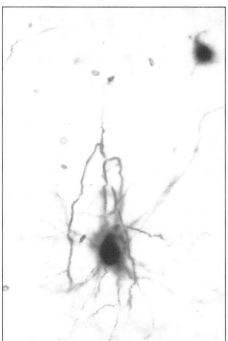

11.47, 11.48 A clone of oligodendrocytes in a 28-day rat brain that derived from a single cell infected by BAG retroviral vector injected into the cerebral vesicles of the 16dpc embryo *in utero* (**11.47**, ×100). A pair of cortical pyramidal neurons from a similar animal (**11.48**, ×330). (The sections were stained blue with X-gal and brown with a monoclonal antibody that marks astrocytes. Micrographs courtesy of Jack Price.)

Organ culture and mouse kidney development

One area where the mouse is particularly useful for studying developmental mechanisms is the epithelial-mesenchymal inductive interactions that give rise to many of the tissues that form during the organogenesis phase of development (for general review, see Gilbert, 1991). These interactions are between domains of mesenchymal cells and specific epithelial-cell buds, and tissues that can be studied relatively easily *in vitro* include the salivary and mammary glands, lung and tooth (where the mesenchyme derives from the neural crest), and, most recently, the gut where a mixed rat-mouse system has been used and the cell origins distinguished by *in situ* hybridization (**11.49**; Del Buono *et al.*, 1992). The richest

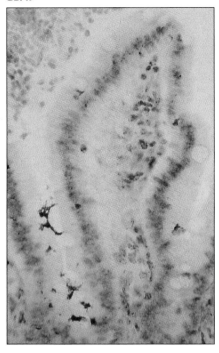

11.49 A section showing crypt differentiation. Here, the intestinal epithelium is from undifferentiated rat intestine placed in a collagen gel and xenografted into a nude mouse. As mouse mesenchyme invades the graft, the tissue differentiates to give villus. The rat DNA is marked with a purple stain, while the mouse DNA is brown (×270). (Micrograph courtesy of Raffaele Del Buono. For further details, see Del Buono et al., 1992.)

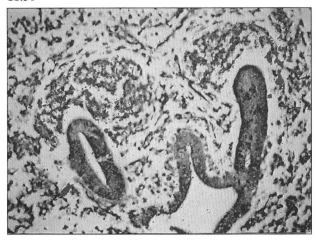

11.50 A section of an 11dpc embryo through the two metanephric buds. The Wilms' tumour gene WT1 (red) is expressed in the metanephric blastema (×120). (For technical details, see caption to **11.55** and **11.56**. Micrograph courtesy of Jane Armstrong.)

system, however, is the kidney and it is discussed here to illustrate some of the technical possibilities of mammalian organ culture.

The kidney found in the adult, the metanephros, is the third that forms from nephrogenic cord tissue (intermediate-plate mesoderm) and follows the production of the very rudimentary pronephros and the transitory mesonephros. It is initiated around 11dpc when a small bud (the ureteric bud) off the mesonephric (Wolffian) duct invades a domain of mesenchyme (the metanephric blastema) and a reciprocal inductive interaction takes place (**11.50**; for review, see Bard, 1992). Although the details remain to be elucidated, it seems that the ureter induces the metanephric mesenchyme to form nephrous, a process with an unexpected neuronal involvement (see Sariola et al., 1991). At the same time, the mesenchyme induces the proximal part of the bud to form the renal pelvis and the ureter, while its distal part extends and bifurcates to form the collecting ducts.

Nephrogenesis starts when some of the mesenchymal stem cells aggregate to form condensations, with the remainder initially staying in their stem-cell state. These condensations first epithelialize and then take on a comma shape, with blood vessels invading the tail which will become the glomerulus, before forming an **S**-shaped tubular structure. This tube then extends and undergoes a complex morphogenetic change so that the centre forms the loop of Henle, the domain near the glomerulus forms the proximal convoluted tubule while the part adjacent to the future collecting duct forms the distal convoluted tubule. As the kidney grows, successive groups of undifferentiated stem cells continue to enter the nephrogenic pathway to form successive tiers of nephrons. The developing metanephric kidney at 13–18dpc thus represents a time course of nephrogenesis with a rind of stem cells at the periphery (subjacent to the kidney capsule), primitive nephrons underlying them and more advanced, functional nephrons nearer to its centre (**11.51**).

The ability of kidney rudiments to develop and form nephrons in culture has facilitated the study of nephrogenesis and has shown, for example, that mesenchyme differentiation can be blocked by LIF, the glycoprotein which blocks ES-cell

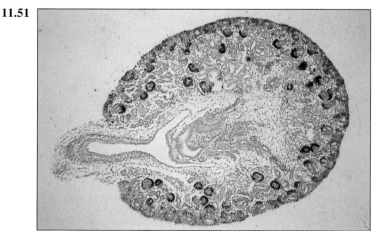

11.51 A 16dpc kidney showing WT1 expression in the peripheral stem cells, the early nephrons near the surface and the glomeruli nearer the centre (×42). (Micrograph courtesy of Jane Armstrong.)

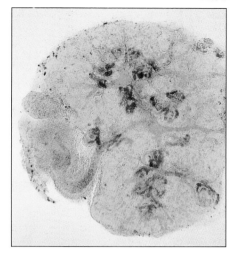
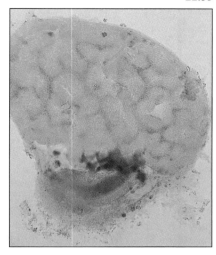

11.52–11.53 A wholemount of an 11-day kidney rudiment cultured for 5 days *in vitro* and stained with X-gal is seen in **11.52**. The stain shows the ducting system and well-differentiated nephrons (dark blue). **11.53** shows a similar rudiment cultured in LIF. The ducting system forms normally, but there are no nephrons (×45).

differentiation (**11.52, 11.53**; Bard and Ross, 1991). *In vitro* work has, however, mainly been used to study mesenchyme induction, and the most recent area of interest is the Wilms' tumour gene, WT1, which helps mediate the inductive response and the mesenchyme-to-epithelium transition (for review, see van Heyningen and Hastie, 1992). Mutation or loss of this gene is associated with the development of Wilms' tumour, which characteristically displays uncontrolled growth of the kidney stem cells. WT1 is expressed at low level in the uninduced blastema and at higher levels in the nephrons. The function of the gene may be clarified by culturing kidneys in the presence of the appropriate anti-sense oligonucleotides which should suppress its transcription (see Sariola *et al.*, 1991).

Developmental genetics

It is not possible to discuss mouse embryogenesis without touching on the major thrust of current research on the organism, elucidating the molecular basis of the various control mechanisms that regulate its development. As space precludes any attempt to review the literature here, we will merely summarize some data on the mouse genome and the procedures being used to analyse gene regulation in development (for review see Wilkins, 1993).

The normal laboratory mouse (always an inbred or random-bred strain of *Mus musculus*) has 38 autosomes and two sex chromosomes (**11.54**) which together contain approximately 3×10^9 bp (base pairs), or about 20 times the number in *Drosophila*. Of the DNA, about 70% is single copy and the remainder repeated, with some 10% being a 14bp, AT-rich repeat, located in the centromeric heterochromatin and repeated about 10^6 times. Estimates of gene number indicate a figure of about 40,000, give or take a factor of two (see Joyner, 1991). One way of facilitating such genetic analysis is to use crosses between inbred strains and *Mus spretus*, another related species of mouse, as this increases the number of polymorphisms. Southern analysis of the crosses can, for example, be used to look for mutant genes (see Hogan *et al.*, 1986).

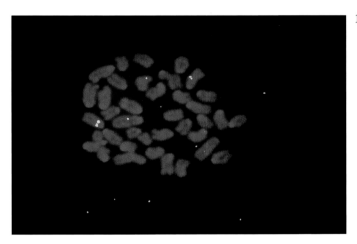

11.54 A confocal micrograph of the chromosomes of a mouse ES cell into which a dominant selectable marker (hygromycin resistance) has integrated. The probe for the marker was labelled with biotin and Avidin DCS-FITC and the chromosome spread counterstained with propidium iodide/DAPI (×1500, micrograph courtesy of John Gosden).

Finding new regulatory genes is not easy and the simplest way is to use known genes from, say, *Drosophila* to probe mouse cDNA libraries under low stringency conditions and so identify their mouse homologues (e.g. Davidson and Hill, 1991). Such has been the success of this procedure that it is now hard to count the number of such mouse control genes, and it is sufficient to mention here that genes have been identified with all of the known DNA-binding motifs (homeobox, POU, zinc-finger, helix-turn-helix, leucine-zipper etc.).

Identifying these genes is, however, very much easier than elucidating their function. For this, the first step is to identify when and where they are expressed by using *in situ* hybridization to reveal sites of expression in sectioned tissue (**11.55, 11.56**). If transgenic mice are then made that incorporate one or another regulatory sequences of the gene together with a reporter gene such as β-galactosidase (which converts the substrate X-gal to a blue compound), one can then identify which of these sequences is an appropriate promoter for the gene in a particular position (**11.57**). This technique can be extended to look for promoter sequences by randomly inserting DNA sequences. If the probe inserts near such a site, production of the reporter gene can be upregulated and the resulting expression pattern can then be investigated and analysed (Joyner, 1991). Alternatively, if the probe contains a weak promoter, the method can be used to look for other regulatory or enhancer sites.

The third step in this long process is to make a transgenic mouse in which one copy of the gene is either not expressed or is present in a non-functional form or under the control of a selectable promoter. Following mating between transgenic siblings, deformities in the homozygous offspring together with local abnormalities in the expression of other genes indicate tissues where the molecule plays a controlling role. Unfortunately, there are good reasons for supposing that several genes play such roles in each tissue and it is going to take some time for us to untangle the tiers of regulation in the formation of even simple tissues (if such tissues exist).

11.55
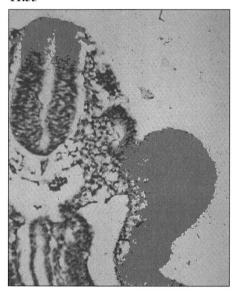

11.56
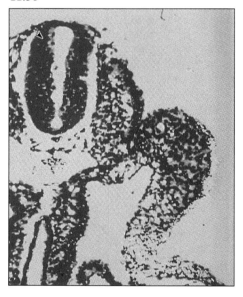

11.57
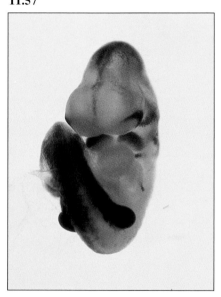

11.55, 11.56 The expression of the homeodomain genes, hox7 (red) and hox8 (yellow) in the limb-bud region of a 9.5dpc embryo (TS15). Digital images were made of the sections under bright field and dark field optics and these images were then electronically colour-enhanced and superimposed (×140, micrographs courtesy of Duncan Davidson and Richard Baldock).

11.57 A 9.5dpc transgenic mouse (TS15) carrying a reporter gene under the influence of the hox7 promoter and stained with X-gal (×23, photograph courtesy of Bob Hill and Duncan Davidson).

Transgenic mice

In a more general context, the mouse embryo is becoming as important for its use in making transgenic adult mice as for its developmental interest and we now consider how these mice are made (for detailed reviews of the area, see Jaenisch, 1988; Joyner, 1991).

A transgenic mouse is one whose germ line contains DNA exogenously added so that its offspring will inherit the added genes. This result can be achieved in three distinct ways, of which the easiest is to use retroviruses to infect a fertilized egg or a later-stage embryo (in which case there is a low chance of the genes being taken up into the germ cells) and such infections can be done with either intact or recombinant retroviruses. The proviral DNA will of course insert randomly, but, if the point of insertion is in the middle of a gene whose function is then destroyed, the phenotype can be recognized and the disrupted gene identified by its proximity to the proviral insert. This was the first technique to be used and one success of the method was the making of the Mov13 mouse when a Moloney leukaemia provirus inserted into the first intron of the collagen α1 gene (Schnieke *et al.*, 1983); the resulting mice failed to lay down collagen I (see Bard and Kratochwil, 1987). The use of the method is, however, limited partly by the fact that insertion is random and partly because of the small amount of genetic material that can be introduced.

The latter set of problems is overcome in the second technique, that of injecting DNA of up to 50kb or more directly into the pronucleus of the fertilized egg (**11.58**) where it may insert itself into the host genome (see Hogan *et al.*, 1986). The insertion process is complex and usually leads to many copies of the injected DNA in end-to-end array inserting themselves into the host genome at a random site which is often rearranged or suffers a deletion. Although these rearrangements can make it difficult to identify the insertion site in a transgenic mouse, this method is an efficient and common way of inserting the upstream sequence of a gene to which has been attached a lac-Z reporter gene and, more recently, for making 'dominant negatives'. These transgenic mice incorporate at a random site a sequence whose expression can negate the effect of a particular gene, perhaps by overexpressing a mutant subunit that becomes incorporated into the normal protein, rendering it non-functional. This technique has, for example, been used to show that a mutant in the collagen pro-αI(1) gene can mimic the phenotype of the disease *osteogenesis imperfecta* (Stacey *et al.*, 1988).

For investigating the function of normal genes, however, a recent method of making transgenic mice has been developed that is based on *gene targeting* in ES cells. This involves transfecting (usually by electroporation) ES cells with DNA coding a mutated gene which recognizes and homologously recombines with the wild-type gene and so mutates it. This is, of course, a rare event, but, with sufficient number of ES cells and appropriate assay and selection techniques (Joyner, 1991), recombinants can be isolated and inserted into blastocysts where they become incorporated into the developing mice (**11.59**). Those chimeras which incorporate the differentiating ES cells into their germ cells will go on to form transgenic animals (e.g., Dorin *et al.*, 1992).

The use of ES cells thus allows us, in principle at least, to destroy any gene function and also to repair mutant genes. Despite the technique being slow and labour intensive, it is nevertheless being used to investigate the roles of transcription factors in early development. One example is the transgenic mouse which fails to make Hox-1.5, a homeobox-containing gene which is widely expressed in the embryo. The pattern of abnormalities in the mouse is, however, but a small subset of its expression pattern: although there are defects in glands and skeletal tissues in the anterior part of the mouse, the posterior region appears to be normal (Chisaka and Capecchi, 1991). This unexpected result points to a measure of functional redundancy in the gene's activity.

Transgenic mice now have an important role in contexts other than elucidating genetic control mechanisms, as the following examples show. First, mice can be made that allow specific cell types to be ablated by inserting toxin genes such as ricin or the A chain of diphtheria toxin that are under the control of tissue-specific promoters or drug administration (Bernstein and Breitman, 1989). Second, transgenic mice are being used to analyse the role of molecules such as growth factors in development and the extent to which they are affected by imprinting (e.g. DeChiara, *et al.*, 1990, 1991). Third, mouse models can be made of such human diseases as cystic fibrosis (Dorin *et al.*, 1992). Finally, and of particular interest in the context of cancer, is a transgenic mouse which includes a temperature-sensitive transforming gene (Jat *et al.*, 1991) and which should allow permanent cell lines to be made of almost any tissue. Over the next few years, the use of transgenic mice can be expected to become widespread throughout mammalian research as such mice will be designed to model a very wide range of phenotypes.

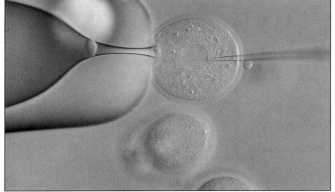

11.58 Micro-injection of DNA into the pronucleus of a fertilized mouse oocyte which is held at the end of a holding pipette by gentle suction. The photograph illustrates the oocyte after the injection of the DNA and before the needle is removed. (×360, Nomarski optics. Micrograph courtesy of John Mullins.)

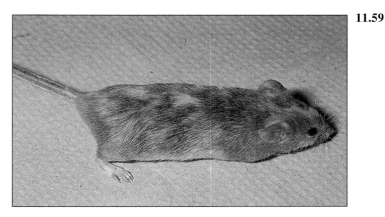

11.59 A chimeric mouse made by inserting an ES cell from a 129 mouse homozygous for *chinchilla* (C^{ch}) and *pink-eye* (*p*), both of which lighten coat colour, into the blastocyst of a CBAxC57BL/6 mouse. Hence the striped appearance. (Photograph courtesy of Julia Dorin.)

The future

The last decade has seen very great advances in the techniques for investigating mouse development and in the uses to which they can be put, and it seems unlikely that there will be the same level of technical progress in the near future. Instead, we can expect an explosion of data that will come from using the techniques for growing, culturing and manipulating mouse embryos as we investigate the expression and function of the genes that control the emergence of the developmental phenotype. This ever-increasing amount of expression, lineage and functional data will bring its own problems as it will not be easy to keep track of data which will span all the tissues of the mouse as it develops from a blastocyst to a functioning animal.

One solution to the problem of handling all this information may lie in making databases and there are two obvious needs. The first is to list all the transgenic mice that are being made and the second is to store the tissue-specific and time-specific data on mouse expression that has been published. The former will be relatively simple, but the latter will not as it will require both the complete atlas of mouse development and the expression data to be stored digitally. The expression database will then have to be regularly updated in a manner analagous to that used for DNA sequences in the human genome database. Such a project is not, however, impossible: the embryonic sections and the detailed developmental anatomy are now available (Kaufman, 1992a), as are many of the programs required for setting up a computer-based system (Baldock et al., 1992). If this system turns out to be feasible, it will be possible to interrogate the database to ask about which genes are expressed as a tissue develops or where a particular gene is expressed. Such a system should also be useful for lineage and cell movement studies, and will allow us to integrate much that is known of mRNA and protein expression as the mouse develops.

At the beginning of this chapter, it was pointed out that the mouse embryo has come a long way from its relative obscurity in the '60s to its central position today. This importance is likely to increase over the next few years as its use as a model system for studying all aspects of development and the relative ease of making transgenic mice become more widely appreciated. It is, indeed, not unrealistic to expect that, in a decade or two, we will know almost as much about the genetic base of mammalian development as we will about that of *Drosophila* and *C. elegans*.

Acknowledgements

We thank Duncan Davidson and Bob Hill for discussion.

References

For many years, the *Mouse Newsletter* (now renamed the *Mouse Genome*) has permitted the rapid publication of mouse data, mainly with reference to genetics, and it is a useful source of information.

Austin, C.R. (1965). *Fertilization*. Prentice-Hall, New Jersey.

Baldock, R., Bard, J., Kaufman, M. and Davidson, D. (1992). A real mouse for your computer. *BioEssays*, **14**, 501–502.

Balinski, B.I. (1981). *An Introduction to Embryology*. W.B. Saunders Co., Philadelphia.

Bard, J.B.L. (1992). The development of the kidney: embryogenesis writ small. *Curr. Opin. Genet. Devel.*, **2**, 589–586.

Bard, J.B.L. and Kratochwil, K. (1987). Corneal morphogenesis in the Mov13 mutant mouse is characterised by normal cellular organization but disordered and thin collagen. *Development*, **101**, 547–555.

Bard, J.B.L. and Ross, A.S.A. (1991). The blocking of mouse nephron development *in vitro* by DIA/LIF, the ES-cell differentiation inhibitor. *Development*, **113**, 193–198.

Beddington, R.S.P and Lawson, K.A. (1990). Clonal analysis of cell lineages. In *Postimplantation Mammalian Embryos. A Practical Approach* (A.J. Copp, and D.L. Cockroft, eds) pp. 267–292, I.R.L. Press, Oxford.

Beddington, R.S.P. and Martin, P. (1989). An *in situ* marker to monitor migration of cells in the mid-gestation mouse embryo: somite contribution to the early forelimb bud. *Mol. Biol. Med.*, **6**, 263–274.

Bernstein, A. and Breitman, M. (1989). Genetic ablation in transgenic mice. *Mol. Biol. Med.*, **6**, 523–530.

Bradley, A. (1987). Production and analysis of chimaeric mice. In *Teratocarcinomas and Embryonic Stem Cells. A Practical Approach* (E.J. Robertson, ed.) pp. 113–151, I.R.L. Press, Oxford.

Bradley, A., Evans, M., Kaufman, M.H. and Robertson, E. (1984). Formation of germ-line chimaeras from embryo-derived teratocarcinoma cell lines. *Nature*, **309**, 255–256.

Chiquoine, A.D. (1954). The identification, origin, and migration of the primordial germ cells in the mouse embryo. *Anat. Rec.*, **118**, 135–146.

Chisaka, O. and Capecchi, M.R. (1991). Regionally restricted developmental defects resulting from target disruption of the mouse homeobox gene *hox-1.5*. *Nature*, **350**, 473–479.

Cockroft, D.L. (1990). Dissection and culture of postimplantation embryos. In *Postimplantation Mammalian Embryos. A Practical Approach* (A.J Copp, and D.L. Cockroft, eds) pp. 15–40, I.R.L. Press Oxford.

Copp, A.J. and Cockroft, D.L. (eds) (1990). *Postimplantation Mammalian Embryos. A Practical Approach*. I.R.L. Press, Oxford.

Davidson, D.R. and Hill, R.E. (1991). *Msh-*

like genes: a family of homeobox genes with wide-ranging expression during vertebrate development. *Sem. Dev. Biol.*, **2**, 405–412.

DeChiara, T.M., Efstratiadis, A. and Robertson, E.J. (1990). A growth deficiency phenotype in heterozygous mice carrying an insulin-like growth factor II gene disrupted by gene targeting. *Nature*, **345**, 78–80.

DeChiara, T.M., Robertson, E.J., and Efstratiadis, A. (1991). Parental imprinting of the mouse insulin-like growth factor II gene. *Cell*, **64**, 849–860.

Del Buono, R., Fleming, K.A., Morey, A.L., Hall, P.A., and Wright, N.A. (1992). A nude mouse xenograft model of fetal intestine development and differentiation. *Development*, **114**, 67–73.

Dorin, J.R., et al. (1992). Cystic fibrosis in the mouse by targetted insertional mutagenesis. *Nature,* **359**, 211–215.

Ducibella, T. (1977). Surface changes of the developing trophoblast cell. In *Development in Mammals*, **vol. 1** (M.H. Johnson, ed.) pp. 5–30, Elsivier, Amsterdam, North Holland.

Edwards, R.G. and Gates, A.H. (1959). Timing of the stages of the maturation divisions, fertilization and the first cleavage of eggs of adult mice treated with gonadotrophins. *J. Endocr.*, **18**, 292–304.

Evans, M.J. and Kaufman, M.H. (1981). Establishment in culture of pluripotential cells from mouse embryos. *Nature*, **292**, 154–156.

Fitzgerald, L.W., Muller, K.J., Glick, S.D., Ratty, A.K., Elsworth, M.K. and Gross, K.W. (1991). Ontogeny of hyperactivity and circling behaviour in a transgenic insertional mutant mouse. *Behav. NeuroSci.,* **105**, 755–763.

Gardner, R.L. and Papaioannou, V.E. (1975). Differentiation in the trophectoderm and inner cell mass. In *The Early Development of Mammals* (M. Balls,. and A.E. Wild, eds) pp. 107–132, Cambridge University Press.

Gilbert, S.F. (1991). *Developmental Biology*, 3rd Edition. Sinauer Press, Mass., U.S.A.

Gressens,P., Hill, J.M., Gozes, I., Fridkin, M. and Brenneman, D.E. (1993). Growth factor function of vasoactive intestinal peptide in whole cultured mouse embryos. *Nature,* **362**, 155–158.

Hamilton, W.J. and Mossman, H.W. (1972). *Hamilton, Boyd and Mossman's Human Embryology. Prenatal Development of Form and Function*. W. Heffer and Sons Ltd, Cambridge.

Henery, C.C., Bard, J.B.L. and Kaufman, M.H. (1992). Tetraploidy in mice, embryonic cell number and the grain of the developmental map. *Developmental Biology* **152**, 233–241.

Hogan, B., Constantini, F., and Lacy, E. (1986). *Manipulating the Mouse Embryo.* Cold Spring Harbor Laboratory.

Jaenisch, R. (1988). Transgenic animals. *Science*, **240**, 1468–1474.

Jat, P.S., Noble, M.D., Ataliotis, P., Tanaka, Y., Yannoutsos, N., Larsen, L., and Kioussis, D. (1991). Direct derivation of conditionally immortal cell lines from an H-$2K^b$ -tsA58 transgenic mouse. *Science*, **88**, 5096–5100.

Johnson, M.H. and Ziomek, C.A. (1981). The foundation of two distinct cell lineages within the mouse morula. *Cell*, **24**, 71–80.

Joyner, A.L. (1991). Gene targeting and gene trap screens using embryonic stem cells: new approaches to mammalian development. *BioEssays*, **13**, 649–656.

Kaufman, M.H. (1982). The chromosome complement of single-pronuclear haploid mouse embryos following activation by ethanol treatment. *J. Embryol. Exp. Morph.*, **71**, 139–154.

Kaufman, M.H. (1983). *Early Mammalian Development: Parthenogenetic Studies.* Cambridge University Press.

Kaufman, M.H. (1990). Morphological stages of postimplantation embryonic development. In *Postimplantation Mammalian Embryos. A Practical Approach* (A.J. Cop and D.L. Cockroft, eds) pp. 81–91, I.R.L. Press, Oxford.

Kaufman, M.H. (1991a). New insights into triploidy and tetraploidy, from an analysis of model systems for these conditions. *Human Reproduction*, **6**, 8–16.

Kaufman, M.H. (1991b). Histochemical identification of primordial germ cells and differentiation of the gonads in homozygous tetraploid mouse embryos. *J. Anat.*, **179**, 169–181.

Kaufman, M.H. (1992a). *The Atlas of Mouse Development*. Academic Press, London.

Kaufman, M.H. (1992b). Postcranial morphological features of homozygous tetraploid mouse embryos. *J. Anat.*, **180**, 521–534.

Kaufman, M.H., Lee, K.K.H. and Speirs, S. (1989). Influence of diandric and digynic triploid genotypes on early mouse embryogenesis. *Development*, **105**, 137–145.

Kaufman, M.H and Speirs, S. (l987). The postimplantation development of spontaneous digynic triploid embryos in LT/Sv strain mice. *Development*, **101**, 383–391.

Kaufman, M.H. and Webb, S. (1990). Postimplantation development of tetraploid mouse embryos produced by electrofusion. *Development*, **110**, 1121–1132.

Lawson, K.A., Meneses, J.J., and Pedersen, R.A. (1991). Clonal analysis of epiblast fate during germ layer formation in the mouse embryo. *Development*, **113**, 891–911.

Lyon, M.F. and Searle, A.G. (eds) (1989). *Genetic Variants and Strains of the Laboratory Mouse*, 2nd Edition. Oxford University Press.

Moore, K.L. (1982). *The Developing Human. Clinically Oriented Embryology,* 3rd Edition. W.B. Saunders Company, Philadelphia.

Morriss–Kay, G. and Tan, S.-S. (1987). Mapping cranial neural crest cell migration pathways in mammalian embryos. *Tr. Genet.,* **3**, 257–261.

Murphy, P. and Hill, R.E. (1991). Expression of the mouse homeobox-containing genes, *Hox 2.9* and *Hox 1.6*, during segmentation of the hindbrain. *Development*, **111**, 61–74.

New, D.A.T. (1978). Whole-embryo culture and the study of mammalian embryos during organogenesis. *Biol. Rev.*, **53**, 81–122.

Niemierko, A. and Opas, J. (1978). Manipulation of ploidy in the mouse. In *Methods in Mammalian Reproduction* (J. C. Daniel, Jr., ed.) pp. 49–65, Academic Press, New York.

O'Neill, G.T. and Kaufman, M.H. (1987). Cytogenetic analysis of first cleavage fertilized mouse eggs following *in vivo* exposure to ethanol shortly before and at the time of conception. *Development*, **100**, 441–448.

Ozdzenski, W. (l967). Observations on the origin of primordial germ cells in the mouse. *Zool. Polon.*, **17**, 367–379.

Pedersen, R.A. (1986). Potency, lineage and allocation in preimplantation mouse embryos. In *Experimental Approaches to Mammalian Embryonic Development.* Cambridge University Press.

Price, J. (1987). Retroviruses in the study of cell lineages. *Development*, **101**, 409–419.

Rashbass, P., Cooke, L.A., Herrmann, B.G., and Beddington, R.S.P. (1991). A cell autonomous function of *Brachyury* in *T/T* embryonic stem cell chimaeras. *Nature*, **353**, 348–351.

Resnick, J.L., Bixler, L.S., Cheng, L. and Donovan, P.J. (1992). Long-term proliferation of mouse primordial germ cells in culture *Nature*, **359**, 550–551.

Robertson, E.J. (ed.) (1987). *Teratocarcinomas and Embryonic Stem Cells. A Practical Approach*. I.R.L. Press, Oxford.

Rogers, D. (1988). Inhibition of

pluripotential embryonic stem cell differentiation by purified polypeptides. *Nature*, **336**, 688–690.

Rossant, J. and Papaioannou, V.E. (1977). The biology of embryogenesis. In *Concepts in Mammalian Embryogenesis* (M.I. Sherman, ed.) pp. 1–36. MIT Press, Cambridge, Ma, USA.

Rossant, J. and Pedersen, R.A. (eds) (1986). *Experimental Approaches to Mammalian Embryonic Development*. Cambridge University Press.

Sariola, H., Saarma, M., Sainio, K., Arumae, U., Palgi, J., Vaahtokari, A., Thesleff, I. and Karavanov, A. (1991). Dependence of kidney morphogenesis on the expression of nerve growth factor receptor. *Science*, **254**, 571–573.

Schnieke, A., Harbers, K., and Jaenisch, R. (1983). Embryonic lethal mutant in mice induced by retrovirus insertion into the alpha 1(I) collagen gene. *Nature*, **304**, 315–320.

Smith, A.G., Heath, J.K., Donaldson, D.D., Wong, G.G., Moreau, J., Stahl, M., and Rogers, D. (1988). Inhibition of pluripotential embryonic stem cell differentiation by purified polypeptides. *Nature*, **336**, 688–690.

Smith, R. and McLaren, A. (1977). Factors affecting the time of formation of the mouse blastocoele. *J. Embryol. Exp. Morph.*, **41**, 79–92.

Snell, G.D. and Stevens, L.C. (1966). Early embryology. In *Biology of the Laboratory Mouse*, 2nd Edition. (E.L. Green, ed.) pp. 205–245, McGraw-Hill, New York.

Snow, M.H.L. (1973). Tetraploid mouse embryos produced by cytochalasin B during cleavage. *Nature*, **244**, 513–515.

Stacey, A., Bateman, J., Choi, T., Mascara, T., Cole, W., and Jaenisch, R. (1988). Perinatal lethal *osteogenesis imperfecta* in transgenic mice bearing an engineered mutant pro-alpha-(1)I gene. *Nature*, **332**, 131–136.

Szulman, A.E. and Surti, U. (1978a). The syndromes of hydatidiform mole. I. Cytogenetic and morphologic correlations. *Am. J. Obstet. Gynec.*, **131**, 665–671.

Szulman, A.E. and Surti, U. (1978b). The syndromes of hydatidiform mole. II. Morphologic evolution of the complete and partial mole. *Am. J. Obstet. Gynec.*, **132**, 20–27.

Theiler, K. (1989). *The House Mouse: Atlas of Embryonic Development*. Springer-Verlag, New York.

van Heyningen, V. and Hastie, N.D. (1992). Wilms' tumour: reconciling genetics and biology. *Tr. Genet.*, **8**, 16–21.

Wilkins, A. (1993). *Genetic Analysis of Animal Development*. Wiley-Liss, New York.

12. The Human

Marjorie A. England

Introduction

The earliest written records contain accounts of the development of the human embryo and fetus, albeit relatively infrequently, probably because of the internal and hidden nature of the conceptus' development. Despite their lack of sophisticated equipment and methodology, these early authors were both fascinated by their subject and observant in their records.

Embryos have been described in ancient Sanskrit documents (1416BC) and by Aristotle and Hippocrates, while note was also made of them in the Bible and in the Koran (AD600). Because of the paucity of early embryos and instrumentation for magnification at that time, most authors could only make limited observations on the external features of the human fetus. While Galen (AD100) described the morphology of older fetuses, nothing of substance was done through the Dark Ages and the illustrated accounts in those times contained little that could be considered realistic. Accuracy again returned when, in the fifteenth century, Leonardo da Vinci made measurements of embryonic growth and illustrated fetuses *in utero*. Soon after, another significant treatise entitled *De Formatu Foetu* by Fabricius of Aquapendente (1533–1619) was also published. Its illustrations are a realistic representation of the developing fetus as are those of William Hunter (1774) more than a century later in *The Anatomy of the Human Gravid Uterus* and others during this period (**12.1**).

With the discovery of the microscope, the early stages of embryonic development could be studied with more accuracy. In 1683, Van Leeuwenhoek wrote that "the humane fetus though no bigger then a green pea, yet is furnished with all its parts". The microscope also allowed information about the female and male gametes to become available and the ovarian follicles were first described in the rabbit by de Graaf in 1672 and human sperm by Hamm and Leeuwenhoek in 1677. The true nature of the gametes was not, however, appreciated then and there were strong arguments for the view that a miniature preformed human was in the egg (ovists) or in sperm (homunculists). A century later, Spallanzoni (1775) finally laid such preformation arguments to rest by proving that both the ovum and sperm are necessary to produce a new individual. At around the same time, Wolff (1733–1794) published the concept of epigenesis which proposed that development occurred through differentiation and progressive growth.

It was not, however, until the nineteenth century that major advances were made and the foundations of modern human embryology established. In 1829, von Baer put forward the germ layer theory of embryos and, with the publication of the

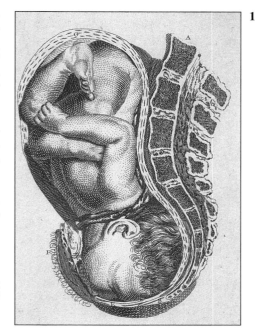

12.1 A late human embryo (reproduced from Smellie's Tables) by Frederick Birnie, anatomical draughtsman to Dr William Hunter and engraved by William Taylor, 1791.

cell theory in 1839 by Schleiden and Schwann, it became clear that the body develops from two single cell gametes, one from each parent. In 1880, Wilhelm His stimulated great interest with the publication of *The Anatomy of Human Embryos* and subsequent workers established the framework of the major morphogenetic events in human embryos and fetuses. Most of these findings have been based on natural and therapeutic abortions and do not include many of the earliest stages. It is only since 1969 that human eggs were successfully fertilized *in vitro* and since 1970 that cleavage was achieved in culture. The practical side of this work was successfully brought to fruition in 1978 when the work of Edwards and Steptoe culminated in the birth of Louise Brown. Since then, human embryologists have had access to the earliest stages of embryonic development *in vitro* (fertilization to blastula stages).

Sources of human embryos

While animal embryos can be obtained systematically in a scientifically controlled environment, human embryos only become available in a haphazard way when development is interrupted by natural causes or by accidental or deliberate intervention. In these cases, more often than not, only minimal information is available about the embryo. Large numbers of embryos of any designated age group are similarly difficult to obtain for study, while, because of moral and legal restraints, certain age groups (e.g. late blastulae) are very rarely if ever obtainable.

To make matters worse, only one conceptus is normally produced in humans and multiple births are the exception, while the low frequency of twinning precludes easy comparison of pairs of such embryos. Moreover, the parental contribution varies with each specimen and they are not, therefore, genetically comparable for experimental observations.

Many workers, therefore, combine the observations on embryos that they have obtained with those made on existing museum specimens which include both wholemount and histological preparations. Measurements on such specimens may, however, be unreliable as, if a period of time has elapsed between the discovery of the embryo and its subsequent fixation, there will be a degree of shrinkage and other post mortem changes, and the embryo may even have been damaged during handling.

Those embryos and fetuses in recent museum collections have often been fixed in formal saline, sometimes supplemented with sodium hydrosulphite to help maintain a pink colour in the skin. If these specimens are used in subsequent investigations, these chemical additions should be noted. Older fetal specimens prepared in the 1920–1950s were frequently fixed in their entirety by perfusion techniques, but this form of preservation is no longer acceptable. Moreover, because of their relative rarity, many suboptimal specimens have been retained which would have been discarded in a study of another species. Today, specimens for light microscopy are often fixed in formal saline, Bouin's fluid, or Carnoy's solution, and those for electron microscopy in gluteraldehyde-formaldehyde solutions. Standard procedures are then used for preparing sections.

Ageing human embryos

The normal gestational period of human development is from fertilization to approximately week 38. In the past, many different methods of assessing age have been used that include crown-rump (CR) length, lunar months and 'last menstrual period' (LMP) dates. Clinicians routinely use the LMP for ageing and normally divide the nine months' gestation into trimester periods each of three months' duration. Embryologists have, however, divided the span of gestation into two major periods, that of embryogenesis (from fertilization to the end of week 8) and that of fetal development (from week 9 to birth). In addition, the early part of the embryonic period, before the appearance of separate groups of cells for the embryo and its membranes, is now usually referred to as the pre-embryonic period (up to about Stage 6, about day 14, when there is the first evidence of the primitive streak). The primordia of all the tissues of the body are established during the embryonic period. While some major morphogenetic events occur during the fetal period, it is primarily a period of differentiation, growth and the onset of function in the various organs and systems. There is also a manifold increase in size and body weight.

While all of the different methods of ageing have proved to be of use, the most accurate method of dating in the pre-embryonic and embryonic periods is based on a series of progressively developing features. The original stageing series introduced by Streeter (1942, 1945, 1948, 1951) for human embryos was based on primate development, but O'Rahilly (1973) and O'Rahilly and Müller (1987) have recently introduced a system based on human embryonic development, using an age based on days post ovulation (dpo) rather than days post conception as it is not easy to tell the exact time of conception. They have revised, updated and modified Streeter's system with the introduction of the Carnegie Stages of human development.

This internationally agreed and accepted method of ageing human embryos is thus based on a series of developmental stages which incorporates both external and internal features. It commences at the stage of the single fertilized cell (approximately one post-ovulatory day) and ends at Stage 23 which marks the end of the embryonic period (approximately 56–57 days). The older Streeter system of ageing has been retained during the fetal period of development (9 post-ovulatory weeks to birth). In addition to the external and internal features, each Carnegie Stage specifies the length of the embryo in millimetres as well as the appropriate number of post-ovulatory days. From Carnegie Stage 6 it is practicable to use the greatest length (GL) measurement of the embryo.

Since experimental studies are normally conducted on various laboratory animals it is necessary to compare their developmental stages with those of the human. As gestational periods differ among these animals, the age, both absolute and relative, at which a particular morphogenetic event will occur can vary substantially (*Table 12.1*).

Table 12.1 Developmental ages for some common animals.

	Man	Monkey	Mouse	Rat	Rabbit	Guinea pig	Hamster	Chick
Total Gestational period (days)	approximately 266	166	18–20	21–22	31–34	64–68	15½–16	21
Developmental event								
Blastula	4	5–8 4–9	3–6	3–5	2–6	4–5	3–4½	–
Implantation	7–12	9	5½–7	5–6	6	6	4–5	–
Primitive Streak	13	18–20	8	9	6½	11¾–13½	7	½–1
Neural Plate	16	19–21	8–8½	9½	7½	13½	7½	1–1½
First Somite	20±1	20–21	8+	10	–	14½	7+	1–1½
Ten Somites	22±1	23–24	8½	10½	9–	15½	8	1½–2
Anterior Neuropore Closes	24±1	23–25	9	10½	9½	15½	8+	2–2½
Upper Limb Bud	26±1	25–26	9+	10½	10½	16½	8¼	2+
Twenty Somites	24±1	26	9+	11+	10–	16½	8½	2
Posterior Neuropore Closes	26±1	–	9½	11+	–	16+	8½	3¾
Lower Limb Bud	28	26	10+	11+	11	18½	9¾	2½
Gut Herniation	41	–	12+	12½	14½	23¾	–	19
Digital Rays	41	34	12+	13+	14½	23¾	10¼	4¾
Testis: Histological Diff.	44	–	–	14½	20	26+	12	5½

Normal development

The embryonic staging, descriptions and terminology given here are based on the series published by O'Rahilly and Müller (1987), but, in using them, one should be aware that each stage incorporates a range of development rather than an exact description. Times are given as days post ovulation (dpo), lengths are crown-rump and descriptions are based on gross observations and light microscopy. The summaries given below are necessarily brief and focus on major features alone; for greater detail, the reader should consult O'Rahilly and Müller (1987). One further point: the reader should note that the terminology used to describe human embryogenesis is not always the same as that used in non-medical branches of development. Here, we will mainly use the latter terminology and give the former in brackets.

Stage 1.   Age: about 1dpo.   Diameter: 0.1–0.15mm.

Characteristic feature: unicellular

Fertilization normally occurs in the ampulla of the oviduct (uterine tube) and takes place in three stages. First spermatozoa contact the oocyte, with one or more penetrating the *zona pellucida* of the oocyte, and making contact with the plasma membrane of the egg; the spermatozoan head swells and the second polar body is extruded shortly afterwards; next female and male pronuclei form; and then some 24–36hr later, the first mitotic division or zygote cleavage begins.

Stage 2.   Age: 1.5–3dpo.   Diameter: 0.1–0.2mm.

Characteristic features: two or more cells, but no blastocyst cavity

This stage includes specimens from two blastomeres up to when the embryo compacts, but stops before the appearance of the blastocystic cavity. While this happens, the zygote passes from the ampulla of the uterine tube to the uterine cavity (3–4dpo, 8–12 cells). Up to the 16-cell stage, blastomeres may form either inner cell mass (ICM) or trophoblast cells (**12.2–12.5**).

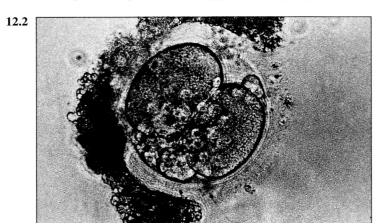

12.2 Stage 2. The 3-cell stage *in vitro* by light microscopy (×260; courtesy of Dr Carla Mills, Hallam Medical Centre).

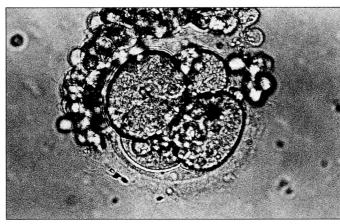

12.3 Stage 2. The 4-cell stage *in vitro* by light microscopy (×260; courtesy of Dr Carla Mills, Hallam Medical Centre.)

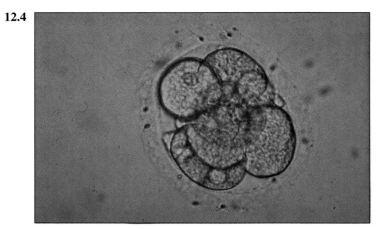

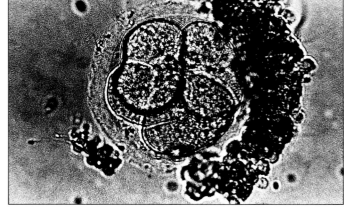

12.4, 12.5 Stage 2. The 6-cell stage *in vitro* by light microscopy (12.4 ×250, **12.5** ×290; courtesy of Dr Carla Mills, Hallam Medical Centre).

Stage 3.   Age: 4dpo.   Diameter: 0.1–0.2mm.

Characteristic feature: free blastocyst

By light microscopy, the conceptus is a free, *zona*-intact blastocyst of about 32 cells, with an outer trophectodermal layer of cells surrounding a blastocystic (or segmentation) cavity (although this cavity can form in cultured 16–20 cell embryos; Edwards, 1972) with a recognizable ICM. The *zona pellucida* may or may not be lost in more advanced embryos of this stage, although the blastocyst 'hatches' *in vitro* at 6–7 days (**12.6**).

Blastomeres lying peripherally form the trophoblastic cells, with those covering the ICM being referred to as 'polar' and the remainder as 'mural'. The embryo will later implant at the embryonic pole. The ICM will form the epiblast layer (which in turn forms the 'germ disc'), the hypoblast layer, and may add cells to the trophoblast. A polarity is evident here: the ICM facing the blastocystic cavity represents the future ventral surface, while that portion adjacent to the polar trophoblast represents the future dorsal surface.

During Stages 1–3, there is a general correlation between the number of cell divisions and the age of the embryos as the cleavage rate during the first week is not usually more than one per day.

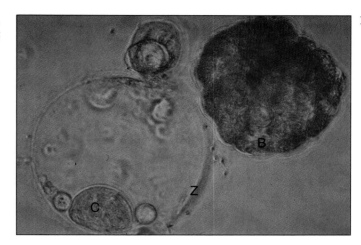

12.6 Stage 3. The 'hatching' blastocyst stage. (Key: **B**, blastocyst; **C**, retained cells; **Z**, zona pellucida; ×245; courtesy of Dr Carla Mills, Hallam Medical Centre.)

Stage 4.   Age: 5–6dpo.   Diameter: 0.1–0.2mm.

Characteristic feature: attachment of the blastocyst
The blastocyst begins implantation (completed at Stage 5). This starts with the dissolution of the *zona pellucida* and is followed by the attachment of the blastocyst to the maternal endometrium, through which it penetrates and migrates. No satisfactory reports of this stage have been recorded in humans and many workers rely on descriptions based on the development of the macaque monkey which is believed to be similar.

Stage 5.   Age: 7–12dpo.   Diameter: 0.1–0.2mm.

Characteristic feature: the trophoblast develops lacunae which subsequently fill with maternal blood
Implantation continues, most commonly on the posterior uterine wall, and the amniotic cavity and extra-embryonic mesoblast form. The blastocoelic surface of the epiblast is covered by primary embryonic endoderm (the embryo is now bilaminar, **12.7**) and extra-embryonic mesoderm (mesoblast) is forming in more advanced embryos. The chorion is beginning to form (0.3–1mm), but still lacks villi.

Implantation is an interactive process between the embryo and the maternal endometrium. During Stage 5, a decidual reaction occurs in the maternal endometrium which is characterized by the presence of endometrial connective tissue cells that enlarge due to cytoplasmic accumulation of glycogen or lipoids. The trophoblast, initially known as the cytotrophoblast, develops to form the syncytiotrophoblast and Stage 5 has been divided into three sub-stages which are based on the appearance of the trophoblast and its vascular relationships.

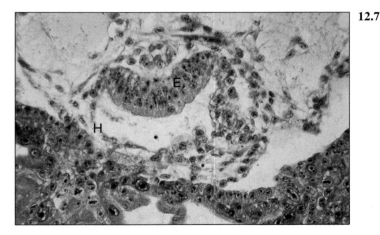

12.7 Stage 5. The bilaminar embryonic disc. (Key: **E**, ectoblast; **H**, hypoblast; ×270.)

Stage 5a.   Age: 7–8dpo.   Embryonic diameter: 0.1mm.   Chorion diameter: less than 0.5mm.

In this rarely described stage, definitive lacunae are not apparent and the trophoblast appears uniformly dense; extra-embryonic mesoderm (mesoblast) cells and the amniotic cavity begin to appear. The embryonic disc consists of the epiblast and primary endoderm layers.

Stage 5b. Age: 9dpo. Embryonic diameter: 0.1mm. Chorion diameter: 0.6mm.

The embryonic disc is incompletely covered by uterine epithelium and a decidual reaction is present. Definitive lacunae appear in the trophoblast which communicate with endometrial blood vessels and the syncytiotrophoblast is penetrated by clumps of cytotrophoblast. The chorion is about 0.6mm in diameter, the primary yolk sac cavity (umbilical vesicle) is developing and the amniotic cavity is almost closed by amniogenic cells. The embryonic disc is little changed from Stage 5a, while the mesoderm continues to develop.

Stage 5c. Age: 11–12dpo. Embryonic diameter: 0.15–2mm. Chorion diameter: 0.75–1mm.

The implantation site has a closing plate or operculum which occludes the site of entry into the maternal endometrium and there is an early decidual reaction in the endometrial stroma around the conceptus. The trophoblastic lacunae intercommunicate and contain enough maternal blood on the endometrial surface to be seen by the naked eye. The extra-embryonic mesoderm continues to develop. The amniotic cavity is almost completely closed and is smaller than the yolk sac.

Stage 6. Age: 13dpo. Diameter: 0.2mm.

Characteristic features: formation of chorionic villi and the primitive streak

Axial differentiation is apparent at the midpoint of the stage (defined as Stage 6a before and Stage 6b after the primitive streak can be seen). This is the stage at which the earliest prechordal plate and cloacal membrane have been recorded and at which gastrulation starts (**12.8**). This occurs as some of the epiblast cells join the primitive streak, move ventrally through it and migrate away as mesodermal cells between the superficial epiblast (ectoderm) and the underlying hypoblast (endoderm). The primitive node may appear before the primitive streak (see Stage 7).

Early on in this stage, the chorionic villi appear and start branching. The chorion is now 1–4.5mm in diameter and its greatly expanding cavity is 0.6–4.5mm in diameter, while the disc and its cloacal membrane is 0.15–0.5mm across. The secondary yolk sac (umbilical vesicle) develops as the site of origin of primordial germ cells and as a centre of haemopoietic activity. The amnion is well developed and its cavity may be smaller or larger than that of the yolk sac.

The decidual reaction is maintained at the site of implantation and three regions can now be recognized: the *decidua basalis* situated adjacent to the embryonic pole of the conceptus; the *decidua capsularis* reflected over the rest of the chorionic sac; and the *decidua parietalis* which lines the uterine cavity except at the implantation site. The connecting stalk consists of two portions, the amnio-embryonic stalk and the umbilical stalk, with the latter later forming part of the umbilical cord.

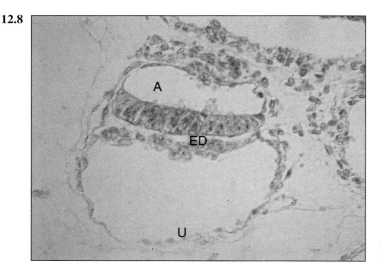

12.8 Stage 6. The embryonic disc, amnion, and umbilical vesicle. (Key: **A**, amnion; **ED**, embryonic disc; **U**, umbilical vesicle; ×310.)

Stage 7.   Age: 16dpo.   Diameter: 0.4mm.

Characteristic feature: notochordal process forms

The notochord process can now be seen rostral to the primitive streak and the node and is up to 300μm in length. It fuses with the endoderm layer and may have lateral mesodermal wings. The embryonic disc is typically 0.3–0.7mm long but may extend to 1mm. The node can now clearly be seen, although it may have formed a little earlier (Jirasek, 1983). The mesodermal cells spread laterally and rostrally to the primitive streak, while their ventral and caudal movements may form a temporary end node.

The diameter of the chorion is 4–9mm and its cavity 1.5–8mm. Primary and secondary villi are 0.5–1mm in length and may be present over the entire chorionic surface (see **12.21**). The amnion has an inner ectoderm layer, some mesenchyme in its middle layer and an outer mesothelial layer.

The yolk sac is likewise trilaminar in many areas and comprises an internal endoderm, a middle layer of mesoderm and an outer layer of mesothelium. Angioblastic tissue is differentiating in the ventral wall and its cavity is slightly larger than the amniotic cavity. The cloacal membrane is clearly defined and has extended (30–85μm). The allantoic (allanto-enteric) primordium now appears and the presence of primordial germ cells has been reported in this region.

Stage 8.   Age: 18 ±1dpo.   Length: 1.0–1.5 mm.

Characteristic features: neural groove, primitive pit and notochordal canal form

The embryonic disc is 0.5–2mm in length. Much of gastrulation has now taken place and endoderm is apparent, although incomplete, subjacent to the columnar ectoderm. There is evidence of neural groove formation and an underlying notochordal process (0.01–0.8mm) of much the same size which is fused to the endoderm. The primitive streak continues to extend.

The intra-embryonic mesoderm (mesoblast) layer is now essentially complete and has fused with the extra-embryonic mesoderm (mesoblast) layer at the margin of the disc. In the pericephalic region, isolated spaces are forming in the mesoderm (mesoblast) layer and mark the beginning of the intra-embryonic coelomic cavity. The uteroplacental circulation is now well established. The chorion is now 9–15mm in diameter while its cavity is 3–10mm.

Stage 9.   Age: 20 ±1dpo.   Length: 1.5–2.5mm.

Characteristic features: 1–3 somites, three identifiable brain regions.

The embryo now has well-defined head and tail folds and the primitive streak extends to the cloacal membrane. There is clear head organization with the prosencephalon, mesencephalon and rhombencephalon regions of the brain and the otic discs being readily identifiable. The cloacal membrane and allantoic diverticulum can also be seen.

In sections, the three germ layers can now clearly be seen, with the mesoderm layer having two main regions separated by intermediate mesoderm, a paraxial band that forms somites and lateral plate whose somatopleuric and splanchnopleuric layers are separated by the coelomic cavity. Neural crest cells are starting to migrate off the neural tube. The heart is beginning to form, initially within a horseshoe-shaped pericardial cavity, and matures over the stage to form sulci, cardiac jelly and mesocardium, while the septum transversum is also present. In addition, there are blood vessels in the umbilical vesicle, connecting stalk, chorion, amnion and in the embryo itself.

Stage 10.   Age: 22 ±1dpo.   Length: 1.5–3mm.

Characteristic features: 4–12 somites, initial fusion of neural folds and early heart formation

The embryo now has clearly defined neural folds that are beginning to fuse and, by the end of this stage, the neural tube extends from the hind brain (rhombencephalon) to beyond the last somite (**12.9, 12.10**). The head has a characteristic flexure at the mesencephalon, and the mandibular and hyoid arches are present; the notochordal plate, but not the notochord, can also be seen. In addition, the first pharyngeal arches are apparent, the primitive streak is at the caudal end of the embryo and the eye rudiment is beginning to form.

Externally, the heart is particularly prominent; internally, it has formed its myocardial mantle, reticulum (cardiac jelly) and endocardium and, as it develops, it folds to form an S-shaped curve with a ventricular loop. Lung and liver primordia are present, the mesonephric cord is beginning to form and the midgut is being delimited.

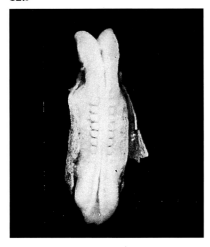
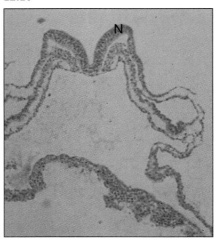
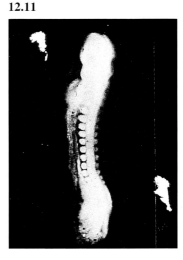

12.9 Stage 10. An embryo with 10 pairs of somites (×25).

12.10 Stage 10. The neural folds (N) in transverse section (×75).

12.11 Stage 11. An embryo with 13 pairs of somites (×20).

Stage 11.   Age: 24 ±1dpo.   Length: 2.5–4.5mm.

Characteristic features: 13–20 somites, initial stages of eye, ear and mesonephros formation

The rostral neuropore closes and neural crest cells migrate off the neural tube into the mesenchyme of the head, whose brain now possesses many well-defined anatomical features (**12.11**), while several cranial nerve primordia are present. The most prominent organ is the heart, now beating, while a network of endothelial channels forms the primitive vascular system. The lung primordium is present and the hepatic diverticulum is growing into the septum transversum. The urogenital system is beginning to differentiate as the mesonephros and mesonephric duct form, and the primordial germ cells are moving from the wall of the yolk sac (umbilical vesicle) to the hindgut.

Stage 12.   Age: 26 ±1dpo.   Length: 3–5mm.

Characteristic features: 21–29 somites, arm or upper limb buds appear, embryo is C-shaped

Externally, the embryo now has a pronounced curvature and there is the first sign of the upper limb buds. The three pharyngeal arches are visible and the caudal neuropore and otic vesicles are closing.

Internally, the gut and the epithelium of the alimentary canal are differentiating, with the latter forming, for example, the thyroid. The single lung bud and the liver rudiment are apparent, the central nervous system is well developed and internal organs such as the dorsal pancreas and gallbladder can all be readily identified. The heart and blood system are now connected and closed, and vascularization of the embryo is proceeding.

Stage 13.   Age: 28dpo.   Length: 4–6mm.

Characteristic features: leg or lower limb buds appear, lung buds enlarge.

After a month of development, most of the essential tissues are present and there are some 30 somites. In particular, the heart and liver are together as large as the head (**12.12, 12.13**). Much of the heart structure has now formed (except for valves) and a co-ordinated contractile beat has been established. There are circulating blood cells and gaseous interchange takes place with the maternal blood in the intervillar spaces. Such alimentary organs as the thyroid and thymus are particularly prominent and the spleen and ventral pancreas are apparent. The leg buds appear, the two lung rudiments are beginning to enlarge, the trachea is developing, the mesonephros is differentiating and the germ cells have reached the mesonephric ridges.

The umbilical cord is established as amnion covers the slender connecting stalk, the coelomic cavity and the stalk of the yolk sac.

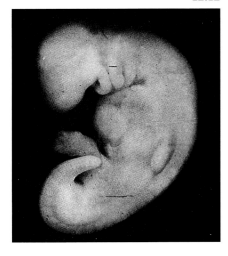

12.12 Stage 13. The 'C' shaped 4mm CR embryo with 30 or more pairs of somites.

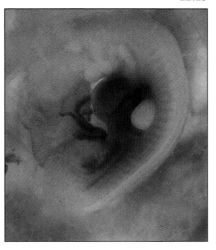

12.13 Stage 13. All four limb buds are present at this stage. The heart and liver are evident (×9).

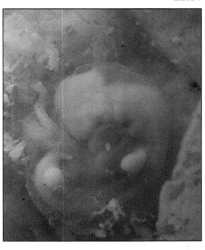

12.14 Stage 14. The elongated upper and lower limb buds are evident (×6.6).

Stage 14.   Age: 32dpo.   Length: 5–7mm.

Characteristic features: limb buds elongate, brain develops, metanephros forms

It is now becoming difficult to stage embryos from their external appearance, other than by their limb development, as most of the external features are present (**12.14**). On the head, the lens of the eye has formed a pit and the nasal plate is flattened. The heart has its four major subdivisions, the bifurcating trachea and its two lung buds can be seen, and the mesonephros extends down the abdomen. The metanephros is beginning to develop as the ureteric bud meets the metanephrogenic cap. The primordial germ cells have now migrated into the gonadal ridges and substantial innervation is taking place as spinal nerves attach to muscle primordia and extend into the upper limbs.

Stage 15.   Age: 33dpo.   Length: 7–9mm.

Characteristic features: hand plate and nasal pits form, retinal pigment

The embryo is now beginning to enlarge due to an increase in body mesenchyme, the differentiation of the muscular plates and the growth of spinal ganglia in the trunk region. Externally, the Stage 15 embryo is characterized by closed lens vesicles, the formation of nasal pits, hyoid arch differentiation, and the segmentation of the upper limb into shoulder, arm, forearm and handplate (**12.15**), while the muscular plates, somites and spinal ganglia produce

12.15 Stage 15. The upper limb bud is regionally subdivided along its proximo-distal axis into the shoulder, arm, forearm and hand plate (×6.8).

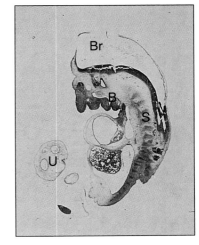

12.16 Stage 15. The 9mm CR embryo sectioned longitudinally and stained with haematoxylin and eosin. (Key: **B**, branchial arches; **Br**, brain; **S**, somites; **U**, umbilical stalk; ×5.)

identifiable elevations on the outer surface of the embryo.

Internally (**12.16**), the heart begins to develop fine structures such as the foramen secundum and the semilunar cusps, the midgut region of the intestine forms a loop and the ureteric bud within the metanephric kidney is expanding to form the renal pelvis. Innervation of the leg buds is initiated at this stage, while the gonadal ridges contain large numbers of germ cells. In the eye, the retinal pigment is evident.

Stage 16.   Age: 37dpo.   Length: 11–14mm.

Characteristic features: face begins to acquire form, lower limbs segment.

The face is now developing and, as it widens, some lateral structures such as the nasal pits become relatively closer to the midline. External ear structures are beginning to form and there are internal thickenings that will become the semicircular canals. The vertebral column has 36 skeletal elements, the gut has a definitive mesentery and is beginning to rotate, the ureter is elongating and the gonads are beginning to differentiate. The lower limbs are segmenting into thigh, leg and foot while the hand now has a digital plate (**12.17**).

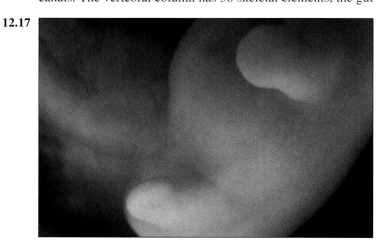

12.17 Stage 16. The leg bud is subdividing along its proximo-distal axis into the thigh, the leg, and foot plate (×8).

Stage 17.   Age: 41dpo.   Length: 11–14mm.

Characteristic features: face forms, slight lumbar curvature.

At this stage (**12.18, 12.19**), the face is becoming evident and the palate is appearing. The auditory ossicles but not the semicircular canals are present. Indifferent external genitalia are apparent as are the buds that will form nipples. Ossification is beginning in the limb bones and some vertebrae. The chondrocranium, within which the brain is differentiating, begins at this stage. The various lung lobes are now visible. A slight lumbar curvature is evident.

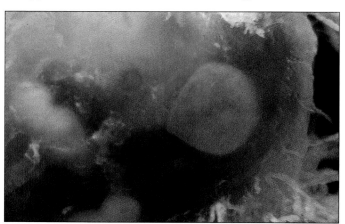

12.18 Stage 17. The hand plate has digital rays while the foot plate is rounded (×8).

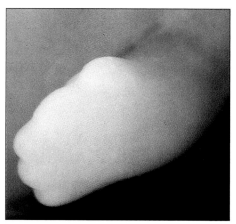

12.19 Stage 17. The digital rays of the hand and the arm (×17).

Stage 18.   Age: 44dpo   Length: 13–17mm.

Characteristic features: elbows are forming, cervical and lumbar flexures are present and nasal passages develop.

On the head, eyelids and ears are forming, the semicircular ducts are differentiating and the nasal passages developing. Most limb structures are defined and the elbows are forming. The heart has largely completed its development. In the metanephric kidney, collecting tubules are extending and nephrons differentiating, while testicular cords may appear in male gonads. The cervical and lumbar flexures are present.

Stages 19–23.

These stages are as follows: Stage 19, age 47–48 dpo, length 17–20mm (12.20, 12.21); Stage 20, age 50–51dpo, length 21–23mm; Stage 21, age 52dpo, length 22–24mm; Stage 22, age 54dpo, length 25–27mm; Stage 23, age 56–57dpo, length 23–32mm (12.22, 12.23). Characteristic features: organ development.

The 10 days during which the embryo progresses from Stages 19–23 is a period of consolidation when the tissues that are present continue to mature and few new structures form (e.g., at Stage 19, the limbs extend forward (**12.20**), while at Stage 23, the head becomes erect and the eyes open (**12.22**); in addition the fingers and toes become evident (**12.23**)).

An example of the fine level of development now happening is provided by the cornea: at Stage 20, it is merely an epithelium anterior to the lens; at Stage 21, neural crest cells invade the space between this epithelium and the lens and start to lay down the organized collagen fibrils responsible for its transparency. By Stage 23, the posterior layer of neural crest cells is differentiating into the endothelium.

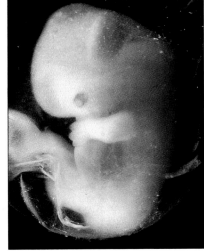

12.20 Stage 19. The 20mm CR embryo. The limbs extend forward and the pre-axial and post-axial borders are established (×3.3).

12.21 An SEM micrograph of a Stage 19 embryo showing the chorionic villi (×18,000).

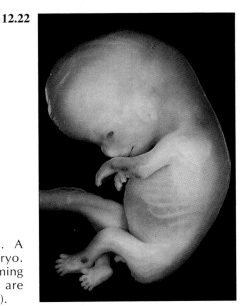

12.22 Stage 23. A 27mm CR embryo. The head is becoming erect. The eyes are largely open (×2.6).

12.23 Stage 23. A 30mm CR embryo. The limbs have increased in length and both fingers and toes are evident (×2.5).

Later development

Stage 23 marks the transition between the end of embryogenesis and the beginning of the fetal period, although in practice it is hard to point to a sharp division between these two phases of development. In essence, the former period is characterized by the generation of tissue structure and the latter by their growth and acquisition of function. It should therefore be noted that, by Stage 23, over 90% of the 4500 or so named structures in the human body are now present and recognizable.

Traditionally, fetal development was studied using normal anatomical techniques, but, with the advent of ultrasound, the development, growth and functioning of the embryo and fetus have been clarified further. Large gaps do, however, remain in our knowledge of the formation of the human embryo and fetus.

Experimentation

Embryos from animal models were initially studied for the value they might impart to the knowledge of the human embryo. While many investigators then studied animal models for their intrinsic value, they also wished to understand human development, even though there was sometimes only a tenuous connection between the animal model and the human embryo.

There is, however, a legal aspect to such work. In the case of animal embryos that can be readily obtained, the legislation controlling their use in experimentation is well established. The law governing the use of human embryos in experimental studies is a recent development and there are particularly strict limits on the use of material. The field is, however, constantly evolving and, within the permitted parameters, new techniques are developing very rapidly. The innovations in this field, which are almost exclusively confined to the period between conception and 14 days of development, are being closely monitored and reviewed by several elements in society (i.e. government, religious leaders and the general public) and the moral and religious implications of human experimentation will ensure that this close monitoring will continue.

In Britain, experimentation on human embryos and their associated tissue (e.g. cumulus cells) is very strictly controlled by local Ethical Committees in approved establishments and by a national body, the Human Fertilisation and Embryology Authority (HFEA). The Committees are the arbiters of experimentation as based on the Human Fertilisation and Embryology Act, 1990, which limits studies on gametes and from fertilization to Day 14 (up to primitive streak formation).

Much of embryonic experimentation involves modifying media and identifying the conditions necessary for improving the relatively low success rate (0–26%) for *in vitro* fertilization. The requirements for early growth *in vitro* have therefore been studied extensively in co-cultures of oocytes and other tissues (e.g. uterine tube cells; Bongso *et al.*, 1989) and these systems have added valuable information on the growth factors needed for the *in vitro* systems. Other experiments have dealt more directly with embryonic development and, for example, Handyside *et al.* (1990) were recently able to sample a single cell from a pre-embryo and assess genetically whether a sex-linked defect had been inherited. Experimentation has also

Table 12.2 Therapeutic techniques for female infertility with gamete and embryological applications.

Named procedure	Methodology
In vitro fertilization (IVF) and embryo transfer (ET)	Follicular oocytes obtained, inseminated and cultured *in vitro*. Tranferred to uterus.
Pronuclear stage tubal transfer (PROST) or Zygote intrafallopian tube transfer (ZIFT)	Follicular oocytes obtained, inseminated and cultured *in vitro*. Transferred to uterine tube.
Tubal embryo stage transfer (TEST)	As for PROST but embryo cultured *in vitro* an additional day.
Gamete intra-fallopian transfer (GIFT)	Follicular oocytes placed with sperm in uterine tubes.
Peritoneal oocyte and sperm transfer (POST)	Follicular oocytes obtained, inseminated and injected near fimbria of uterine tube (and ovary).
Direct intraperitoneal insemination (DIPI)	Sperm injected into peritoneal cavity near fimbria of uterine tube (and ovary).
Oocyte donation	Used in IVF, PROST, POST, GIFT or TOT.
Surrogacy	Donor oocyte and/or sperm placed in surrogate mother.
Hormone treatment	Hormone deficiency or imbalance leading to ovulation failure.
Cryopreservation	Pre-embryos.

shown that, as the *zona pellucida* 'toughens' in culture, blastocyst 'hatching' is often inhibited. If, however, a hole is made in the zona mechanically *in vitro*, blastocyst hatching is greatly enhanced (Cohen *et al.*, 1990).

This experimental work has resulted in several investigative and therapeutic techniques with routine applications in treating female and male infertility and in assisting pregnancy. In addition, they have yielded information concerning gamete and early embryonic development. Techniques which help overcome female and male infertility are given in *Tables 12.2* and *12.3*.

Experimentation on developing fetuses is particularly restricted and most studies are, of course, non-invasive and designed to elucidate function. It has, for example, been established that, from the sixth month, fetuses normally hear *in utero* and that the new-born infant can recognize sounds such as the heart beat or the mother's voice heard prior to birth. Ultrasound has been used to show that a predominant right- or left-handedness is established before birth (Hepper *et al.*, 1990) and this laterality will be reflected in the development of the brain. In addition, fetal surgery (mainly on the heart) has been carried out in a few centres in the world and here the techniques were of course established using animal models.

Table 12.3 Therapeutic techniques for male infertility with gamete and embryological applications.

Named procedure	*Methodology*
Motile sperm selection or IVF sperm preparation	Concentrate motile sperm by centrifugation or swim-up or gradient separation for IVF, or IUI, or uterine tubes (PROST or ZIFT), or peritoneal insemination.
Intrauterine insemination (IUI)	
Sperm freeze	Cryopreserved sperm.
Donor insemination (DI) or donor (IUI)	Donor sperm.
Microsurgical epididymal sperm aspiration (MESA)	Sperm recovered surgically from junction of epididymis and vas deferens for IVF.

The future

By even the most optimistic expectations of the last generation of human embryologists, the future has arrived! We can initiate human life *in vitro* and allow an individual to develop before re-implantation into the natural or a surrogate mother. Furthermore, it is now possible to operate on a fetus and to correct certain physical anomalies so that it subsequently continues normal gestation. We are also able to save early premature babies by enabling them to continue development in specialist units under carefully controlled conditions and these infants afford us a unique opportunity to study normal development. Future research should allow even younger babies to survive.

The other area where we can look forward to significant progress is in correcting defective genes using the technology developed with mouse embryos, and cystic fibrosis is the obvious first candidate. Moreover, with our increasing understanding of the molecular basis of differentiation, it may prove possible in the longer term to put early biopsied cells into culture and then 'bank' them in case a person later needs a stock of his or her own early cells, which can then be modulated and used for repair (e.g. skin) or for replacing tissues affected by tumours (e.g. bone marrow).

Acknowledgements

I am very grateful to Dr Carla Mills, Director of Embryology and Andrology, Hallam Medical Centre, London, who kindly and generously supplied **12.2–12.6** and read and corrected the sections on IVF. I am also grateful to Professor K. Shiota, Department of Anatomy, Kyoto University, Japan, for allowing me to use **12.9–12.12** from their Department's Embryology Collection. **12.6** and **12.7** were reproduced from *A Colour Atlas of Life Before Birth*, Wolfe Medical Publications (1983, p. 216) and were originally photographed at the Anatomy Department, Charing Cross and Westminster Hospital Medical School. Professor K.E. Carr and The Anatomy Department, The Queen's University, Belfast, kindly gave permission for **12.15** to be reproduced. Dr E.C. Blenkinsopp, Consultant Pathologist, Watford General Hospital, also provided generous help.

References

Bongso, A., Soon-Chye, M., Sathananthan, H., Pohlian, N.G., Rauf, M. and Ratnam, S. (1989). Improved quality of human embryos when co-cultured with human ampullary cells. *Hum. Reprod.*, **4**, 706–713.

Cohen, J., Elsner, C., Kort, H., Malter, H., Massey, J., Maver, M.P. and Wiemer, K. (1990). Impairment of the hatching process following *in vitro* fertilisation in the human and improvement of implantation by assisted hatching using micromanipulation. *Hum. Reprod.*, **5**, 7–13.

Edwards, R.G. (1972). Fertilization and cleavage *in vitro* of human ova. In *Biology of Mammalian Fertilization and Implantation* (K.S. Moghissi and E.S.E. Hafez, eds) pp. 263–278. Thomas, Springfield, Illinois.

Handyside, A.H., Kontogianni, E.H., Hardy, K. and Winston, R.M.L. (1990). Pregnancies from biopsied human pre-implantation embryos sexed by Y-specific DNA amplification. *Nature*, **344**, 768–770.

Hepper, P.G., Shahidullah, S. and White, R. (1990). Origins of fetal handedness. *Nature*, **347**, 731.

Her Majesty's Stationary Office, London (1990) *Human Fertilisation and Embryology Act*. Chapter 37. pp. 1–39.

Jirasek, J.E. (1983). *Atlas of Human Prenatal Morphogenesis*. Martinus Nijhoff, Hague.

O'Rahilly, R. (1973). Developmental Stages in Human Embryos including a Survey of the Carnegie Collection, Part A: Embryos of the first three weeks (Stages 1–9). *Carnegie Inst. Wash. Publ.* 631, Washington D.C.

O'Rahilly, R. and Muller, F. (1987). *Developmental Stages in Human Embryos including a revision of Streeter's "Horizons" and a survey of the Carnegie Collection*. *Carnegie Inst. Wash. Publ.*, 637, 1–306.

Rosenfeld, M.A. *et al.* (1992). *In vivo* transfer of the human cystic-fibrosis transmembrane conductance regulator gene to the airway epithelium. *Cell*, **68**, 143–155.

Streeter, G.L. (1942). Developmental horizons in human embryos. Description of age group XI, 13–20 somites, and age group XII, 21–29 somites. *Carnegie Inst. Wash. Publ.* 541. *Contrib. Embryol.*, **30**, 211–245.

Streeter, G.L. (1945) Developmental horizons in human embryos. Description of age group XIII, embryos of about 4 or 5 mm long, and age group XIV, period of indentation of the lens vesicle. Carnegie Inst. Wash. Publ. 557. *Contrib. Embryol.*, **31**, 27–63.

Streeter, G.L. (1948) Developmental horizons in human embryos. Description of age groups XV, XVI, XVII and XVIII, being the third issue of a survey of the Carnegie Collection. Carnegie Inst. Wash. Publ. 575. *Contrib. Embryol.*, **32**, 133–203.

Streeter, G.L. (1951) Developmental horizons in human embryos. Description of age groups XIX, XX, XXI, XXII and XXIII, being the fifth issue of a survey of the Carnegie Collection (prepared for publication by Heuser, C.H. and Corner, G.W.). *Carnegie Inst. Wash. Publ.* 592. *Contrib. Embryol.*, **34**, 165–196.

Index

A
activin
  chick 177
  xenopus 164
Amphibia *see* molluscs
Aplysia see molluscs
Apoptosis *see* cell death
Arabidopsis thaliana
  callus 17
  developmental stages 10
    octant 11
    dermatogen 12
    globular 12
    triangular 13
    heart 14
    torpedo 15
    bent cotyledon 15
    dessication 16
  fertilization 8
  gamete formation 8
  in vitro work 17
  life cycle 8
  normal development 8
  ovary 9
  pattern mutants
    fackel 18, 19
    fass 18, 19
    gnom 18, 19
    gurke 18, 19
    keule 18, 19
    knolle 18, 19
    knopf 18, 19
    mickey 18, 19
    monopteros 18, 19
  pollen 8
  pollen tube 9
  RFLP maps 8, 20
  seed germination 16
  self-fertilization 7
  transgenic plants 17
  vegetative phase 8

B
Bithynia see molluscs
block to polyspermy
  mouse 184
  sea urchin 39

C
C. elegans see nematode
cell adhesion
  sea urchin 49
cell communication
  Dictyostelium discoideum 26
  mollusc 87
  nematode 68
cell cycle
  chick 170, 177
  leech 102
  mollusc 80
  mouse 186
  Xenopus 151
  zebrafish 136
cell death (apoptosis)
  leech 108
  nematode 69
cell interactions in specification
  molluscs 86
cell lineage *see* lineage analysis
cell migration
  chick 168
  Dictyostelium discoideum 24–29
  leech 99, 109
  mouse 186, 187, 189, 199
  nematode 72
  sea urchin 46
  Xenopus 151–153
  zebrafish 137, 138, 144
cell rearrangement
  sea urchin 47, 48
chemotaxis
  Dictyostelium discoideum 24
  sea urchin 39
chick
  activin 177
  axis formation 168
  cell fate 168
  cell markers
    Chick-en 178
    HNK-1 178, 180
    L5 178
  cell movement 168
  cleavage 168
  egg organization 167
  epiblast formation 168
  fertilization 168
  gastrulation 168
  germ cells 168
  gut formation 168
  Henson's node 168, 174
  hypoblast formation 168
  induction 174
  in vitro development 181
  limb formation 175
    apical ectodermal ridge (AER) 175
    duplications 176
    zone of polarizing activity (ZPA) 175
  lineage markers 177
  manipulation of embryo 174
  mesoderm formation 168
  nervous system formation
    neurulation 174
    origins of neurons 171
    segmentation 171, 173
  neural crest cells 171, 180, 181
  quail
    development 167
    grafts 179, 181
    HNK marker 179
  retinoic acid 175, 177
  rhombomeres 173
  somitogenesis 169
  teratological techniques 177
  transgenic birds 167, 181
  twin formation 175
cleavage
  Arabidopsis 11
  chick 168
  human 210
  leech 95
  mollusc 80, 83
  mouse 186, 195
  nematode 57, 63, 66
  sea urchin 38, 39
  Xenopus 150, 150, 160
  zebrafish 136

D
databases
  mouse 204
  nematode 70
Dictyostelium discoideum
  cAMP and chemotaxis 24, 25
  Dictyostelids 32
  D. lacteum 33
  D. mucoroides 32
  DIF 25, 31, 35
  differentiation 30
  genes and markers
    α-actin 31
    ecmA,B 28
    fdgA 31
    G protein 31
    myosin2 31
    rdeA 32
    rdeC 31
    Slugger 31
    streamer 31
  migration 24–29
  mutants 24,31
  normal development 24
  Polysphondylids 32
  P. violaceum 32
  spore-stalk patterning 28, 29
  transformation 23
DIF 25, 31, 35
differentiation *in vitro*
  Arabidopsis 17
  chick 181
  human 218
  mouse 199–201
  nematode 66
  Xenopus 161
Drosophila
  axis formation 129
  balancer chromosomes 128
  experimental manipulations 126
  fate map 115–118
  fertilization 114
  gastrulation 118
  genes
    abdominal-A (Abd-A) 119, 120
    achaete-scute 123
    actin 114
    Antp 119

bicoid 129–132
delta 123
dorsal 123, 128, 130
dpp 123, 130
engrailed 131, 132
even-skipped 124, 131, 132
fushi-tarazu 131
giant 129, 132
hairy 131, 132
hunchback 129–132
knirps 129, 132
Krüppel 128–132
MyoD 115
myosin 114
nanos 129–132
Notch 123
runt 132
snail 115, 123
S59 122
Toll 129
torso 129
twist 115, 122, 130
Ultrabithorax (Ubx) 120, 121
vasa 125
wingless 132
genetic manipulations 127
genetic screens 127
germ-band development 117
gonad 125
gradient 131
gut 115, 120
homeotic genes 120
imaginal discs 114, 127
mesoderm development 114
muscle differentiation 122
mitosis 120
morphogenesis 117
nervous system 123
    identified neurons 123
    mesectoderm cells 123
    ventral nerve cord 123
normal development 113
parasegments 117
pole cells 115
salivary gland 120
segmentation 118, 132
tracheae 125, 126

E
echinoderms *see* sea urchin
enhancer trap embryo
  Drosophila 121
epiboly
  sea urchin 47
  zebrafish 137
exogastrulation
  sea urchin 48

F
fate map and details (also *see* lineage analysis)
  chick 168
  Drosophila 115–118
  Dictyostelium discoideum 25–28
  leech 96
  molluscs 85
  mouse 198
  nematode 58
  sea urchin 41

Xenopus 158, 159, 162
zebrafish 144
fertilization
  Arabidopsis 8
  chick 168
  Drosophila 115
  leech 95
  molluscs 80
  mouse 184
  nematode 56
  Xenopus 150
  zebrafish 136
flowering plants *see Arabidopsis*
follicle cells
  molluscs 79
  mouse 184

G
gastrulation
  Drosophila 118
  human 212
  molluscs 87
  mouse 187
  nematode 57
  sea urchin 43
  Xenopus 151–164, 164
  zebrafish 137
genes — *see* under specific organisms
germ band, *Drosophila* 118
germ cells
  chick 168
  Drosophila 115
  mouse 189
gradient
  chick 167
  Dictyostelium discoideum 23, 26
  Drosophila 131
  leech 99
  sea urchin 37, 45
gut development
  chick 168
  Drosophila 120
  human 214
  leech 98
  mouse 188, 193, 199
  nematode 56
  sea urchin 47
  Xenopus 161

H
haploid embryos
  zebrafish 140
hemisomites
  molluscs 82
homeotic genes
  Drosophila 121
  nematode 67
human development
  ageing of embryos 208
  blastocyst formation 210
  chorionic villi 212
  cleavage 210
  comparisons among mammals 209
  corneal development 217
  epiblast formation 210
  experimental techniques 218
  gastrulation 212
  gut development 214
  heart formation 213

implantation 211
infertility, female 218
infertility, male 219
inner cell mass 210
in vitro fertilization 207, 218
kidney formation 214
legal restrictions 218
mesoderm formation 212
neural crest cell migration 213, 214
neural development 213
normal development 209
polar body extrusion 209
primitive streak formation 212
somitogenesis 213, 214
spinal ganglion formation 215
trophoblast differentiation 211
umbilical cord formation 214
zonar pellucida 210

I
imaginal discs
  Drosophila 114, 127
induction
  chick 174
  mollusc 86
  mouse 199
  Xenopus 163, 165
  zebrafish 145

K
kidney
  chick 168
  human 214
  leech 108
  mouse 199–201
  Xenopus 163

L
leech
  bandlets 94, 107, 108
  blast cells
    classes 99, 103
    cycle 102
    death 108
    divisions 103
  cleavage 95, 96
  comb cells 109, 110
  epiboly 97
  experimental techniques 99
  families
    glossiphoniidae 94, 109
    hirudinidae 94, 99, 109, 110
  fertilization 95
  ganglia
    formation 99, 105
    segmental 93, 100, 101, 109
    supraoesophageal 98
  genes
    ht-en 105, 106
    lox2 105
  genital primordia 100
  germinal bands 94, 97, 108
  germinal plate 97
  gradient 99
  gut formation 98
  kinship groups 101
  lineage 96, 102
  muscles 100, 109, 110
  nephridial duct 108

nerve cord formation 98
neurons 99, 108
    identifiable 101, 109
    Retzius 109
neural and muscle antigens 100
    Lan 3-14 109, 110
    SCP-like 109
    serotonin 93, 108
normal development 95
RNA synthesis 102
segmentation 97, 98, 106, 108
size 94
teloblasts 94, 96, 97, 99, 101, 103, 107
teloplasm 96, 104
terminal differentiation 107
lineage analysis
    chick 177
    leech 96, 102
    molluscs 85
    mouse 198
    nematode 59
    sea urchin 41
    Xenopus 157, 162
    zebrafish 143
lithium
    sea urchin 45, 46
    Xenopus 160
Lymnea see molluscs

M

mammalian development comparisons
    human 209
metamorphosis
    molluscs 90
mesoderm formation
    chick 168
    human 212
    molluscs 86
    mouse 187
midblastula transition
    Xenopus 151
    zebrafish 136
molluscs
    Aplysia 80, 81
    Bithynia 82, 85
    blastula formation 87
    cell cycle 80
    cell interactions in specification 86
    cell lineage 85
    cleavage pattern 80, 86
    fate map 85
    fertilization 80
    follicle cells 79
    gastropod morphology 78
    gastrulation 87
    hemisomites 82
    larva 77
    Lymnea 77, 79, 80, 81, 89
    Nassarius 77, 84, 85
    macromeres 83, 86
    mesoderm stem cells 86
    metamorphosis 90
    micromeres 83
    molluscan cross 87, 88
    neural development 88
    nomenclature of blastomeres 83
    ooplasmic segregation 80, 81
    quadrant specification 85
    Patella 77, 79, 80, 84, 85, 87, 88, 89

phylum members 77–79
polar lobe formation 80, 82, 85
prototroch 77
spiral cleavage 77, 80, 83, 84
trochoblasts 87
trochophore 77, 89
veliger 77
morphogenesis
    chick 170
    Drosophila 118
    Dictyostelium discoideum 33
    leech 98
    mollusc 80
    sea urchin 43, 47
mouse
    block to polyspermy 184
    brain development 195, 198
    compaction 186
    culturing embryos 197, 199
    databases 204
    dominant negatives 203
    ES (embryonic stem) cells 194, 198, 203
    erythropoiesis 188
    embryonic manipulations 194
    extraembryonic membranes 189, 190
    fertilization 184
    gastrulation 187
    gene targeting 203
    genetics 183, 201
    genes
        Brachyury (T) 187
        chinchilla 203
        cystic fibrosis 203
        Hox1.5 202
        Hox7 202
        Hox8 202
        pink-eye (p) 202
        osteogenesis imperfecta 203
        WT1 200, 201
    germ cells 189
    growth 193
    gut formation 188, 193, 199
    heart formation 188, 191
    implantation 186
    induction 199
    imprinting 195, 203
    inner cell mass 185, 186
    kidney development 199–201
    LIF 194, 200
    lineage studies 198
    mesoderm formation 188
    Mus spretus 201
    neural crest cells 193, 198
    neurulation 188, 198
    normal development 183
    parthenogenesis 194, 195
    rhombomeres 191
    somites 188, 198
    tetraploids 195
    transgenic mice 202–204
    triploids 195
    trophectoderm 185, 186
muscle development
    Drosophila 122
    leech 100, 109, 110
mutant screens
    Drosophila 127
    nematode 58, 67
    zebrafish 140

N

Nassarius see molluscs
nematode (*C. elegans*)
    Ascaris 71
    anatomy 56
    cell death 69
    cell lineage 57, 59
    cell lineage mutations 67
    cell types
        Z1ppp 65
        Z4aaa 65
    database 70
    dauer 63
    experimental manipulations
        laser microsurgery 63, 65
        micromanipulation 64
    gastrulation 58
    genes
        bli-4(e937) 67
        dpy-2(e8) 67
        lin family 66, 68
        lon-2l(e678) 67
        tra family 68
        unc family 66, 67, 68, 72
    genome map 70, 73
    gonad development 63, 67
    granules
        gut 61
        P 57, 60
        rhabditin 58
    gut 56
    in vitro development 66
    hermaphrodite 63, 65
    homeotic mutations 67
    muscle development 72, 73
    nervous system 71
    normal development 57
    pattern mutants 67
    phylogenetic tree 55
    post-embryonic development 62
    sex determination 68
    transformation 70
neural crest
    chick 171, 180, 181
    human 213, 214
    mouse 193, 198
    Xenopus 153
neural development
    chick 171, 173, 174
    Drosophila 123
    human 213
    leech 98, 108
    mollusc 79
    mouse 188, 198
    nematode 71
    sea urchin 44
    Xenopus 153
    zebrafish 140–142
normal development
    Arabidopsis 8
    Dictyostelium discoideum 24
    Drosophila 113
    human 209
    leech 95
    mouse 183
    Xenopus 150
    zebrafish 135

O
ooplasmic segregation
  molluscs 80, 81

P
parthenogenetic activation
  mouse 194, 195
  sea urchin 39
pattern mutations
  Arabidopsis 18
  Dictyostelium discoideum 24, 31
  Drosophila 130–132
  nematode 67
  zebrafish 144
Patella see molluscs
plants *see Arabidopsis*
polar lobe formation
  molluscs 80, 82, 85

Q
quail *see* chick

R
retinoic acid
  chick 175, 177
RFLP maps
  Arabidopsis 7
rhombomeres
  chick 173
  mouse 191

S
saturation screen
  Drosophila 127
sea urchin
  archenteron formation 43, 47, 50
  block to polyspermy 39
  cell adhesion 49
  cell migration 45
  cell rearrangement 47, 48
  chemotaxis 39
  contact interactions 45
  epiboly 47
  exogastrulation 48
  extracellular matrix 45
  fate map 41
  gradient 37, 45
  gastrulation 43
  granules
    cortical 37
    extracellular matrix 37
    pigment 37
  hatching enzyme 41
  lithium 45, 46
  macromeres 39, 40
  markers
    BP-10 43
    Ecto V 43
    Endo 1 43
    LvN1.2 43
    LvS1 43
    Meso 1 43
    msp130 43
    Spec 1 43
  mesenchyme
    primary 40, 41, 42, 46
    secondary 40, 41, 49
  mesomeres 39, 40, 45
  micromeres 39, 40, 42, 45
  microtubules 40
  nerves 44
  nickel, effect of 46
  nomenclature of blastomeres 40
  parthenogenetic activation 39
  phylogenetic differences 50
  pluteus larva 44
  polar body 37
  spicule formation 43, 46, 50
  timing effects 50
  vegetalization 45, 45, 46
self-fertilization
  Arabidopsis 7
  nematode 56
segmentation
  chick 169, 171, 173
  Drosophila 118, 132
  leech 98
  zebrafish 138
somitogenesis
  chick 169
  human 213, 214
  molluscs (hemisomites) 82
  mouse 189, 191, 199
  Xenopus 153, 162
  zebrafish 138
spiral cleavage
  molluscs 77, 80, 83, 84

T
transgenic animals
  chick 167, 181
  mice 202–204
  zebrafish 140
transgenic plants
  Arabidopsis 17

V
vegetalization
  sea urchin 45, 46

W
worm *see* nematode

X
Xenopus and other amphibians
  activin 164, 165
  Ambystoma mexicanum (axolotl) 149, 152,153, 154
  Cynops pyrrhogaster (Japanese newt) 149
  blood cell development 156
  differentiation *in vitro* 161
  dorsal lip of blastopore 151
  dorsalization 162
    hyperdorsalization 160
  experimental methods
    fate mapping 157
    grafting 162
    lithium 160
    microsurgery 155
    ultraviolet irradiation 159
  fate map 158, 159, 162
  fertilization 150
  fibroblast growth factor (FGF) 164, 165
  gastrulation 151–153, 164
    in *Ambystoma* 153
    convergence extension 152
    in *Pleurodeles* 153
  grey crescent 150
  holoblastic development 149
  induction
    mesoderm 163
    neural (primary) 163
    overview 165
  markers
    alkaline phosphatase 156
    C6 155
    cytokeratin 155
    fibronectin 155
    HNK-1 155
    keratin sulphate 155
    myosin 155
  mid-blastula transition 151
  neural crest 153
  neurulation 153
  normal development 150
  organizer 163, 164
  Pleurodeles (ribbed newt) 153
  somites 153, 162
  specification map 162
  subcortical rotation 150, 162
  ventralization 160

Z
zebrafish
  cleavage 136
  deep cells 137
  enveloping layer 137
  epiboly 137
  experimental studies 141
  fate map 144
  gastrulation 137
  genes
    albino 139
    brass 139
    cyclops 145
    golden 136, 139
    Pax-6 146
    spadetail 139, 140, 144, 145
    sparse 139
  haploid embryos 140
  hemisegments 142
  induction 145
  lateral line system 138
  lineage analysis 143
  marker, antibody zn-12 141
  mutant screens 140
  mutational analysis 144
  neurons
    auditory 141
    identified 141
    lateral-line nerve 141
    Mauthner 142
    motoneurons 141, 142, 143
    primary 141
    reticulospinal 142
    Rohon-Beard 140, 141
    sensory 141
    trigeminal 141
  normal development 135
  pattern mutations 144
  segmentation 138
  somitogenesis 138
  staging series 136
  transgenic fish 140